T0180987

Applied Mathematical Sciences

Volume 189

For further volumes:
www.springer.com/series/34

Juncheng Wei · Matthias Winter

Mathematical Aspects of Pattern Formation in Biological Systems

 Springer

Juncheng Wei
Department of Mathematics
The Chinese University of Hong Kong
Hong Kong, China

Matthias Winter
Department of Mathematical Sciences
Brunel University
Uxbridge, UK

ISSN 0066-5452   ISSN 2196-968X (electronic)
Applied Mathematical Sciences
ISBN 978-1-4471-7261-1   ISBN 978-1-4471-5526-3 (eBook)
DOI 10.1007/978-1-4471-5526-3
Springer London Heidelberg New York Dordrecht

Mathematics Subject Classification: 35B25, 35B36, 35B40, 35J47, 35K40, 92B05, 92C15

Key words and phrases. Biological systems, Reaction-diffusion systems, Mathematical biology, Pattern formation, Mathematical analysis, Localised structures, Singular perturbation, Stability.

Printed on acid-free paper

Springer is part of Springer Science+Business Media (www.springer.com)

Preface

This monograph is concerned with the mathematical analysis of patterns which are encountered in biological systems. It serves a number of purposes. First, we summarise, expand and relate the results we have obtained over the last fifteen years. Secondly, we link these results to biological applications and highlight their relevance to phenomena in nature. Thirdly, we conclude that with these studies we have initiated a systematic programme of rigorous mathematical investigation into pattern formation for large-amplitude patterns far from equilibrium in biologically relevant models. The work is organised as follows:

(i) Our goal is to consider the existence of spiky steady states in reaction-diffusion systems and select as observable patterns only those which are stable.

(ii) The investigation begins by considering a spatially homogeneous two-component activator-inhibitor system in one or two space dimensions (Chaps. 2–6).

(iii) We extend our study of these systems by adding several extra effects or by considering related systems, each motivated by their specific roles in developmental biology. This extended investigation includes the following features: the study of precursor gradients (smooth inhomogeneous coefficients) or discontinuous diffusivities; the study of finite activator diffusivity; an investigation of the effect of large reaction rates; the effect of altering the boundary conditions (from Neumann to Robin type); an investigation of the system on manifolds; the effect of adding saturation terms to the system; and replacing Gierer-Meinhardt kinetics of activator-inhibitor type by kinetics of activator-substrate type (also called Gray-Scott or Schnakenberg kinetics). In our study of reaction-diffusion systems with many components we consider the kinetics given by the hypercycle of Eigen and Schuster, introduce convection into the model, consider a system allowing for the mutual exclusion of spikes, investigate the interaction of many activators and substrates and, finally, establish exotic spiky patterns for a consumer chain model.

The existence of solutions is proved using methods from nonlinear functional analysis in either of the following settings: (i) when the linearised operator about an approximate solution is invertible in a suitable function space; or (ii) when the linearised operator becomes invertible under suitable finite-dimensional projections. Case (ii) has become well-known as the Liapunov-Schmidt reduction method, and, using that approach, in a second step a finite-dimensional nonlinear problem has to be solved to establish the existence of solutions.

Concerning the stability of solutions, the approach we take allows for a rigorous analysis of the stability of large-amplitude patterned states. Large eigenvalues are investigated by nonlocal eigenvalue problems (NLEPs) which are derived by taking the limit of a rescaled eigenvalue problem and analysed using bilinear forms. While previous work introduced the NLEP approach in essentially one-dimensional settings (real line, radially symmetric, axial), we extend it to general partial differential equations without any symmetry assumptions, only requiring a certain smoothness of the boundary. The NLEP approach only provides an answer for the instabilities caused by large eigenvalues. Since there are only finitely many small eigenvalues, this reduces the stability problem to finite dimensions. Thus stability can be derived explicitly by computing the eigenvalues and eigenvectors of certain matrices.

We introduce a new approach to pattern formation in reaction-diffusion systems which differs from previous work on Turing instability (bifurcation of patterned states from unstable homogeneous steady states) [232]. These techniques also go beyond the order parameter approach reviewed in [34], which was introduced to investigate the behaviour of a system after the initial bifurcation has occurred, by including the leading nonlinear terms in the approximation, and for system parameters near the bifurcation point small-amplitude, or, more generally, slowly modulated spatially patterned states have been derived. However, both approaches are only valid for system parameters near the bifurcation point and for small-amplitude, or, more generally, slowly modulated patterns, since otherwise the nonlinear contributions or higher-order nonlinear contributions, respectively, cannot be neglected. Our approach is valid for large-amplitude patterns in the case of multiple spikes which concentrate at certain points of the underlying domain.

Juncheng Wei thanks his collaborators Manuel del Pino, Mike Kowalczyk, Theodore Kolokolnikov, David Iron, S.-I. Ei, Wei-Ming Ni, Xiaofeng Ren and Michael J. Ward for their encouragement, stimulating ideas and generous contributions on this subject. In particular, he thanks Professor Wei-Ming Ni for introducing him to the subject and for his invaluable guidance. Last but not least, he thanks his family, in particular his wife, for their support and understanding. Matthias Winter thanks the Department of Mathematics and the Institute of Mathematical Sciences at The Chinese University of Hong Kong as well as the Institut für Analysis, Dynamik und Modellierung at the Universität Stuttgart for their kind hospitality during the writing of this book. He is deeply grateful for the regular visits he has made to the Department of Mathematics at The Chinese University of Hong Kong over many years while this research has been carried out. He acknowledges the support of Brunel University under the Research Leave Scheme. He thanks Professors Philip Maini and Hans Meinhardt for very interesting discussions which shaped his view of mathematical biology. We thank the advisors and reviewers for valuable suggestions which helped improve the manuscript. We acknowledge the assistance of Lynn Brandon and Catherine Waite at Springer London during the publication process. This book would not have been written without the encouragement and help of Manuel del Pino and Michael J. Ward. This work is supported by an Earmarked Research Grant of RGC of Hong Kong.

Hong Kong, China   Juncheng Wei
London, UK   Matthias Winter
Stuttgart, Germany

Contents

Chapter 1
Introduction

1.1 Reaction-Diffusion Systems

We begin with a classification of reaction-diffusion systems. A two-component reaction-diffusion system in a bounded, smooth domain $\Omega \subset \mathbb{R}^n$, $n = 1, 2, \ldots$, is given by

$$u_t = \epsilon^2 \Delta u + f(u, v), \qquad \tau v_t = D\Delta v + g(u, v), \quad x \in \Omega. \tag{1.1}$$

We mostly consider Neumann (no-flux) boundary conditions

$$\frac{\partial u}{\partial v} = \frac{\partial v}{\partial v} = 0, \quad x \in \partial\Omega. \tag{1.2}$$

The unknowns $u = u(x, t)$ and $v = v(x, t)$ represent the concentrations of two biochemicals at a point $x \in \Omega$ and at a time $t > 0$. Further, $\Delta := \sum_{j=1}^{2} \frac{\partial^2}{\partial x_j^2}$ is the Laplace operator in $\mathbb{R}^n$; $v(x)$ is the outer normal vector at $x \in \partial\Omega$ and $\frac{\partial}{\partial v_x}$ denotes the normal derivative at $x \in \partial\Omega$ (for simplicity we will usually drop the index x); ϵ^2 and D are positive diffusion coefficients; $\tau \geq 0$ is a non-negative time relaxation constant. Note that for the special case $\tau = 0$ we have a mixed parabolic-elliptic system, otherwise it is of parabolic type. The functions $f(u, v)$ and $g(u, v)$ describe nonlinear reaction terms which will be specified below. The terms with second-order spatial derivatives in the system represent its diffusion part.

It is well-known that for two-component reaction-diffusion systems the Turing instability, which leads to pattern formation [232], can occur for exactly *two types* of systems. We first compute the Jacobian matrix J of the reaction-terms in (1.1) at a homogeneous, positive steady state (u_0, v_0) and get

$$J = \begin{pmatrix} \frac{\partial f}{\partial u}(u_0, v_0) & \frac{\partial f}{\partial v}(u_0, v_0) \\ \frac{\partial g}{\partial u}(u_0, v_0) & \frac{\partial g}{\partial v}(u_0, v_0) \end{pmatrix}.$$

J. Wei, M. Winter, *Mathematical Aspects of Pattern Formation in Biological Systems*,
Applied Mathematical Sciences 189, DOI 10.1007/978-1-4471-5526-3_1,
© Springer-Verlag London 2014

Then we introduce two types of sign combinations for the entries in the Jacobian matrix: Firstly, we define *Type 1*:

$$\begin{pmatrix} + & - \\ + & - \end{pmatrix}$$

which, swapping the unknown functions, is equivalent to

$$\begin{pmatrix} - & + \\ - & + \end{pmatrix}.$$

Secondly, we have *Type 2*:

$$\begin{pmatrix} + & + \\ - & - \end{pmatrix}$$

which, swapping the equations, is equivalent to

$$\begin{pmatrix} - & - \\ + & + \end{pmatrix}.$$

This provides a natural classification of the pattern-forming reaction-diffusion systems into two types. Now we consider some important examples of reaction-diffusion systems for both types.

Activator-inhibitor systems which have been suggested in (12) of [73] to have the reaction terms

$$f(u, v) = -u + \frac{u^p}{v^q}, \qquad g(u, v) = -v + \frac{u^r}{v^s},$$

$$\text{where } 1 < \frac{qr}{(p-1)(s+1)}, \quad 1 < p \tag{1.3}$$

(after rescaling) belong to *Type 1*. It is commonly assumed that there is a fast-diffusing inhibitor, v, which inhibits the production of a slowly-diffusing activator, u, and possibly itself. On the other hand, u activates itself and the inhibitor. Finally, both components decay due to the negative linear terms. This mechanism drives spatially localised activator peaks in combination with more shallow inhibitor peaks which are in the same location.

This activator-inhibitor system is now commonly called the *Gierer-Meinhardt system*. An interesting special case is given by the parameter choice $(p, q, r, s) = (2, 1, 2, 0)$. Frequently this case captures the typical behaviour for general exponents. Thus we will often restrict our attention to this case in order to simplify the notation and make the computations easier to follow.

Activator-substrate systems which are suggested in (11) of [73] to have the reaction terms

$$f(u, v) = -u + vu^2, \qquad g(u, v) = A - \mu v - vu^2, \quad \text{where } A > 0 \tag{1.4}$$

(after rescaling) are of *Type 2*. It is commonly assumed that there is a fast-diffusing substrate, v, which is consumed by a slowly-diffusing activator, u, and supplied to the system at a constant rate. On the other hand, u activates itself. Again, both components decay due to the negative linear terms. This mechanism drives spatially localised activator peaks coupled with more shallow substrate dips which are in the same location.

Particular examples of activator-substrate systems are (1.4) with $\mu = 1$ which is now commonly called the *Gray-Scott* system [75, 76] and $\mu = 0$ which is now commonly called the *Schnakenberg* model [214]. In both cases the reaction kinetics can be derived from simple chemical reactions using the mass balance law.

Note that there is a marked difference between activator-inhibitor and activator-substrate systems: At activator peaks the inhibitor forms peaks, whereas the substrate develops dips. This behaviour is easy to understand at an intuitive level, since in the first case high values of activator lead to strong activation of inhibitor which results in an inhibitor peak. However, in the second case high values of activator lead to fast consumption of substrate which causes a substrate dip. This different behaviour is clearly reflected in our analytical results and numerical simulations for multi-spot patterns.

For most of the monograph we will focus on activator-inhibitor systems. In particular, we will study the Gierer-Meinhardt system as a prototype. We will present analytical methods to give a rigorous treatment of the existence and stability of spiky patterns in one and two space dimensions (Chaps. 1–6). Many of our results can be extended to other systems.

In the later chapters we consider extensions of the standard Gierer-Meinhardt system which are all motivated by their important roles in modelling certain processes in developmental biology. We first study the Gierer-Meinhardt system with inhomogeneous coefficients in two specific situations: precursor gradients (smooth, spatially inhomogeneous coefficients) and piecewise constant inhibitor diffusivities (Chap. 7). We investigate the effect of finite activator diffusivities, study concentration phenomena for large reaction rates, explore the influence of the boundary conditions by changing them from Neumann to Robin type and analyse the Gierer-Meinhardt system on manifolds (Chap. 8). Then we consider the Gierer-Meinhardt system with saturation (Chap. 9). Next we study activator-substrate systems, represented by the models of Schnakenberg and Gray-Scott. For the former we also investigate flow-distributed spikes and show the destabilising effect of convection (Chap. 10). Finally, we consider systems with many components, namely the hypercycle of Eigen and Schuster for which we establish the result that it can only have stable spots if the number of components is small enough, a mutual exclusion model of Meinhardt for which we show that there are stable steady states for which spikes for different components are located in nearby positions, a model with multiple activators and substrates for which we derive novel stability conditions, and exotic spiky patterns for a consumer chain model for which we show the existence of a new type of spike cluster and show that it is stable (Chap. 11).

1.2 The Gierer-Meinhardt System for Hydra

The most fundamental phenomenon in pattern-forming activator-inhibitor systems is that a small deviation from spatial homogeneity has a strong positive feedback leading to its further increase. An example from erosion is the following: The more water collects at an initial depression, the faster the erosion proceeds, further deepening the depression. However, self-enhancement is not sufficient. On its own, it would lead to unlimited increase and spreading. Pattern formation requires in addition a long-distance limitation of the locally self-enhancing process which can result from a long-ranging inhibitor that spreads rapidly from such a region.

Gierer and Meinhardt formalised this observation and proposed a molecularly plausible model for pattern formation. The Gierer-Meinhardt model is a typical activator-inhibitor system and has now become one of the most famous models in biological pattern formation.

Gierer and Meinhardt used this model to explain the results of experiments on head regeneration and transplantation in the freshwater polyp *hydra*. It provides a favourable setting for biological experiments since its body parts regenerate quickly, often within hours. In the model it is assumed that the two biochemicals (or morphogens) called the activator and inhibitor are first produced by an external source, then they interact as represented by the nonlinear terms in the system coupled with the decay of both and diffusion at different rates. We will first consider the special case that these reaction and diffusion mechanisms are spatially homogeneous and later will relax these assumptions, for example some spatially inhomogeneous extensions will be considered in Chap. 7. In particular, these inhomogeneities together with the initial conditions are able to select and establish patterns more reliably than before. This model was able to explain the head formation in different situations by following the paradigm that activator peaks provide a signal to trigger change on a cellular level, leading to cell differentiation and finally deciding where a new head will form. A more detailed discussion of *hydra* experiments and their mathematical modelling will be given in Sect. 12.2. In our mathematical investigation we will particularly focus on spiky patterns whose peaks are located at interior or boundary points in a bounded smooth domain.

Thus we consider the Gierer-Meinhardt system (see [73]) as a prototype of a deterministic reaction-diffusion system which is able to model pattern formation. As such it is able to capture the essential behaviours encountered in developmental biology without being too complex to render possible a rigorous and explicit mathematical analysis. After a suitable rescaling, it can be stated as follows:

$$(GM) \quad \begin{cases} A_t = D_a \Delta A - A + \frac{A^2}{H}, & A > 0 \text{ in } \Omega, \\ \tau H_t = D_h \Delta H - H + A^2, & H > 0 \text{ in } \Omega, \\ \frac{\partial A}{\partial \nu} = \frac{\partial H}{\partial \nu} = 0 & \text{on } \partial\Omega. \end{cases} \quad (1.5)$$

The unknowns $A = A(x, t)$ and $H = H(x, t)$ represent the concentrations of two biochemicals, called the activator and inhibitor, respectively, at a point $x \in \Omega \subset \mathbb{R}^n$ and for a time $t > 0$; the positive diffusion constants D_a, D_h and the nonnegative time relaxation constant τ are all independent of $x \in \Omega$ and $t > 0$ with $D_a \ll D_h$, and τ is independent of D_a or D_h; we will also use the notation $D_a = \epsilon^2$ and $D_h = D$. We will consider space dimensions $n = 1$ and $n = 2$. We will mostly study Neumann type boundary conditions as in (1.5), but occasionally we will investigate other types as well.

The meaning of activator and inhibitor can clearly be seen from their effect in the reaction terms: The activator appears in the numerator of the reaction terms in both equations and thus increases the production of both activator and inhibitor. The inhibitor appears in the denominator of the reaction term in the first equation and thus decreases the production of activator. Further, both morphogens decay due to the negative linear terms. It is this type of activator-inhibitor interaction in combination with different diffusivities which leads to stable patterns in the Gierer-Meinhardt system.

The onset of pattern formation can be understood in the context of classical Turing instability [232] which explains how patterns arise from a homogeneous solution (see Sect. 1.3). Pattern formation can also be analysed by studying the existence and stability of the patterns themselves. This approach will be the focus of this research monograph. We will perform an analysis in the neighbourhood of certain inhomogeneous steady states, which are far away from homogeneous states, and rigorously prove their existence and (linearised) stability.

After introducing the system (1.5) to model the formation of new heads for *hydra*, A. Gierer and H. Meinhardt used it to study the orientation of the head and of arms/legs in embryonic growth. More recently, a huge variety of beautiful patterns on sea shells have been explained by it [145, 152, 155]. It can safely be said that the model has been successfully used to predict a huge variety of patterns and to elucidate the mechanisms of their formation in many different biological settings. Typically, in these applications, the problem is posed in a two- or three-dimensional domain (except in the study of sea shells which is considered as a system on a one-dimensional interval with the time axis corresponding to the direction of shell growth). The analysis for the one- and two-dimensional setting is considered in this monograph. The three-dimensional case leaves many challenges for future work.

1.3 Turing's Diffusion-Driven Instability

To start the discussion of the mathematical behaviour of (1.5), we first recall Turing's concept of diffusion-driven instability [232] which considers instabilities of a positive and spatially homogeneous steady state and requires (i) stability with respect to spatially homogeneous perturbations and (ii) instability with respect to some spatially inhomogeneous perturbations. The solution is then expected to follow the spatial instability and develop a spatially inhomogeneous profile which can explain the onset of pattern formation.

This behaviour fits into the general framework of pattern formation by symmetry breaking: We consider an initial state pattern which has a certain symmetry. This could be a steady state like here the spatially homogeneous steady state, but more generally we could begin with a spatially inhomogeneous steady state or even a spatio-temporal pattern. Then we require that (i) there is no instability in the symmetry class of the pattern. On the other hand, however, we assume that (ii) there is an instability which does not belong to the symmetry class of the pattern and therefore is less symmetric. It is expected that the solution follows close to this instability which leads to an explanation of the onset of a pattern which is less symmetric than the original one.

Now we use this important, well-known and popular concept and apply it to the Gierer-Meinhardt system (1.5). For other introductions to Turing instability we refer to [17, 166].

To understand the stability of homogeneous perturbations, we begin by studying the kinetic system associated with (1.5) which is obtained by dropping the diffusion terms:

$$\begin{cases} A_t = -A + \frac{A^2}{H}, \\ \tau H_t = -H + A^2. \end{cases} \tag{1.6}$$

Both the systems (1.5) and (1.6) have the unique positive homogeneous steady state $A \equiv 1$, $H \equiv 1$. Firstly, we consider the (linearised) stability of this steady state for the kinetic system (1.6) which is equivalent to the stability with respect to homogeneous perturbations for the full system (1.5). More precisely, for (1.6) and equivalently for (1.5), we study the initial value problem with initial condition

$$A(t) = 1 + \phi e^{\lambda t}, \qquad H(t) = 1 + \psi e^{\lambda t}, \tag{1.7}$$

where $(\phi, \psi) \in \mathbb{R}^2$ and ϕ, ψ are small and independent of the spatial variable. The linearisation of (1.6) around the steady state $(A, H) = (1, 1)$ gives

$$\begin{cases} \lambda \phi = \phi - \psi, \\ \lambda \psi = \frac{2}{\tau} \phi - \frac{1}{\tau} \psi. \end{cases} \tag{1.8}$$

The steady state $(1, 1)$ for system (1.6) is called (linearly) stable if $\mathrm{Re}(\lambda) < 0$ for both eigenvalues $\lambda \in \mathbb{C}$. The meaning of this criterion is that time-dependent, small perturbations in (1.7) are exponentially decreasing. In general, the parameter $\tau \geq 0$ will determine whether this condition is satisfied. To determine the range of τ for which this is the case, we recall the following criterion from linear algebra: For a real 2×2 matrix M, the eigenvalues are given by

$$\lambda_{1,2} = \frac{1}{2} \left(\mathrm{tr}(M) \pm \sqrt{\mathrm{tr}^2(M) - 4 \det(M)} \right).$$

This implies that both eigenvalues satisfy $\mathrm{Re}(\lambda_i) < 0$, $i = 1, 2$, if and only if $\mathrm{tr}(M) < 0$ and $\det(M) > 0$.

Applying this criterion to (1.8) with

$$M = \begin{pmatrix} 1 & -1 \\ \frac{2}{\tau} & -\frac{1}{\tau} \end{pmatrix},$$

we compute $\text{tr}(M) = 1 - \frac{1}{\tau}$ and $\det(M) = \frac{1}{\tau} > 0$. This implies that $\text{Re}(\lambda_i) < 0$ for both eigenvalues if $0 < \tau < 1$. For $\tau > 1$ both eigenvalues satisfy $\text{Re}(\lambda_i) > 0$. At $\tau = 1$ we have eigenvalues $\lambda_{1,2} = \pm\sqrt{-1}$ and there is a Hopf bifurcation. At this bifurcation point spatially homogeneous oscillations start to appear. To summarise, for $0 < \tau < 1$ the steady state $(A, H) = (1, 1)$ is stable under spatially homogeneous perturbations.

Secondly, to understand the (linearised) instability of the steady state $(A, H) = (1, 1)$ under spatially inhomogeneous perturbations, we consider the full Gierer-Meinhardt system (1.5). As initial condition we choose the perturbation

$$A(x, t) = 1 + \phi(x)e^{\lambda t}, \qquad H(x, t) = 1 + \psi(x)e^{\lambda t} \qquad (1.9)$$

around the steady state $(A, H) = (1, 1)$, where $\phi \in H_N^2(\Omega)$, $\psi \in H_N^2(\Omega)$ and $\|\phi\|_{H_N^2(\Omega)}$ and $\|\psi\|_{H_N^2(\Omega)}$ are both small. Here we have used the notation

$$H_N^2(\Omega) = \left\{ u \in H^2(\Omega) : \frac{\partial u}{\partial \nu} = 0 \text{ on } \partial\Omega \right\}.$$

(For the definitions of Lebesgue spaces, Sobolev spaces and their norms, we refer to Sect. 13.1 in the Appendix.)

We represent the perturbations ϕ and ψ as linear combinations of the eigenfunctions of the Laplace operator $-\Delta$ in Ω with Neumann boundary conditions. Then the instability is equivalent to instability with respect to any of these eigenfunctions. Thus we linearise the system (1.5) around the steady state $(1, 1)$ and get the eigenvalue problem

$$\begin{cases} \lambda\phi = -\epsilon^2 \mu\phi + \phi - \psi, \\ \lambda\psi = -\frac{D}{\tau}\mu\psi + \frac{2}{\tau}\phi - \frac{1}{\tau}\psi, \end{cases} \qquad (1.10)$$

where ϕ and ψ are both eigenfunctions with eigenvalue $\mu \geq 0$ of the Laplace operator on Ω with Neumann boundary conditions.

Next we compute the eigenvalues λ_i in terms of μ and determine their stability. We will get a Turing instability for some $0 < \tau < 1$ if there exists an eigenvalue λ of (1.10) with $\text{Re}(\lambda) > 0$ for some $\mu \leq 0$. This implies that we will have an instability of the steady state $(A, H) = (1, 1)$ for system (1.5) with spatially inhomogeneous perturbation but not with spatially homogeneous perturbations.

By the criterion from linear algebra given above, we compute for

$$M = \begin{pmatrix} 1 - \epsilon^2\mu & -1 \\ \frac{2}{\tau} & -\frac{1}{\tau} - \frac{D}{\tau}\mu \end{pmatrix}$$

that

$$\det(M) = \frac{1}{\tau} + \frac{\epsilon^2 - D}{\tau}\mu + \frac{\epsilon^2 D}{\tau}\mu^2,$$

$$\text{tr}(M) = 1 - \frac{1}{\tau} - \left(\epsilon^2 + \frac{D}{\tau}\right)\mu.$$

This implies that

$$\text{tr}(M) < 0 \quad \text{if } 0 < \tau < 1.$$

For $0 < \tau < 1$, the diffusion-driven instability arises if in addition $\det(M) < 0$ for some $\mu \geq 0$, where μ is an eigenvalue of the Laplace operator with Neumann boundary conditions, resulting in a positive eigenvalue λ. We determine

$$\min_{\mu \geq 0} \det(M) = \frac{1}{\tau} - \frac{1}{4\epsilon^2 D\tau}\left(D - \epsilon^2\right)^2 \quad \text{if } D > \epsilon^2$$

and

$$\min_{\mu \geq 0} \det(M) = \frac{1}{\tau} \quad \text{if } D \leq \epsilon^2.$$

Thus

$$\min_{\mu \geq 0} \det(M) < 0$$

if and only if

$$\frac{D}{\epsilon^2} > 3 + 2\sqrt{2}. \tag{1.11}$$

By the diagonal form of the diffusion matrix in system (1.5), it is clear that diffusion-driven instability is possible only if $\frac{D}{\epsilon^2} \neq 1$. The condition (1.11) specifies how much $\frac{D}{\epsilon^2}$ has to exceed unity for Turing instability to happen. Further, we need to ensure that for some eigenvalue $\mu \geq 0$ of the Laplace operator with Neumann boundary conditions we have $\det(M) < 0$. This will imply a further restriction for (1.11) and so Turing instability occurs only after $\frac{D}{\epsilon^2}$ has exceeded $3 + 2\sqrt{2}$ by a certain amount unless the parameter $\mu \geq 0$ for which $\det(M) = 0$ is initially achieved happens to be an eigenvalue of the Laplace operator with Neumann boundary conditions.

For the special case $\tau = 1$ the Gierer-Meinhardt system (1.5) has very interesting properties in terms of the instabilities just considered. The steady state has homogeneous eigenfunctions which are marginally stable (corresponding to the eigenvalues $\pm i$) and so are at the borderline to Hopf bifurcation, but, depending on the diffusion constants may also allow for spatially varying steady states. Therefore, depending on the initial conditions and the choice of diffusion constants, spatially homogeneous oscillations (due to Hopf bifurcation) or spatial patterns (due to Turing instability) or a combination of both are observed.

Turing's instability seems to contradict the commonly held mathematical intuition that diffusion is a smoothening and trivialising process. We remark that this behaviour is in fact widely observed for single partial differential equations. However, as we have seen, for certain systems of partial differential equations diffusion may destabilise homogeneous steady states and thus lead to the emergence of complex patterns.

It is important to point out that for the Gierer-Meinhardt system there is no variational structure and neither are energy methods available for the analysis. Similarly, the maximum principle is not valid either [168]. Therefore analytical methods have to be used which are independent of these properties, and thus for example the contraction mapping principle, fixed point methods or degree theory play a prominent role.

1.4 Amplitude Equations and Order Parameters

The diffusion-driven instability is able to explain how pattern formation is initiated. As the pattern develops but is still in a small neighbourhood of the homogeneous equilibrium $(A, H) = (1, 1)$, the dynamics of the solution can be described by amplitude equations, or, more generally, modulation equations, whose solutions are the order parameter functions. Here the behaviour of the system is considered beyond the linearisation, and so the influence of the nonlinear terms is taken into account in leading order but higher order nonlinear terms are neglected. In particular, as for Turing instability analysis, the system is considered *near equilibrium*. Using this approach it is possible to compute the interaction between multiple spatial eigenfunctions for the same critical eigenvalue. This approach has been applied to general two-component reaction diffusion systems by Kuramoto and Tsuzuki [122] and to (1.5) by Haken and Olbricht [84]. The latter authors show that the interaction of critical eigenfunctions is able to produce stripy, rectangular or hexagon-like patterns. In [13] this analysis using amplitude equations has been extended to study the Gierer-Meinhardt system on a sphere. A comprehensive account and revision of amplitude equation methods applied to pattern formation problems arising in such diverse areas as fluid mechanics, biology and many other fields is given in [34]. They consider the main theoretical approaches including different classes of amplitude equations, e.g. for stationary periodic, oscillatory uniform or oscillatory periodic patterns. Further they study phase equations which describe spatial modulation and are valid not only near threshold but also far away from it. Elements of bifurcation theory and the effects of boundary conditions are included in their treatment. This is complemented by pattern selection, both in the linear and nonlinear setting, and pathways to chaos. Rigorous justification of the approximation of the full solution by the corresponding modulation equations on unbounded domains has been shown by Mielke and Schneider [157, 158].

1.5 Analytical Methods for Spiky Patterns

In this monograph we consider the limit in which the activator diffusivity is "small enough" and the inhibitor diffusivity is of order unity (in one space dimension) or "large enough" (in two space dimensions). This condition implies that (1.11) holds, so we are within the parameter regime of Turing instability (if the time rate constant

τ in (1.5) or other systems is small enough). We study spiky patterns for which the activator peaks have very small diameter (of order square root of activator diffusivity) and the profile of the inhibitor has a spatial scale of order unity and can often be described by certain Green's functions. By a suitable scaling of kinetic constants in (1.5) we can achieve that the amplitudes are of order unity. In particular, the patterns considered are *far-from-equilibrium* steady states. The problem has been studied in a mathematically rigorous way and local stability of bifurcation solutions has been proved in [159, 225] using the classical results on bifurcation theory by Crandall and Rabinowitz [30, 31].

We will show their existence in a rigorous way. The main strategy is as follows: We first make a good guess of the spiky steady state: For the activator we take the sum of finitely many spikes with given profile but arbitrary amplitudes and positions. By a leading order calculation we compute the amplitudes first. The positions are determined as part of the existence proof: We first reduce the existence problem of spiky patterns to finite dimensions by applying the Liapunov-Schmidt reduction method. Then we solve a finite-dimensional reduced problem which effectively determines the positions of the spikes.

For the stability we use the approach of nonlocal eigenvalue problems (NLEPs) which are derived rigorously by taking the limit of a rescaled eigenvalue problem and analysed using bilinear forms. The NLEP method implies results on the stability or instability due to large eigenvalues. Since there are only finitely many small eigenvalues, we have reduced the stability problem to finite dimensions. Thus stability results can be derived explicitly by computing the eigenvalues and eigenvectors of certain matrices directly.

The NLEP approach to stability has previously been considered by Doelman, Eckhaus, Gardner and Kaper [47–49, 51] and applied to essentially one-dimensional settings (real line, radially symmetric, axial symmetry). We extend it to general partial differential equations without any symmetry assumptions, only requiring certain smoothness assumptions for the boundary.

An alternative approach to stability is the singular limit eigenvalue problem (SLEP) method introduced and applied by Nishiura and Fujii [183]. To summarise their strengths and weaknesses, the formulation of the NLEP problem is more explicit than SLEP and can be treated rather transparently using bilinear forms. On the other hand, SLEP is able to treat small eigenvalues directly, whereas for NLEP they have to be studied separately in a second step. This makes NLEP well suited for problems for which there are only finitely many small eigenvalues such as in the analysis of spikes (i.e. concentrations at points) but less so for problems which involve concentrations or sharp layers along curves, surfaces and other finite-dimensional geometric objects.

Convergence of the SLEP method for bounded domains has been shown by Nishiura and Fujii [183]. Uniform convergence of the SLEP method for unbounded domains has been derived by Taniguchi [228].

Often our analytical results have been confirmed by numerical computations based on the finite-element method. In these computations we have chosen parameters within the same regime as required for the theoretical results, usually some of the diffusivities are very small and the other parameters are of order unity.

Fig. 1.1 A symmetric
two-spike steady state is
displayed (activator *A solid
line*, inhibitor *H dashed line*)
for parameters $\epsilon = 0.02$ and
$D = 0.10$

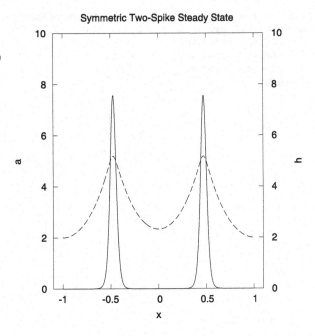

Fig. 1.2 A symmetric
five-spike steady state is
displayed (activator *A solid
line*, inhibitor *H dashed line*)
for parameters $\epsilon = 0.02$ and
$D = 0.04$

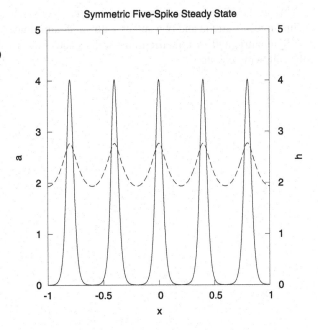

We conclude the introduction by presenting some figures of multiple spike patterns. Figures 1.1 and 1.2 show symmetric two-spike and five-spike steady states for
the reaction-diffusion system (1.5). Here all spikes have the same amplitude.

Fig. 1.3 An asymmetric four-spike steady state is displayed (activator *A solid line*, inhibitor *H dashed line*) for parameters $\epsilon = 0.02$ and $D = 0.0465$

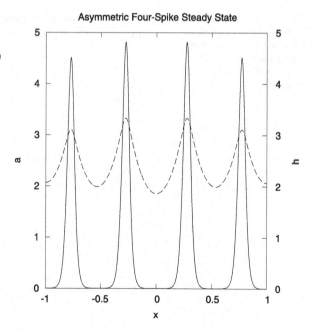

Asymmetric Four-Spike Steady State

Figure 1.3 shows an asymmetric four-spike steady state consisting of two spikes with large amplitude and two spikes with small amplitude.

Symmetric and asymmetric multi-spike steady states will be studied in detail in the following chapters.

Chapter 2
Existence of Spikes for the Gierer-Meinhardt System in One Dimension

In this chapter we give a full account of the existence of multiple spikes for the Gierer-Meinhardt system in an interval on the real line. Without loss of generality, we assume that this interval is $(-1, 1)$. We will construct the solution rigorously by (i) reducing the problem to finite dimensions by applying the Liapunov-Schmidt reduction and (ii) employing a fixed-point argument (e.g. using the mapping degree) to solve this finite-dimensional problem. Here we present a simplification in the case of two spikes. The time-dependent problem consists of a two-component reaction-diffusion system of activator-inhibitor type with no-flux (Neumann) boundary conditions and can be stated as follows:

$$\begin{cases} A_t = \epsilon^2 A'' - A + \frac{A^2}{H}, & x \in (-1, 1), t > 0, \\ \tau H_t = D H'' - H + A^2, & x \in (-1, 1), t > 0, \\ A'(\pm 1, t) = H'(\pm 1, t) = 0, & t > 0. \end{cases} \tag{2.1}$$

This system has three parameters, namely two diffusivities and one time relaxation constant which satisfy

$$0 < \epsilon \ll 1, \qquad 0 < D < \infty, \qquad \tau \geq 0.$$

Throughout this chapter, we assume that

$$D \text{ and } \tau \text{ are real constants which are independent of } \epsilon.$$

The corresponding stationary problem is given by

$$\begin{cases} \epsilon^2 A'' - A + \frac{A^2}{H} = 0, & x \in (-1, 1), \\ D H'' - H + A^2 = 0, & x \in (-1, 1), \\ A'(\pm 1) = H'(\pm 1) = 0. \end{cases} \tag{2.2}$$

Before giving a full discussion of the problem, let us summarise what we mean by a spike. It is a pattern which is narrowly concentrated near a point in the domain

J. Wei, M. Winter, *Mathematical Aspects of Pattern Formation in Biological Systems*,
Applied Mathematical Sciences 189, DOI 10.1007/978-1-4471-5526-3_2,
© Springer-Verlag London 2014

which is characterised by its profile, amplitude and position. The pattern is observed in the activator component and the inhibitor component plays a stabilising role. A multi-spike steady state consisting of N spikes satisfies

$$a_\epsilon \sim \xi_{i,\epsilon} w\left(\frac{x - t_i^\epsilon}{\epsilon}\right) \quad \text{as } \epsilon \to 0, i = 1, \ldots, N,$$

where the profile function w is the unique solution of the problem

$$\begin{cases} w'' - w + w^2 = 0 & \text{in } \mathbb{R}, \\ w > 0, \quad w(0) = \max_{y \in \mathbb{R}} w(y), \\ w(y) \to 0 & \text{as } |y| \to \infty \end{cases} \tag{2.3}$$

and the notation $A(\epsilon) \sim B(\epsilon)$ means that $\lim_{\epsilon \to 0} \frac{A(\epsilon)}{B(\epsilon)} = c_0 > 0$, for some positive number c_0. Existence and uniqueness have been proved in Sect. 13.2. Below we will first compute the amplitudes $\xi_{i,\epsilon}$. We will also determine the positions t_i^ϵ of the spike centres which will approach certain limiting locations t_i^0 as $\epsilon \to 0$.

Problem (2.2) has been studied by numerous authors. Let us first recall several important results on the formation of spiky patterns.

2.1 Symmetric Multi-spike Solutions: A Rigorous Proof of Existence

I. Takagi [226] first established the existence of N-spike steady-state solutions with maxima located exactly at

$$t_j^0 = -1 + \frac{2j - 1}{N}, \quad j = 1, \ldots, N, \tag{2.4}$$

for $\epsilon \ll 1$ and $\epsilon^2 \ll D$. These solutions are periodic and they are obtained by first constructing a single spike on a smaller interval and then using periodic continuation to extend the solution to multiple spikes on the original domain. We call them *symmetric N-spike solutions* since all the spikes have the same amplitudes, and in this special case they are exact copies of each other. Takagi's proof uses the implicit function theorem to construct single spikes. The implicit function theorem is applied in a suitable functional space of even functions. The argument of the existence proof is very elegant since the linearised operator around a suitable approximate solution restricted to this space of even functions is invertible, see Sect. 3.1. These solutions are close to the approximate solution

$$A \sim \xi_\epsilon w\left(\frac{x}{\epsilon}\right), \qquad H(0) = \xi_\epsilon$$

for suitable amplitude ξ_ϵ and given shape function w solving (2.3). Note that w is given by

$$w(y) = \frac{3}{2\cosh^2(y/2)}. \tag{2.5}$$

Therefore single-spike solutions are even functions defined on the interval $(-\frac{1}{N}, \frac{1}{N})$ with the boundary conditions

$$A'\left(-\frac{1}{N}\right) = A'\left(\frac{1}{N}\right) = H'\left(-\frac{1}{N}\right) = H'\left(\frac{1}{N}\right) = 0.$$

Because of these properties, the periodic continuation of the solution from the interval $(-\frac{1}{N}, \frac{1}{N})$ to the whole domain $(-1, 1)$ is a symmetric multi-spike solution.

2.2 Asymmetric Multi-spike Solutions: A Formal Derivation

Using matched asymptotic expansions, Ward and Wei in [240] showed that for any given positive integer N and under the condition $D < D_N$, where the sequence $D_1 > D_2 > \cdots > D_N > \cdots > 0$ has been stated explicitly, problem (2.2) has *asymmetric N-spike solutions* if ϵ is small enough. These asymmetric solutions are generated by two different types of spikes which we call type **A** or type **B**, respectively. For any given order, e.g.

ABAABBB...ABBBA...B,

there is a corresponding asymmetric N-spike solution such that the two types of spikes follow this order. In each of the resulting subintervals the two types of spikes are given asymptotically by the formula

$$A \sim \xi_{i,\epsilon} w\left(\frac{x - t_i^\epsilon}{\epsilon}\right), \qquad H\left(t_i^\epsilon\right) = \xi_{i,\epsilon}, \tag{2.6}$$

where t_i^ϵ is the centre of the spike, $\xi_{i,\epsilon}$ its amplitude and w, given by (2.5), is its shape function. Here ξ_i can either be small (for type **A** spikes) or large (for type **B** spikes). It is also seen that for the small spikes the subinterval is small and for the large spikes the subinterval is large. Further, the small spikes are exact copies of each other as are the large spikes.

First we start from (2.2) in a small interval $(-l, l)$: Let A and H be even functions in the set (compare Sect. 13.1)

$$H_N^2(-l, l) := \left\{v \in H^2(-1, 1) : v'(-1) = v'(1) = 0\right\}$$

satisfying

$$\begin{cases} \epsilon^2 A'' - A + \frac{A^2}{H} = 0 & \text{in } (-l, l), \\ DH'' - H + A^2 = 0 & \text{in } (-l, l), \\ A(x) > 0, \quad H(x) > 0 & \text{in } (-l, l). \end{cases} \tag{2.7}$$

Consider the single-spike solution which was constructed by I. Takagi [226]. By some simple computations based on (2.13) below, it follows that we have

$$H(l) = c(D)b\left(\frac{l}{\sqrt{D}}\right) + o(1) \quad \text{as } \epsilon \to 0, \tag{2.8}$$

where $c(D)$ is some positive constant depending on D only and the function $b(z)$ is given by

$$b(z) := \frac{\sinh(z)}{\cosh^2(z)}. \tag{2.9}$$

We note that $b(0) = \lim_{z \to \infty} b(z) = 0$ and b has a unique maximum for which $\cosh^2 z_{\max} = 2$ and $b(z_{\max}) = \frac{1}{2}$.

The approach now is to fix l and find a positive number $\bar{l}$ such that

$$b\left(\frac{l}{\sqrt{D}}\right) = b\left(\frac{\bar{l}}{\sqrt{D}}\right), \quad 0 < l < \bar{l} < 1, \tag{2.10}$$

which will imply that $H(l) = H(\bar{l}) + o(1)$. This shows that if there exists a solution to (2.10), then $H(l)$ and $H(-\bar{l})$ can be matched. Using the fact that w decays exponentially, solutions of (2.7) in different subintervals can be connected.

It turns out that for D small enough (2.10) is always solvable. Now (2.10) has to be solved together with the following constraint to fit N spikes into the interval $(-1, 1)$:

$$N_1 l + N_2 \bar{l} = 1, \quad N_1 + N_2 = N. \tag{2.11}$$

For a solution l of (2.10) and (2.11) and $j = 1, \ldots, N$ we set

$$l_j = l \quad \text{or} \quad l_j = \bar{l}, \tag{2.12}$$

where the order of the l's and $\bar{l}$'s can be chosen arbitrarily under the constraint that the number of j's such that $l_j = l$ is N_1 and the number of j's such that $l_j = \bar{l}$ is N_2.

Finally, we compute $t_j^0 = \lim_{\epsilon \to 0} t_j^\epsilon, j = 1, \ldots, N$, so that

$$t_{j+1}^0 - t_j^0 = l_j + l_{j+1}, \quad j = 1, \ldots, N - 1$$

and $t_1^0 = -1 + l_1, t_N^0 = 1 - l_N$. This concludes the formal construction of asymmetric multi-spike solutions.

2.3 Existence of Symmetric and Asymmetric Multiple Spikes: A Unified Rigorous Approach

In this section and Chap. 4, we present a unified approach to a rigorous theoretical treatment of the existence and stability of general N-spike (symmetric or asymmetric) solutions for the Gierer-Meinhardt system (2.2) on the interval $(-1, 1)$. The existence proof firstly uses Liapunov-Schmidt reduction to deduce a finite-dimensional problem and secondly a fixed-point argument (e.g. using the mapping degree) to solve the finite-dimensional problem.

The stability is shown by first separating the problem into the case of large eigenvalues which tend to a nonzero limit and the case of small eigenvalues which tend to zero as $\epsilon \to 0$. Large eigenvalues are then explored by studying nonlocal eigenvalue problems. Small eigenvalues are calculated explicitly by an asymptotic analysis with rigorous error estimates. It turns out that for the case of symmetric N-spike solutions the instability always arises first from the small eigenvalues.

Finally, in Sect. 2.4 we state results on the existence of multiple clusters for (2.2) for which different spikes may approach the same point.

We note also that in [50] an alternative dynamical systems approach is used to study the stability of symmetric spikes.

Before stating our main results, we introduce some notation. Let $L^2(-1, 1)$ and $H^2(-1, 1)$ be the usual Lebesgue and Sobolev spaces (see Sect. 13.1). Let $\Omega = (-1, 1)$ and $G_D(x, z)$ be the following Green's function:

$$\begin{cases} DG_D''(x, z) - G_D(x, z) + \delta_z(x) = 0 & \text{in } (-1, 1), \\ G_D'(-1, z) = G_D'(1, z) = 0. \end{cases} \tag{2.13}$$

We can calculate explicitly

$$G_D(x, z) = \begin{cases} \frac{\theta}{\sinh(2\theta)} \cosh[\theta(1 + x)] \cosh[\theta(1 - z)], & -1 < x < z < 1, \\ \frac{\theta}{\sinh(2\theta)} \cosh[\theta(1 - x)] \cosh[\theta(1 + z)], & -1 < z < x < 1, \end{cases} \tag{2.14}$$

where

$$\theta = \frac{1}{\sqrt{D}}. \tag{2.15}$$

We set

$$K_D(|x - z|) = \frac{1}{2\sqrt{D}} e^{-(1/\sqrt{D})|x-z|} \tag{2.16}$$

to be the non-smooth part of $G_D(x, z)$, and we define the regular part H_D of G_D by $H_D = K_D - G_D$. Note that G_D is C^∞ for $(x, z) \in \Omega \times \Omega \setminus \{x = z\}$ and H_D is C^∞ for all $(x, z) \in \Omega \times \Omega$.

We use the abbreviation e.s.t. to denote an exponentially small term of order $O(e^{-d/\epsilon})$ for some $d > 0$ in a suitable norm. By C we denote a generic constant which may change from line to line.

For simplicity of presentation, we will assume that the number of spikes is given by

$$N = 2. \tag{2.17}$$

Let $\mathbf{t}^0 = (t_1^0, t_2^0)$ be fixed, where $-1 < t_1^0 < t_2 < 1$ are two different points in $(-1, 1)$. Using the unique solution w of (2.3), we set

$$\xi_\epsilon := \left(\epsilon \int_{\mathbb{R}} w^2(z) dz \right)^{-1} \tag{2.18}$$

and define

$$\hat{\xi}_i = \xi_{i,\epsilon} (\xi_\epsilon)^{-1}. \tag{2.19}$$

Next we introduce several matrices for later use: Let

$$\mathcal{G}_D(\mathbf{t}) = \big(G_D(t_i, t_j) \big), \tag{2.20}$$

where the Green's function G_D has been defined in (2.14) and $\mathbf{t} = (t_1, t_2) \in (-1, 1)^2$ with $-1 < t_1 < t_2 < 1$ is arbitrary. Then we introduce the matrix

$$\mathcal{B} = (b_{ij}), \quad \text{where } b_{ij} = G_D(t_i^0, t_j^0) \hat{\xi}_j^0. \tag{2.21}$$

Remark 2.1 Since the matrix $\mathcal{B}$ is of the form $\mathcal{G}_D \mathcal{D}$, where $\mathcal{G}_D$ is a symmetric and $\mathcal{D}$ is a diagonal matrix, the eigenvalues of $\mathcal{B}$ are real.

Next we compute the partial derivatives of $G_D(t_i, t_j)$. Recall that

$$G_D(t_i, t_j) = K_D\big(|t_i - t_j|\big) - H_D(t_i, t_j).$$

For $i \neq j$, we define

$$\nabla_{t_i} G(t_i, t_j) := \frac{\partial}{\partial x} G(x, t_j) \bigg|_{x=t_i^0}.$$

For $i = j$, we have that $K_D(|t_i - t_j|) = K_D(0) = \frac{1}{2\sqrt{D}}$ is a constant and we set

$$\nabla_{t_i} G_D(t_i, t_i) = -\nabla_{t_i} H_D(t_i, t_i) := -\frac{\partial}{\partial x} H_D(x, t_i) \bigg|_{x=t_i}.$$

Similarly, we define

$$\nabla_{t_i} \nabla_{t_j} G_D(t_i, t_j) = \begin{cases} -\frac{\partial}{\partial x}\big|_{x=t_i} \frac{\partial}{\partial y}\big|_{y=t_i} H_D(x, y) & \text{if } i = j, \\ \frac{\partial}{\partial x}\big|_{x=t_i} \frac{\partial}{\partial y}\big|_{y=t_j} G_D(x, y) & \text{if } i \neq j. \end{cases} \tag{2.22}$$

Then the following two matrices of first and second order derivatives of $\mathcal{G}_D$ can be introduced:

$$\nabla \mathcal{G}_D(\mathbf{t}) = \big(\nabla_{t_i} G_D(t_i, t_j) \big), \qquad \nabla^2 \mathcal{G}_D(\mathbf{t}) = \big(\nabla_{t_i} \nabla_{t_j} G_D(t_i, t_j) \big). \tag{2.23}$$

In order to guarantee the existence of two-spike solutions, we make the following three assumptions. The first two assumptions will ensure that we can find suitable amplitudes for the spikes. This can be seen by the following leading-order computation: Substituting (2.6) into (2.2) and using (2.19), we compute

$$\hat{\xi}_i \sim \sum_{j=1}^{2} G(t_i^\epsilon, t_j^\epsilon)(\hat{\xi}_j)^2$$

which gives condition (H1). Then the nondegeneracy with respect to the amplitudes $(\hat{\xi}_1, \hat{\xi}_2)$ is given by

$$\det\left(\nabla_{(\hat{\xi}_1, \hat{\xi}_2)} \left(\sum_{j=1}^{2} G(t_i^\epsilon, t_j^\epsilon)(\hat{\xi}_j)^2 - \hat{\xi}_i \right)_{i=1,2} \right) \neq 0$$

which implies condition (H2). We now state these two conditions.

(H1) *For given* $\mathbf{t}^0 \in (-1, 1)^2$ *with* $-1 < t_1^0 < t_2^0 < 1$, *there exists a solution* $(\hat{\xi}_1^0(\mathbf{t}), \hat{\xi}_2^0(\mathbf{t}))$ *of the equation*

$$\sum_{j=1}^{2} G_D(t_i^0, t_j^0)(\hat{\xi}_j^0)^2 = \hat{\xi}_i^0, \quad i = 1, 2. \tag{2.24}$$

(H2) *We have*

$$\frac{1}{2} \notin \sigma(\mathcal{B}), \tag{2.25}$$

where $\sigma(\mathcal{B})$ *is the set of eigenvalues of* $\mathcal{B}$.

Applying the implicit function theorem to equation (2.24) for $\mathbf{t} = (t_1, t_2)$ in a small neighbourhood of $\mathbf{t}^0 = (t_1^0, t_2^0)$, the linearised operator is invertible by assumption (H2). Thus there exists a unique solution $\hat{\boldsymbol{\xi}}(\mathbf{t}) = (\hat{\xi}_1(\mathbf{t}), \hat{\xi}_2(\mathbf{t}))$ of the equation

$$\sum_{j=1}^{2} G_D(t_i, t_j)\hat{\xi}_j^2 = \hat{\xi}_i, \quad i = 1, 2, \tag{2.26}$$

if $\mathbf{t}$ is sufficiently close to $\mathbf{t}^0$.

Our final assumption assures us that we will be able to choose suitable positions for the spikes. It is stated in terms of a vector field $F(\mathbf{t})$ which is defined by

$$F(\mathbf{t}) = \big(F_1(\mathbf{t}), F_2(\mathbf{t}) \big),$$

where

$$F_i(\mathbf{t}) = \sum_{l=1}^{2} \big(\nabla_{t_i} G_D(t_i, t_l)\big)\hat{\xi}_l^2$$

$$= -\big(\nabla_{t_i} H_D(t_i, t_i)\big)\hat{\xi}_i^2 + \big(\nabla_{t_i} G_D(t_i, t_{3-i})\big)\hat{\xi}_{3-i}^2, \quad i = 1, 2, \tag{2.27}$$

and a matrix $\mathcal{M}(\mathbf{t})$ which is given by

$$\mathcal{M}(\mathbf{t}) = \big(\hat{\xi}_i^{-1}\nabla_{t_j} F_i(\mathbf{t})\big). \tag{2.28}$$

In Sect. 3.4 we will see how these expressions for $F(\mathbf{t})$ and $\mathcal{M}(\mathbf{t})$ will naturally appear in an explicit calculation. We now state our third condition:

(H3) *We assume that at the point* $\mathbf{t}^0 = (t_1^0, t_2^0)$ *given in* (H1) *we have*

$$F(\mathbf{t}^0) = 0, \tag{2.29}$$

$$\det\big(\mathcal{M}(\mathbf{t}^0)\big) \neq 0. \tag{2.30}$$

Let us now calculate the matrix $\mathcal{M}(\mathbf{t})$. We first compute the derivatives of $\hat{\boldsymbol{\xi}}(\mathbf{t})$. Using (2.26), it follows that locally around $\mathbf{t}^0$ the function $\hat{\boldsymbol{\xi}}(\mathbf{t})$ is C^1 and we can calculate

$$\nabla_{t_j}\hat{\xi}_i = 2\sum_{l=1}^{2} G_D(t_i, t_l)\hat{\xi}_l \nabla_{t_j}\hat{\xi}_l + \sum_{l=1}^{2}\left(\frac{\partial}{\partial t_j} G_D(t_i, t_l)\right)\hat{\xi}_l^2.$$

For $i \neq j$, we have

$$\nabla_{t_j}\hat{\xi}_i = 2\sum_{l=1}^{2} G_D(t_i, t_l)\hat{\xi}_l \nabla_{t_j}\hat{\xi}_l + \nabla_{t_j} G_D(t_i, t_j)\hat{\xi}_j^2,$$

where

$$\nabla_{t_j} G_D(t_i, t_j) = \frac{\partial}{\partial t_j} G_D(t_i, t_j).$$

For $i = j$, we get

$$\nabla_{t_i}\hat{\xi}_i = 2\sum_{l=1}^{2} G_D(t_i, t_l)\hat{\xi}_l \nabla_{t_i}\hat{\xi}_l + \sum_{l=1}^{2}\frac{\partial}{\partial t_i}\big(G_D(t_i, t_l)\big)\hat{\xi}_l^2$$

$$= 2\sum_{l=1}^{2} G_D(t_i, t_l)\hat{\xi}_l \nabla_{t_i}\hat{\xi}_l + \nabla_{t_i} G_D(t_i, t_i)\hat{\xi}_i^2 + \sum_{l=1}^{2}\nabla_{t_i} G_D(t_i, t_l)\hat{\xi}_l^2.$$

Note that by (H3) we have

$$\sum_{l=1}^{2} \nabla_{t_i} G_D(t_i, t_l) \hat{\xi}_l^2 = 0.$$

Then, recalling that

$$\left(\nabla_{t_j} G_D(t_i, t_j) \right) = \left(\nabla \mathcal{G}_D(\mathbf{t}) \right)^T,$$

setting

$$\mathcal{H}(\mathbf{t}) = \left(\hat{\xi}_i(\mathbf{t}) \delta_{ij} \right) \tag{2.31}$$

and using matrix notation

$$\nabla \xi(\mathbf{t}) = \left(\nabla_{t_j} \hat{\xi}_i(\mathbf{t}) \right), \tag{2.32}$$

we have

$$\nabla \xi(\mathbf{t}) = \left(I - 2\mathcal{G}_D(\mathbf{t}) \mathcal{H}(\mathbf{t}) \right)^{-1} \left(\nabla \mathcal{G}_D(\mathbf{t}) \right)^T \left(\mathcal{H}(\mathbf{t}) \right)^2. \tag{2.33}$$

Let

$$\mathcal{Q} = (q_{ij}) = \left(\left(-\frac{\theta^2}{\hat{\xi}_i} + \frac{\theta^3}{2} \right) \delta_{ij} \right). \tag{2.34}$$

Using (2.33), we now compute $\mathcal{M}(\mathbf{t})$. First note that for $i \neq j$ we have

$$\left(\sum_{l=1}^{2} \frac{\partial}{\partial t_j} \nabla_{t_i} G_D(t_i, t_l) \right) \hat{\xi}_l^2 = \left(\nabla_{t_i} \nabla_{t_j} G_D(t_i, t_j) \right) \hat{\xi}_j^2$$

and for $i = j$ we get

$$\left(\sum_{l=1}^{2} \frac{\partial}{\partial t_i} \nabla_{t_i} G_D(t_i, t_l) \right) \hat{\xi}_l^2$$

$$= \left(\frac{\partial^2}{\partial t_i^2} G_D(t_i, t_{3-i}) \right) \hat{\xi}_{3-i}^2 - \frac{\partial^2}{\partial x^2} \bigg|_{x=t_i} H_D(x, t_i) \hat{\xi}_i^2 + \left(\nabla_{t_i} \nabla_{t_i} G_D(t_i, t_i) \right) \hat{\xi}_i^2$$

$$= \theta^2 \sum_{l=1}^{2} G_D(t_i, t_l) \hat{\xi}_l^2 - \theta^2 K_D(0) \hat{\xi}_i^2 + \nabla_{t_i} \nabla_{t_i} G_D(t_i, t_i) \hat{\xi}_i^2$$

$$= \theta^2 \hat{\xi}_i - \frac{\theta^3}{2} \hat{\xi}_i^2 + \nabla_{t_i} \nabla_{t_i} G_D(t_i, t_i) \hat{\xi}_i^2$$

since

$$\frac{\partial^2}{\partial t_i^2} G_D(t_i, t_{3-i}) - G_D(t_i, t_{3-i}) = 0, \qquad \frac{\partial^2}{\partial x^2} \bigg|_{x=t_i} H_D(t_i, t_i) - H_D(t_i, t_i) = 0.$$

In vector notation we get

$$\mathcal{M}(t) = \left(\mathcal{H}(t)\right)^{-1}\left(\nabla^2\mathcal{G}_D(t) - \mathcal{Q}\right)\left(\mathcal{H}(t)\right)^2$$

$$+ 2\left(\mathcal{H}(t)\right)^{-1}\nabla\mathcal{G}_D(t)\mathcal{H}(t)\left(I - 2\mathcal{G}_D(t)\mathcal{H}(t)\right)^{-1}\left(\nabla\mathcal{G}_D(t)\right)^T\left(\mathcal{H}(t)\right)^2$$

$$= \left(\mathcal{H}(t)\right)^{-1}\left[\nabla^2\mathcal{G}_D(t) - \mathcal{Q} + 2\nabla\mathcal{G}_D(t)\mathcal{H}(t)\left(I - 2\mathcal{G}_D(t)\mathcal{H}(t)\right)^{-1}\left(\nabla\mathcal{G}_D(t)\right)^T\right]$$

$$\times \left(\mathcal{H}(t)\right)^2. \tag{2.35}$$

Our first result can be stated as follows:

Theorem 2.2 *Assume that* (H1), (H2) *and* (H3) *are satisfied. Then for* $\epsilon \ll 1$, *problem* (2.2) *has an N-spike solution which satisfies in the limit* $\epsilon \to 0$:

$$A_\epsilon(x) = \sum_{j=1}^{N} \xi_\epsilon \hat{\xi}_j^0 w\left(\frac{x - t_j^\epsilon}{\epsilon}\right) + o(1) \quad \text{in } H_N^2(-1, 1), \tag{2.36}$$

$$H_\epsilon\left(t_i^\epsilon\right) \sim \xi_\epsilon \hat{\xi}_i^0 + o(1), \quad i = 1, \ldots, N, \tag{2.37}$$

$$t_i^\epsilon \to t_i^0, \quad i = 1, \ldots, N. \tag{2.38}$$

Theorem 2.2 for the case $N = 2$ will be proved in the following subsections.

Remark 2.3 In the case of symmetric N-spike solutions, conditions (H2) and (H3) are not needed for the existence proof since in the construction of solutions one can restrict the function space to the class of symmetric functions (see for example [226]). Note that then for all ϵ small enough the spikes are exact copies of each other and thus they are placed equidistantly. For use in the stability proof the three assumptions (H1), (H2) and (H3) will be computed in Sect. 4.1.3 in the case of symmetric spikes.

Remark 2.4 Our results will provide a *rigorous proof* for the existence and stability of asymmetric N-spike solutions which consist of spikes with different amplitudes after the three assumptions (H1), (H2) and (H3) have been verified.

2.3.1 Some Preliminaries

In this subsection, we consider the following vectorial linear operator:

$$L\Phi := \Phi'' - \Phi + 2w\Phi - 2B\frac{\int_{\mathbb{R}} w\Phi dy}{\int_{\mathbb{R}} w^2 dy}w^2, \tag{2.39}$$

where $\mathcal{B}$ is given by (2.21) and

$$
\Phi = \begin{pmatrix} \phi_1 \\ \phi_2 \\ \vdots \\ \phi_N \end{pmatrix} \in \left(H^2(\mathbb{R}) \right)^N.
$$

In this subsection, we consider a general positive integer N for later use. Set

$$
L_0 u := u'' - u + 2wu, \quad \text{where } u \in H^2(\mathbb{R}). \tag{2.40}
$$

Then, using Remark 2.1, the conjugate operator of L under the scalar product in $L^2(\mathbb{R})$ is given by

$$
L^* \Psi = \Psi'' - \Psi + 2w\Psi - 2\mathcal{B}^T \frac{\int_{\mathbb{R}} w^2 \Psi \, dy}{\int_{\mathbb{R}} w^2 \, dy} w, \tag{2.41}
$$

where

$$
\Psi = \begin{pmatrix} \psi_1 \\ \psi_2 \\ \vdots \\ \psi_N \end{pmatrix} \in \left(H^2(\mathbb{R}) \right)^N.
$$

We obtain the following result.

Lemma 2.5 *Assume that* (H2) *holds. Then*

$$
\mathrm{Ker}(L) = X_0 \oplus X_0 \oplus \cdots \oplus X_0, \tag{2.42}
$$

where

$$
X_0 = \mathrm{span}\{w'(y)\}
$$

and

$$
\mathrm{Ker}(L^*) = X_0 \oplus X_0 \oplus \cdots \oplus X_0. \tag{2.43}
$$

Proof Let us first prove (2.42). Suppose that

$$
L\Phi = 0.
$$

We diagonalise $\mathcal{B}$ so that

$$
P^{-1} \mathcal{B} P = J,
$$

where P is an orthogonal matrix. Note that by Remark 2.1 J has diagonal form, i.e.,

$$J = \begin{pmatrix} \sigma_1 & & & 0 \\ & \sigma_2 & & \\ & & \ddots & \\ 0 & & & \sigma_N \end{pmatrix}$$

with suitable real numbers σ_j for $j = 1, \ldots, N$. Defining

$$\Phi = P\tilde{\Phi},$$

we have

$$\tilde{\Phi}'' - \tilde{\Phi} + 2w\tilde{\Phi} - 2\frac{\int_{\mathbb{R}} w(J\tilde{\Phi})dy}{\int_{\mathbb{R}} w^2 dy}w^2 = 0. \tag{2.44}$$

For $l = 1, \ldots, N$ we consider the l-th equation of system (2.44):

$$\tilde{\phi}_l'' - \tilde{\phi}_l + 2w\tilde{\phi}_l - 2\sigma_l\frac{\int_{\mathbb{R}} w\tilde{\phi}_l dy}{\int_{\mathbb{R}} w^2 dy}w^2 = 0. \tag{2.45}$$

By Theorem 3.1(3) below, the last equation (2.45) implies that

$$\tilde{\phi}_l \in X_0, \quad l = 1, \ldots, N. \tag{2.46}$$

Here we have used condition (H2) which gives $2\sigma_l \neq 1$. Thus (2.42) is proved.

To prove (2.43), we proceed in the same way for L^*. Using $\sigma(\mathcal{B}) = \sigma(\mathcal{B}^T)$, the l-th equation of the diagonalised system is given by

$$\tilde{\psi}_l'' - \tilde{\psi}_l + 2w\tilde{\psi}_l - 2\sigma_l\frac{\int_{\mathbb{R}} w^p \tilde{\psi}_l dy}{\int_{\mathbb{R}} w^r dy}w = 0, \quad l = 1, \ldots, N. \tag{2.47}$$

Multiplying (2.47) by w and integrating over the real line, we obtain

$$(1 - 2\sigma_l)\int_{\mathbb{R}} w^2 \tilde{\psi}_l dy = 0,$$

which implies that

$$\int_{\mathbb{R}} w^2 \tilde{\psi}_l dy = 0,$$

since by (H2) we know that $2\sigma_l \neq 1$. Thus all the nonlocal terms vanish and we have

$$L_0\tilde{\psi}_l = 0, \quad l = 1, \ldots, N. \tag{2.48}$$

This implies that $\tilde{\psi}_l \in X_0$ for $l = 1, \ldots, N$. Now (2.43) follows. $\square$

As a consequence of Lemma 2.5, we have

Lemma 2.6 *The operator*

$$L : \left(H^2(\mathbb{R})\right)^N \to \left(L^2(\mathbb{R})\right)^N$$

is invertible if it is restricted as follows

$$L : (X_0 \oplus \cdots \oplus X_0)^\perp \cap \left(H^2(\mathbb{R})\right)^N \to (X_0 \oplus \cdots \oplus X_0)^\perp \cap \left(L^2(\mathbb{R})\right)^N.$$

Moreover, L^{-1} is bounded.

Proof This follows from the Fredholm Alternative (see Theorem 13.2) and Lemma 2.5. □

2.3.2 Study of the Approximate Solutions

Let $-1 < t_1^0 < t_2^0 < 1$ be two points satisfying the assumptions (H1)–(H3) and again we will use the notation $\mathbf{t}^0 = (t_1^0, t_2^0)$. Let $\hat{\xi}^0 = (\hat{\xi}_1^0, \hat{\xi}_2^0)$ be the unique solution of (2.24).

We first construct an approximate two-spike solution to (2.2) which concentrates near these prescribed two points.

Let $-1 < t_1 < t_2 < 1$ be such that $\mathbf{t} = (t_1, t_2) \in B_{c_0\epsilon}(\mathbf{t}^0)$, where the constant c_0 will be chosen below. Set

$$w_j(x) = w\left(\frac{x - t_j}{\epsilon}\right) \tag{2.49}$$

and

$$r_0 = \frac{1}{10}\left(\min\left(t_1^0 + 1, 1 - t_2^0, \frac{1}{2}|t_2^0 - t_1^0|\right)\right). \tag{2.50}$$

Let $\chi : \mathbb{R} \to [0, 1]$ be a smooth cut-off function such that

$$\chi(x) = 1 \quad \text{for } |x| < 1 \quad \text{and} \quad \chi(x) = 0 \quad \text{for } |x| > 2. \tag{2.51}$$

We now define our approximate solution. Firstly, we set

$$\tilde{w}_j(x) = w_j(x)\chi\left(\frac{x - t_j}{r_0}\right). \tag{2.52}$$

Then it is easy to see that $\tilde{w}_j(x) \in H_N^2(-1, 1)$ satisfies

$$\epsilon^2 \tilde{w}_j'' - \tilde{w}_j + \tilde{w}_j^2 = \text{e.s.t.} \tag{2.53}$$

in $L^2(-1, 1)$ where e.s.t. denotes an exponentially small term.

Secondly, we let $\hat{\boldsymbol{\xi}}(\mathbf{t}) = (\hat{\xi}_1, \hat{\xi}_2)$ be the unique solution of (2.26) and put

$$w_{\epsilon,\mathbf{t}}(x) = \sum_{j=1}^{2} \hat{\xi}_j \tilde{w}_j(x). \tag{2.54}$$

For any function $A \in H^2(-1, 1)$ we define $T[A]$ to be the unique solution of the linear problem

$$\begin{cases} DT[A]'' - T[A] + \xi_\epsilon A^2 = 0, & -1 < x < 1, \\ T[A]'(-1) = T[A]'(1) = 0, \end{cases} \tag{2.55}$$

where ξ_ϵ was defined in (2.18). Then the solution $T[A]$ is unique and positive.

For $A = w_{\epsilon,\mathbf{t}}$, where $\mathbf{t} \in B_{c_0\epsilon}(\mathbf{t}^0)$, we now compute

$$\tau_i := T[A](t_i). \tag{2.56}$$

From (2.55), we have

$$\tau_i = \xi_\epsilon \int_{-1}^{1} G_D(t_i, z) A^2(z) dz$$

$$= \xi_\epsilon \epsilon \sum_{j=1}^{2} \hat{\xi}_j^2 \int_{-1}^{1} G_D(t_i, z) \tilde{w}_j^2(z) dz \big(1 + O(\epsilon)\big)$$

$$= \xi_\epsilon \epsilon \sum_{j=1}^{2} \hat{\xi}_j^2 \left[G_D(t_i, t_j) \int_{-\infty}^{+\infty} w_j^2(y) dy + O(\epsilon) \right] = \sum_{j=1}^{2} G_D(t_i, t_j) \hat{\xi}_j^2 + O(\epsilon),$$

where we have used (2.18). Thus we have derived the following system of algebraic equations:

$$\tau_i = \sum_{j=1}^{2} G_D(t_i, t_j) \hat{\xi}_j^2 + O(\epsilon). \tag{2.57}$$

By the implicit function theorem and assumptions (H1), (H2), the system (2.57) has a unique solution

$$\tau_i = \hat{\xi}_i + O(\epsilon), \quad i = 1, 2.$$

Hence

$$T[A](t_i) = \hat{\xi}_i + O(\epsilon). \tag{2.58}$$

Next for $x = t_i + \epsilon y$ and $A = w_{\epsilon,\mathbf{t}}$ we calculate

$$T[A](x) - T[A](t_i)$$

$$= \xi_\epsilon \int_{-1}^{1} \big[G_D(x, z) - G_D(t_i, z) \big] A^2(z) dz$$

$$= \xi_\epsilon \hat{\xi}_i^2 \int_{-1}^{1} [G_D(x, z) - G_D(t_i, z)] \tilde{w}_i^2(z) dz$$

$$+ \xi_\epsilon \hat{\xi}_{3-i}^2 \int_{-1}^{1} [G_D(x, z) - G_D(t_i, z)] \tilde{w}_{3-i}^2(z) dz$$

$$= \xi_\epsilon \hat{\xi}_i^2 \int_{-1}^{1} [K_D(|x - z|) - K_D(|t_i - z|)] \tilde{w}_i^2(z) dz$$

$$- \xi_\epsilon \hat{\xi}_i^2 \int_{-1}^{1} [H_D(x, z) - H_D(t_i, z)] \tilde{w}_i^2(z) dz$$

$$+ \xi_\epsilon \hat{\xi}_{3-i}^2 \int_{-1}^{1} [G_D(x, z) - G_D(t_i, z)] \tilde{w}_{3-i}^2(z) dz$$

$$= \epsilon^2 \xi_\epsilon \hat{\xi}_i^2 \int_{-\infty}^{+\infty} \left[\frac{1}{2D} |z| - \frac{1}{2D} |y - z| \right] w^2(|z|) dz (1 + O(\epsilon |y|))$$

$$+ \epsilon \hat{\xi}_i^2 [-y \nabla_{t_i} H_D(t_i, t_i) + O(\epsilon y^2)]$$

$$+ \epsilon [y \nabla_{t_i} G_D(t_i, t_{3-i}) \hat{\xi}_{3-i}^2 + O(\epsilon y^2)]$$

$$= \epsilon [\hat{\xi}_i^2 P^i(|y|) - \hat{\xi}_i^2 y \nabla_{t_i} H_D(t_i, t_i)$$

$$+ y \nabla_{t_i} G_D(t_i, t_{3-i}) \hat{\xi}_{3-i}^2 + O(\epsilon y^2)], \tag{2.59}$$

where

$$P^i(|y|) = \left(\int_{-\infty}^{+\infty} w^2 \right)^{-1}$$

$$\times \int_{-\infty}^{+\infty} \left[\frac{1}{2D} |z| - \frac{1}{2D} |y - z| \right] w^2(|z|) dz. \tag{2.60}$$

Note that P^i is an even function.

Let us now define the rescaled domain $\Omega_\epsilon = (-\frac{1}{\epsilon}, \frac{1}{\epsilon})$ and the operator

$$S : H_N^2(\Omega_\epsilon) \to L^2(\Omega_\epsilon), \quad S[A] := A'' - A + \frac{A^2}{(T[A])}, \tag{2.61}$$

where $T[A]$ has been introduced in (2.55). Then, choosing $A = w_{\epsilon, \mathbf{t}}$, we compute $S[w_{\epsilon, \mathbf{t}}]$ as follows:

$$S[w_{\epsilon, \mathbf{t}}] = w_{\epsilon, \mathbf{t}}'' - w_{\epsilon, \mathbf{t}} + \frac{w_{\epsilon, \mathbf{t}}^2}{T[w_{\epsilon, \mathbf{t}}]}$$

$$= \sum_{j=1}^{2} \hat{\xi}_j (\tilde{w}_j'' - \tilde{w}_j) + \frac{w_{\epsilon, \mathbf{t}}^2}{T[w_{\epsilon, \mathbf{t}}]} + \text{e.s.t.}$$

$$= \left[\frac{(\sum_{j=1}^2 \hat{\xi}_j \tilde{w}_j)^2}{T[w_{\epsilon,t}]} - \sum_{j=1}^2 \hat{\xi}_j \tilde{w}_j^2 \right] + \text{e.s.t.}$$

$$= E_1 + E_2 + \text{e.s.t.} \quad \text{in } L^2\left(-\frac{1}{\epsilon}, \frac{1}{\epsilon}\right), \tag{2.62}$$

where

$$E_1 = \left[\frac{(\sum_{j=1}^2 \hat{\xi}_j \tilde{w}_j)^2}{T[w_{\epsilon,t}](t_j)} - \sum_{j=1}^2 \hat{\xi} \tilde{w}_j^2 \right] \tag{2.63}$$

and

$$E_2 = \left[\frac{(\sum_{j=1}^2 \hat{\xi}_j \tilde{w}_j)^2}{T[w_{\epsilon,t}](x)} - \frac{(\sum_{j=1}^2 \hat{\xi}_j \tilde{w}_j)^2}{T[w_{\epsilon,t}](t_j)} \right]. \tag{2.64}$$

For E_1, we calculate using (2.58)

$$E_1 = \frac{(\sum_{j=1}^2 \hat{\xi}_j \tilde{w}_j)^2}{T[w_{\epsilon,t}](t_j)} - \sum_{j=1}^2 \hat{\xi}_j \tilde{w}_j^2$$

$$= \sum_{j=1}^2 \left(\frac{\hat{\xi}_j^2}{\hat{\xi}_j + O(\epsilon)} - \hat{\xi}_j \right) \tilde{w}_j^2 = O(\epsilon) \sum_{j=1}^2 \hat{\xi}_j \tilde{w}_j^2. \tag{2.65}$$

Thus we have

$$\|E_1\|_{L^2(-1/\epsilon, 1/\epsilon)} = O(\epsilon). \tag{2.66}$$

For E_2, we calculate

$$E_2 = -\sum_{j=1}^2 \frac{(\hat{\xi}_j \tilde{w}_j)^2}{(T[w_{\epsilon,t}](t_j))^2} \left(T[w_{\epsilon,t}](x) - T[w_{\epsilon,t}](t_j) \right)$$

$$+ O\left(\sum_{j=1}^2 |T[w_{\epsilon,t}](x) - T[w_{\epsilon,t}](t_j)|^2 \tilde{w}_j^2 \right)$$

$$= -\sum_{j=1}^2 \hat{\xi}_j \tilde{w}_j^2 \frac{T[w_{\epsilon,t}](x) - T[w_{\epsilon,t}](t_j)}{T[w_{\epsilon,t}](t_j)} + O\left(\epsilon^2 y^2 \sum_{j=1}^2 \tilde{w}_j^2 \right)$$

$$= -\epsilon \sum_{j=1}^2 \tilde{w}_j^2 \{ \hat{\xi}_j^2 P^j(|y|) - \hat{\xi}_j^2 y \nabla_{t_j} H_D(t_j, t_j) + y \nabla_{t_j} G_D(t_j, t_{3-j}) \hat{\xi}_l^2 \}$$

$$+ O\left(\epsilon^2 y^2 \sum_{j=1}^2 \tilde{w}_j^2 \right). \tag{2.67}$$

This implies that

$$\|E_2\|_{L^2(-1/\epsilon,1/\epsilon)} = O(\epsilon). \tag{2.68}$$

Combining (2.66) and (2.68), we conclude that

$$\|S[w_{\epsilon,\mathbf{t}}]\|_{L^2(-1/\epsilon,1/\epsilon)} = O(\epsilon). \tag{2.69}$$

The estimates in this subsection show that the approximate solution solves the system up to a small error.

2.3.3 The Liapunov-Schmidt Reduction Method

In this subsection, we use the Liapunov-Schmidt reduction method to solve the problem

$$S[w_{\epsilon,\mathbf{t}} + v] = \sum_{j=1}^{2} \beta_j \frac{d\tilde{w}_j}{dx} \tag{2.70}$$

for real constants β_j and a function $v \in H_N^2(-\frac{1}{\epsilon}, \frac{1}{\epsilon})$ which is small in the H^2-norm, where $\tilde{w}_j$ and $w_{\epsilon,\mathbf{t}}$ are given by (2.52) and (2.54), respectively.

To this end, we need to study the linearised operator

$$\tilde{L}_{\epsilon,\mathbf{t}} : H^2(\Omega_\epsilon) \to L^2(\Omega_\epsilon)$$

defined by

$$\tilde{L}_{\epsilon,\mathbf{t}} := S'_\epsilon[A]\phi = \phi'' - \phi + \frac{2A\phi}{T[A]} - \frac{A^2}{(T[A])^2}(T'[A]\phi),$$

where $A = w_{\epsilon,\mathbf{t}}$, and, for given $\phi \in L^2(\Omega)$, we denote by $T'[A]\phi$ the unique solution of the linear problem

$$\begin{cases} D(T'[A]\phi)'' - (T'[A]\phi) + 2\xi_\epsilon A\phi = 0, & -1 < x < 1, \\ (T'[A]\phi)'(-1) = (T'[A]\phi)'(1) = 0. \end{cases} \tag{2.71}$$

We define the approximate kernel and cokernel, respectively, as follows:

$$\mathcal{K}_{\epsilon,\mathbf{t}} := \mathrm{span}\left\{\frac{d\tilde{w}_i}{dx} : i = 1, 2\right\} \subset H_N^2(\Omega_\epsilon),$$

$$\mathcal{C}_{\epsilon,\mathbf{t}} := \mathrm{span}\left\{\frac{d\tilde{w}_i}{dx} : i = 1, 2\right\} \subset L^2(\Omega_\epsilon).$$

Recall the definition of the following vectorial linear operator introduced in (2.39):

$$L\Phi := \Phi'' - \Phi + 2w\Phi - 2B\frac{\int_{\mathbb{R}} w\Phi}{\int_{\mathbb{R}} w^2}w^2,$$

where

$$\Phi = \begin{pmatrix} \phi_1 \\ \phi_2 \end{pmatrix} \in \left(H^2(\mathbb{R})\right)^2.$$

By Lemma 2.5, we know that

$$L : (X_0 \oplus X_0)^\perp \cap \left(H^2(\mathbb{R})\right)^2 \to (X_0 \oplus X_0)^\perp \cap \left(L^2(\mathbb{R})\right)^2$$

is invertible with a bounded inverse.

We will see that this system is the limit of the linear operator $\tilde{L}_{\epsilon,t}$ as $\epsilon \to 0$. To this end, we introduce the projection $\pi_{\epsilon,t}^\perp : L^2(\Omega_\epsilon) \to \mathcal{C}_{\epsilon,t}^\perp$ and study the linear operator $L_{\epsilon,t} := \pi_{\epsilon,t}^\perp \circ \tilde{L}_{\epsilon,t}$. By letting $\epsilon \to 0$, we will show that $L_{\epsilon,t} : \mathcal{K}_{\epsilon,t}^\perp \to \mathcal{C}_{\epsilon,t}^\perp$ is invertible with a bounded inverse provided ϵ is small enough. This statement is contained in the following proposition.

Proposition 2.7 *There exist positive constants $\bar{\epsilon}$, $\bar{\delta}$, λ such that for all $\epsilon \in (0, \bar{\epsilon})$ and all $\mathbf{t} \in \Omega^2$ satisfying $\min(|1 + t_1|, |1 - t_2|, \frac{1}{2}|t_1 - t_2|) > \bar{\delta}$ we have*

$$\|L_{\epsilon,t}\phi\|_{L^2(\Omega_\epsilon)} \geq \lambda \|\phi\|_{H^2(\Omega_\epsilon)} \quad \text{for all } \phi \in \mathcal{K}_{\epsilon,t}^\perp. \tag{2.72}$$

Further, the linear operator

$$L_{\epsilon,t} = \pi_{\epsilon,t}^\perp \circ \tilde{L}_{\epsilon,t} : \mathcal{K}_{\epsilon,t}^\perp \to \mathcal{C}_{\epsilon,t}^\perp$$

is surjective.

Proof This proof follows the method of Liapunov-Schmidt reduction.

Suppose (2.72) is false. Then there are sequences $\{\epsilon_k\}$, $\{\mathbf{t}^k\}$, $\{\phi^k\}$ such that $\epsilon_k \to 0$, $\mathbf{t}^k \in \Omega^2$ with $\min(|1 + t_1^k|, |1 - t_2^k|, \frac{1}{2}|t_1^k - t_2^k|) > \bar{\delta}$ and $\phi^k = \phi_{\epsilon_k} \in \mathcal{K}_{\epsilon_k,t^k}^\perp$, $k = 1, 2, \ldots$ such that

$$\|L_{\epsilon_k,t^k}\phi^k\|_{L^2(\Omega_{\epsilon_k})} \to 0 \quad \text{as } k \to \infty, \tag{2.73}$$

$$\|\phi^k\|_{H^2(\Omega_{\epsilon_k})} = 1, \quad k = 1, 2, \ldots. \tag{2.74}$$

By using the cut-off function introduced in (2.51), we define $\phi_{\epsilon,i}$, $i = 1, 2, 3$ for $\epsilon > 0$ small enough as follows:

$$\phi_{\epsilon,i}(x) = \phi_{\epsilon}(x)\chi\left(\frac{x - t_i}{r_0}\right), \quad x \in \Omega, i = 1, 2,$$

$$\phi_{\epsilon,3}(x) = \phi_{\epsilon}(x) - \sum_{i=1}^{2}\phi_{\epsilon,i}(x), \quad x \in \Omega. \tag{2.75}$$

After rescaling, first the functions $\phi_{\epsilon,i}$ are defined only on Ω_{ϵ}. Then, by a standard result (see Sect. 7.12 in [74]), they can be extended to $\mathbb{R}$ so that for ϵ small enough their norms in $H^2(\mathbb{R})$ are bounded by a constant independent of ϵ and $\mathbf{t}$. In the following we will study this extension, where for simplicity we use the same notation for the original functions and its extension. Since for $i = 1, 2$ both sequences $\{\phi_i^k\} := \{\phi_{\epsilon_k,i}\}$ ($k = 1, 2, \ldots$) are bounded in $H_{\mathrm{loc}}^2(\mathbb{R})$, they have weak limits in $H_{\mathrm{loc}}^2(\mathbb{R})$ and thus also strong limits in $L_{\mathrm{loc}}^2(\mathbb{R})$ and $L_{\mathrm{loc}}^{\infty}(\mathbb{R})$. We denote these limits by ϕ_i. Further, by a barrier argument, these functions have uniform exponential decay, and the limits are also strong in the $H^2(\mathbb{R})$ and $L^{\infty}(\mathbb{R})$ sense. Thus $\phi = (\phi_1, \phi_2)^T$ solves the system $L\phi = 0$ with the operator L introduced in (2.39). By Lemma 2.5, we know that $\phi \in \mathrm{Ker}(L) = X_0 \oplus X_0$. Since $\phi^k \in \mathcal{K}_{\epsilon_k,\mathbf{t}^k}^{\perp}$, by taking the limit $k \to \infty$ we get $\phi \in \mathrm{Ker}(L)^{\perp}$ and so $\phi = 0$.

By elliptic estimates, we derive $\|\phi_i^k\|_{H^2(\mathbb{R})} \to 0$ as $k \to \infty$ for $i = 1, 2$ and $\phi_3^k \to \phi_3$ in $H^2(\mathbb{R})$, where ϕ_3 satisfies

$$\Delta\phi_3 - \phi_3 = 0 \quad \text{in } \mathbb{R}.$$

Therefore we conclude $\phi_3 = 0$ and $\|\phi_3^k\|_{H^2(\mathbb{R})} \to 0$ as $k \to \infty$. This contradicts $\|\phi^k\|_{H^2(\Omega_{\epsilon_k})} = 1$.

To complete the proof of Proposition 2.7, we just need to show that the conjugate operator of $L_{\epsilon,\mathbf{t}}$ (denoted by $L_{\epsilon,\mathbf{t}}^*$) is injective from $\mathcal{K}_{\epsilon,\mathbf{t}}^{\perp}$ to $\mathcal{C}_{\epsilon,\mathbf{t}}^{\perp}$. Note that $L_{\epsilon,\mathbf{t}}^*\psi = \pi_{\epsilon,\mathbf{t}} \circ \tilde{L}_{\epsilon,\mathbf{t}}^*$ with

$$\tilde{L}_{\epsilon,\mathbf{t}}^*\psi = \epsilon^2\Delta\psi - \psi + \frac{2A\psi}{(T[A])} - T'[A]\frac{A^2\psi}{(T[A])^2}.$$

The proof for $L_{\epsilon,\mathbf{t}}^*$ follows exactly along the same lines as the proof for $L_{\epsilon,\mathbf{t}}$ and is therefore omitted. $\square$

Now we have derived all the technical tools needed to solve the equation

$$\pi_{\epsilon,\mathbf{t}}^{\perp} \circ S_{\epsilon}(w_{\epsilon,\mathbf{t}} + \phi) = 0. \tag{2.76}$$

Since the restriction of the linear operator $L_{\epsilon,\mathbf{t}}$ to $\mathcal{K}_{\epsilon,\mathbf{t}}^{\perp}$ is invertible we can write (2.76) in equivalent form as

$$\phi = -\left(L_{\epsilon,\mathbf{t}}^{-1} \circ \pi_{\epsilon,\mathbf{t}}^{\perp} \circ S_{\epsilon}(w_{\epsilon,\mathbf{t}})\right) - \left(L_{\epsilon,\mathbf{t}}^{-1} \circ \pi_{\epsilon,\mathbf{t}}^{\perp} \circ N_{\epsilon,\mathbf{t}}(\phi)\right) =: M_{\epsilon,\mathbf{t}}(\phi), \tag{2.77}$$

where $L_{\epsilon,t}^{-1}$ is the inverse of $L_{\epsilon,t}$, and the nonlinear operators

$$N_{\epsilon,t}(\phi) = S_\epsilon(w_{\epsilon,t} + \phi) - S_\epsilon(w_{\epsilon,t}) - S_\epsilon'(w_{\epsilon,t})\phi \tag{2.78}$$

and $M_{\epsilon,t}(\phi)$ introduced in (2.77) are both defined for $\phi \in H_N^2(\Omega_\epsilon)$.

Finally, we show that the operator $M_{\epsilon,t}$ is a contraction on

$$B_{\epsilon,\delta} \equiv \left\{ \phi \in H_N^2(\Omega_\epsilon) : \|\phi\|_{H^2(\Omega_\epsilon)} < \delta \right\}$$

if δ and ϵ are suitably chosen. First, from (2.69) and Proposition 2.7 we know that

$$
\begin{aligned}
\left\| M_{\epsilon,t}(\phi) \right\|_{H^2(\Omega_\epsilon)} &\leq \lambda^{-1} \left(\left\| \pi_{\epsilon,t}^\perp \circ N_{\epsilon,t}(\phi) \right\|_{L^2(\Omega_\epsilon)} + \left\| \pi_{\epsilon,t}^\perp \circ S_\epsilon(w_{\epsilon,t}) \right\|_{L^2(\Omega_\epsilon)} \right) \\
&\leq \lambda^{-1} C_0 \left(c(\delta)\delta + \epsilon \right),
\end{aligned}
$$

where $\lambda > 0$ is independent of $\delta > 0$, $\epsilon > 0$ and $c(\delta) \to 0$ as $\delta \to 0$. Similarly, we have

$$\left\| M_{\epsilon,t}(\phi) - M_{\epsilon,t}(\phi') \right\|_{H^2(\Omega_\epsilon)} \leq \lambda^{-1} C_0 \left(c(\delta)\delta \right) \left\| \phi - \phi' \right\|_{H^2(\Omega_\epsilon)},$$

where $c(\delta) \to 0$ as $\delta \to 0$. If we choose

$$\delta = C_1 \epsilon, \quad \text{where } \lambda^{-1} C_0 < C_1 \text{ and } \epsilon \text{ small enough,} \tag{2.79}$$

then $M_{\epsilon,t}$ maps from $B_{\epsilon,\delta}$ into $B_{\epsilon,\delta}$ and is a contraction mapping in $B_{\epsilon,\delta}$. Now the existence of a fixed point $\phi_{\epsilon,t} \in B_{\epsilon,\delta}$ follows from the standard contraction mapping principle. Thus we have rigorously constructed a solution $\phi_{\epsilon,t} \in H_N^2(\Omega_\epsilon)$ of (2.77).

We summarise our result as follows:

Lemma 2.8 *There exist $\bar{\epsilon} > 0$ and $\bar{\delta} > 0$ such that for every pair ϵ, t with $0 < \epsilon < \bar{\epsilon}$ and $t \in \Omega^2$, $1 + t_1 > \bar{\delta}$, $1 - t_2 > \bar{\delta}$, $\frac{1}{2}|t_2 - t_1| > \bar{\delta}$ there is a unique $\phi_{\epsilon,t} \in \mathcal{K}_{\epsilon,t}^\perp$ satisfying $S_\epsilon(w_{\epsilon,t} + \phi_{\epsilon,t}) \in \mathcal{C}_{\epsilon,t}$. Further, we have the estimate*

$$\|\phi_{\epsilon,t}\|_{H^2(\Omega_\epsilon)} \leq C_1 \epsilon, \tag{2.80}$$

where C_1 has been defined in (2.79).

2.3.4 The Reduced Problem

In this subsection, we solve the reduced problem and conclude the proof of our main existence result given in Theorem 2.2.

By Lemma 2.8, for every $t \in B_{c_0\epsilon}(t^0)$, there exists a unique solution $\phi_{\epsilon,t} \in \mathcal{K}_{\epsilon,t}^\perp$ such that

$$S[w_{\epsilon,t} + \phi_{\epsilon,t}] = v_{\epsilon,t} \in \mathcal{C}_{\epsilon,t}. \tag{2.81}$$

Now we are going to determine the position $\mathbf{t}^\epsilon = (t_1^\epsilon, t_2^\epsilon) \in B_{c_0\epsilon}(\mathbf{t}^0)$ such that also

$$S[w_{\epsilon,\mathbf{t}^\epsilon} + \phi_{\epsilon,\mathbf{t}^\epsilon}] \perp \mathcal{C}_{\epsilon,\mathbf{t}^\epsilon}. \tag{2.82}$$

Then, by combining (2.81) and (2.82), we have found $\mathbf{t}^\epsilon = (t_1^\epsilon, t_2^\epsilon) \in B_{c_0\epsilon}(\mathbf{t}^0)$ and $\phi_{\epsilon,\mathbf{t}^\epsilon} \in \mathcal{K}_{\epsilon,\mathbf{t}^\epsilon}^\perp$ such that $S[w_{\epsilon,\mathbf{t}^\epsilon} + \phi_{\epsilon,\mathbf{t}^\epsilon}] = 0$. This means that we have found a solution of (2.2) and Theorem 2.2 follows in the case of two spikes.

To this end, we introduce the vector field

$$W_{\epsilon,i}(\mathbf{t}) := \epsilon^{-1} \int_{-1}^{1} S[w_{\epsilon,\mathbf{t}} + \phi_{\epsilon,\mathbf{t}}] \frac{d\tilde{w}_i}{dx} dx,$$

$$W_\epsilon(\mathbf{t}) := \left(W_{\epsilon,1}(\mathbf{t}), W_{\epsilon,2}(\mathbf{t}) \right) : B_{c_0\epsilon}(\mathbf{t}^0) \to \mathbb{R}^2.$$

Then $W_\epsilon(\mathbf{t})$ is a continuous map in $\mathbf{t}$ and our problem is reduced to finding a zero of the vector field $W_\epsilon(\mathbf{t})$.

Next we explicitly calculate $W_\epsilon(\mathbf{t})$:

$$W_{\epsilon,i}(\mathbf{t}) = \epsilon^{-1} \int_{-1}^{1} S[w_{\epsilon,\mathbf{t}} + \phi_{\epsilon,\mathbf{t}}] \frac{d\tilde{w}_i}{dx} dx$$

$$= \epsilon^{-1} \int_{-1}^{1} S[w_{\epsilon,\mathbf{t}}] \frac{d\tilde{w}_i}{dx} dx + \epsilon^{-1} \int_{-1}^{1} S'_\epsilon[w_{\epsilon,\mathbf{t}}] \phi_{\epsilon,\mathbf{t}} \frac{d\tilde{w}_i}{dx} dx$$

$$+ \epsilon^{-1} \int_{-1}^{1} N_\epsilon(\phi_{\epsilon,\mathbf{t}}) \frac{d\tilde{w}_i}{dx} dx$$

$$= I_1 + I_2 + I_3,$$

where I_1, I_2 and I_3 are defined by the last equality.

The computation of I_3 is the easiest: Note that by Taylor expansion for (2.78), the first term in the expansion of N_ϵ is quadratic in $\phi_{\epsilon,\mathbf{t}}$ and so

$$I_3 = O(\epsilon). \tag{2.83}$$

We will now compute I_1 and I_2. The result will be that I_1 is the leading term and $I_2 = O(\epsilon)$.

For I_1, we have

$$I_1 = \epsilon^{-1} \int_{-1}^{1} (E_1 + E_2) \frac{d\tilde{w}_i}{dx} dx$$

$$= \epsilon^{-1} \int_{-1}^{1} E_2 \frac{d\tilde{w}_i}{dx} dx + O(\epsilon),$$

where E_1 and E_2 were defined in (2.63) and (2.64), respectively. Here we have used that E_1 is an even function.

We calculate by (2.67)

$$\epsilon^{-1} \int_{-1}^{1} E_2 \frac{d\tilde{w}_i}{dx} dx$$

$$= \sum_{j=1}^{2} \nabla_{t_i} G_D(t_i, t_j) \hat{\xi}_j^2 \int_{\mathbb{R}} yw^2(y)w'(y)dy + O(\epsilon)$$

$$= \sum_{j=1}^{2} \nabla_{t_i} G_D(t_i, t_j) \hat{\xi}_j^2 \frac{1}{3} \int_{\mathbb{R}} w^3 dy + O(\epsilon).$$

Thus we have

$$I_1 = \sum_{j=1}^{2} \nabla_{t_i} G_D(t_i, t_j) \hat{\xi}_j^2 \frac{1}{3} \int_{\mathbb{R}} w^3(y)dy + O(\epsilon). \tag{2.84}$$

For I_2, we calculate

$$\epsilon I_2 = \int_{-1}^{1} S'[w_{\epsilon,t}](\phi_{\epsilon,t}) \frac{d\tilde{w}_i}{dx} dx$$

$$= \int_{-1}^{1} \left[\epsilon^2 \Delta \phi_{\epsilon,t} - \phi_{\epsilon,t} + \frac{2w_{\epsilon,t}\phi_{\epsilon,t}}{T[w_{\epsilon,t}]} - \frac{w_{\epsilon,t}^2}{(T[w_{\epsilon,t}])^2} (T'[w_{\epsilon,t}]\phi_{\epsilon,t}) \right] \frac{d\tilde{w}_i}{dx} dx$$

$$= \int_{-1}^{1} \left[\epsilon^2 \Delta \frac{d\tilde{w}_i}{dx} - \frac{d\tilde{w}_i}{dx} + \frac{d\tilde{w}_i}{dx} \frac{2w_{\epsilon,t}}{T[w_{\epsilon,t}]} \right] \phi_{\epsilon,t} dx$$

$$- \int_{-1}^{1} \frac{w_{\epsilon,t}^2}{(T[w_{\epsilon,t}])^2} (T'[w_{\epsilon,t}]\phi_{\epsilon,t}) \frac{d\tilde{w}_i}{dx} dx$$

$$= \int_{-1}^{1} \left(2\frac{\hat{\xi}_i \tilde{w}_i}{T[w_{\epsilon,t}]} - 2\tilde{w}_i \right) \phi_{\epsilon,t} \frac{d\tilde{w}_i}{dx} dx - \int_{-1}^{1} \frac{w_{\epsilon,t}^2}{(T[w_{\epsilon,t}])^2} (T'[w_{\epsilon,t}]\phi_{\epsilon,t}) \frac{d\tilde{w}_i}{dx} dx$$

$$= O(\epsilon^2),$$

since

$$\left\| \left(\frac{2\hat{\xi}_i \tilde{w}_i}{T[w_{\epsilon,t}]} - 2\tilde{w}_i \right) \phi_{\epsilon,t} \right\|_{L^2(\Omega_\epsilon)} = O(\epsilon),$$

$$\|\phi_{\epsilon,t}\|_{H^2(\Omega_\epsilon)} = O(\epsilon),$$

$$\left| T'[w_{\epsilon,t}](\phi_{\epsilon,t})(t_i) \right| = O(\epsilon),$$

$$\left| T'[w_{\epsilon,t}](\phi_{\epsilon,t})(t_i + \epsilon y) - T'[w_{\epsilon,t}](\phi_{\epsilon,t})(t_i) \right| = O(\epsilon^2|y|).$$

Combining I_1 and I_2, we have

$$W_{\epsilon,i}(\mathbf{t}) = \sum_{j=1}^{2} \nabla_{t_i} G_D(t_i, t_j)\hat{\xi}_j^2 \frac{1}{3} \int_{\mathbb{R}} w^3 dy + O(\epsilon)$$

$$= F_i(\mathbf{t})\frac{1}{3} \int_{\mathbb{R}} w^3 dy + O(\epsilon),$$

where $F_i(\mathbf{t})$ was defined in (2.27). By assumption (H3), we have $F(\mathbf{t}^0) = 0$ and

$$\det\left(\nabla_{\mathbf{t}^0} F(\mathbf{t}^0)\right) \neq 0.$$

This implies

$$W_\epsilon(\mathbf{t}) = -c_1 \mathcal{H}(\mathbf{t}^0)\mathcal{M}(\mathbf{t}^0)(\mathbf{t} - \mathbf{t}^0) + O\left(|\mathbf{t} - \mathbf{t}^0|^2 + \epsilon\right),$$

$$c_1 = -\frac{1}{3}\int_{\mathbb{R}} w^3 dy = -2.4. \tag{2.85}$$

Then by standard degree theory (see Sect. 13.1) we conclude that for ϵ small enough there exists a $\mathbf{t}^\epsilon \in B_{c_0\epsilon}(\mathbf{t}^0)$ such that $W_\epsilon(\mathbf{t}^\epsilon) = 0$. Here it is important to note that, by choosing c_0 large enough (independent of ϵ), the leading term in (2.85) dominates the error terms for all $\mathbf{t} \in B_{c_0\epsilon}(\mathbf{t}^0)$ and ϵ small enough. We have thus proved the following proposition:

Proposition 2.9 *For ϵ small enough there exist points $\mathbf{t}^\epsilon$ with $\mathbf{t}^\epsilon \to \mathbf{t}^0$ such that $W_\epsilon(\mathbf{t}^\epsilon) = 0$.*

Finally, we conclude the proof of Theorem 2.2.

Proof of Theorem 2.2 By Proposition 2.9, there exist $\mathbf{t}^\epsilon \to \mathbf{t}^0$ such that $W_\epsilon(\mathbf{t}^\epsilon) = 0$. This implies $S[w_{\epsilon,\mathbf{t}^\epsilon} + \phi_{\epsilon,\mathbf{t}^\epsilon}] = 0$. Let $A_\epsilon = \xi_\epsilon(w_{\epsilon,\mathbf{t}^\epsilon} + \phi_{\epsilon,\mathbf{t}^\epsilon})$ and $H_\epsilon = \xi_\epsilon T[w_{\epsilon,\mathbf{t}^\epsilon} + \phi_{\epsilon,\mathbf{t}^\epsilon}]$. By the maximum principle, it follows that $A_\epsilon > 0$ and $H_\epsilon > 0$. Then (A_ϵ, H_ϵ) satisfies all the properties of Theorem 2.2. $\qquad\square$

2.4   Clustered Multiple Spikes

We first study n-spike clusters on the real line for any positive integer n, i.e. there are n spikes which all approximate the same point on the real line. Secondly, taking multiple spike clusters and placing them near different points in a bounded interval, we will derive results on multiple clusters. This construction is analogous to that of multiple spikes in Sect. 2.3, where each spike is replaced by a cluster.

The result on an n-spike cluster on the real line can be stated as follows:

Theorem 2.10 (*n*-spike cluster on the real line) *Consider the stationary Gierer-Meinhardt system on the real line:*

$$\begin{cases} 0 = \epsilon^2 \Delta A - A + \frac{A^2}{H}, & x \in \mathbb{R}, \\ 0 = \Delta H - H + A^2, & x \in \mathbb{R}. \end{cases} \tag{2.86}$$

Then (2.86) has a solution with which consists of n spikes which all approach the same point, where

$$A_\epsilon(x) \sim \sum_{k=1}^{n} \xi_\epsilon \hat{\xi}^0 w\left(\frac{x - t_k^\epsilon}{\epsilon} \right),$$

$$H_\epsilon\left(t_k^\epsilon\right) \sim \xi_\epsilon \hat{\xi}^0, \quad k = 1, \ldots, n,$$

$$t_k^\epsilon \to t^0 \in \mathbb{R}, \quad k = 1, \ldots, n,$$

where ξ_ϵ has been defined in (2.18). Further, we have

$$t_s^\epsilon - t_{s-1}^\epsilon = \epsilon \log \frac{1}{\epsilon} - \epsilon \log\left[\frac{\hat{\xi}^0}{2D}(s-1)(n+1-s) \right] + o(\epsilon), \tag{2.87}$$

$s = 2, \ldots, n$, *and*

$$\hat{\xi}^0 = \frac{2}{n\theta}.$$

Remark 2.11

1. The spikes are able to stabilise each other even without the presence of a boundary. There is a balance between the mutual attraction of the activators and the mutual repulsion of the inhibitors which leads to the existence of an *n*-spike ground state. Here the activator part of each spike interacts only with its neighbour(s) but the inhibitor parts interact with all other spikes.
2. Note that the distance between neighbouring spikes scales like $O(\epsilon \log \frac{1}{\epsilon})$ as $\epsilon \to 0$, i.e. it tends to 0 and all spikes approach the same point.
3. Note that unlike multiple spikes on the interval which are exact copies of each other this is not the case here. In contrast, for these ground states the distance between neighbouring spikes changes from spike to spike.
4. Due to translation invariance, the point $t_0 \in \mathbb{R}$ can be chosen arbitrarily.

Proof of Theorem 2.10 The proof uses Liapunov-Schmidt reduction similarly to Sect. 2.3.3. Solving the reduced problem consists of finding a zero of the vector field

$$W_{\epsilon,k} = \sum_{l \neq k} \left(\frac{1}{2D} \hat{\xi}^0 - \frac{1}{\epsilon} w\left(\frac{x_k - x_l}{\epsilon} \right) \right) \frac{x_k - x_l}{|x_k - x_l|}$$

$$+ O\left(\epsilon^{3/4}\right), \quad k = 1, \ldots, n. \tag{2.88}$$

Here the positive and negative terms can be balanced for $x_{k+1} - x_k \sim C_k \epsilon \log \frac{1}{\epsilon}$, where C_k are positive constants which depend on k. $\qquad \Box$

Next we consider the existence of multiple clusters in an interval, where each of the clusters consists of multiple spikes approaching the same point. Again, the proof is based on Liapunov-Schmidt reduction.

We need to make three assumptions which are extensions of those given in Sect. 2.3 for multiple spikes. If the clusters are indexed by j, then n_j denotes the number of spikes at cluster j.

Our first assumption is as follows:

(H1a) There exists a solution $(\hat{\xi}_1^0, \ldots, \hat{\xi}_N^0)$ of the equation

$$\sum_{j=1}^{N} G_D(t_i^0, t_j^0) n_j (\hat{\xi}_j^0)^2 = \hat{\xi}_i^0, \quad i = 1, \ldots, N.$$

Next we introduce the matrix

$$b_{ij} = G_D(t_i^0, t_j^0) n_j (\hat{\xi}_j^0), \quad \mathcal{B} = (b_{ij}).$$

Our second assumption is the following:

(H2a) It holds that

$$\frac{1}{2} \notin \sigma(\mathcal{B}),$$

where $\sigma(\mathcal{B})$ is the set of eigenvalues of $\mathcal{B}$.

We define the following vector field:

$$F(\mathbf{t}) = \big(F_1(\mathbf{t}), \ldots, F_N(\mathbf{t})\big),$$

where

$$F_i(\mathbf{t}) = \sum_{l=1}^{N} \nabla_{t_i} G_D(t_i, t_l) n_l \hat{\xi}_l^2$$

$$= -\nabla_{x_i} H_D(t_i, t_i) n_i \hat{\xi}_i^2 + \sum_{l \neq i} \nabla_{t_i} G_D(t_i, t_l) n_l \hat{\xi}_l^2, \quad i = 1, \ldots, N.$$

Set

$$\mathcal{M}(\mathbf{t}) = \big(\nabla_{t_j} F_i(\mathbf{t})\big).$$

Our third assumption concerns the vector field $F(\mathbf{t})$:

(H3a) We assume that at $\mathbf{t}^0 = (x_1^0, \ldots, x_N^0)$ we have

$$F(\mathbf{t}^0) = 0,$$

$$\det(\mathcal{M}(\mathbf{t}^0)) \neq 0.$$

Our first result is about the existence of *symmetric* multiple cluster solution.

Theorem 2.12 (Existence of symmetric multiple clusters) *Let N and n be two positive integers and*

$$t_j^0 = -1 + \frac{2j-1}{N}, \quad j = 1, \ldots, N.$$

Then, for $\epsilon \ll 1$, problem (2.2) has a solution with N equidistant clusters which concentrate at $t_1^0, \ldots, t_N^0$ and each of which consists of n spikes. We have

$$A_\epsilon(x) \sim \sum_{j=1}^{N} \sum_{k=1}^{n} \xi_\epsilon \hat{\xi}^0 w\left(\frac{x - t_{j,k}^\epsilon}{\epsilon}\right),$$

$$H_\epsilon\left(t_{j,k}^\epsilon\right) \sim \xi_\epsilon \hat{\xi}^0, \quad j = 1, \ldots, N, k = 1, \ldots, n,$$

$$t_{j,k}^\epsilon \to t_j^0, \quad j = 1, \ldots, N, k = 1, \ldots, n,$$

where ξ_ϵ has been defined in (2.18). Further,

$$t_{j,s}^\epsilon - t_{j,s-1}^\epsilon = \epsilon \log \frac{1}{\epsilon} - \epsilon \log\left[\frac{\hat{\xi}^0}{2D}(s-1)(n+1-s)\right] + o(\epsilon), \tag{2.89}$$

$j = 1, \ldots, N, s = 2, \ldots, n,$ *and*

$$\hat{\xi}^0 = \frac{2\tanh(\theta/N)}{n\theta}.$$

Our final result concerns the existence of *asymmetric* multiple clusters.

Theorem 2.13 (Existence of asymmetric multiple clusters) *Let $N, n_1, \ldots, n_N$ be $N+1$ positive integers.*
　　Assume that for $(t_1^0, \ldots, t_N^0) \in (-1, 1)^N$ with $t_1^0 < \cdots < t_N^0$ assumptions (H1a), (H2a) and (H3a) are satisfied. Let $(\xi_1^0, \ldots, \xi_N^0)$ be given by (H1). Then for $\epsilon \ll 1$, problem (2.2) has a solution with N clusters which concentrate at $t_1^\epsilon, \ldots, t_N^\epsilon$. Namely, it holds that

$$A_\epsilon(x) \sim \sum_{j=1}^{N} \sum_{k=1}^{n_j} \xi_\epsilon \hat{\xi}_j^0 w\left(\frac{x - t_{j,k}^\epsilon}{\epsilon}\right),$$

$$H_\epsilon\left(t_{j,k}^\epsilon\right) \sim \xi_\epsilon \hat{\xi}_j^0, \quad j = 1, \ldots, N, k = 1, \ldots, n_j,$$

$$t_{j,k}^\epsilon \to t_j^0, \quad j = 1, \ldots, N, k = 1, \ldots, n_j,$$

$$t_{j,s}^\epsilon - t_{j,s-1}^\epsilon = \epsilon \log \frac{1}{\epsilon} - \epsilon \log\left[\frac{\hat{\xi}_j^0}{2D}(s-1)(n_j+1-s)\right] + o(\epsilon), \tag{2.90}$$

$j = 1, \ldots, N, s = 2, \ldots, n_j.$

Remark 2.14 Equations (2.89) and (2.90) express the fact that we have two different scalings in the spike locations: the distance between the centers of clusters which is of order $O(1)$ and the distance between spikes within each cluster which is of order $O(\epsilon \log \frac{1}{\epsilon})$.

2.5 Notes on the Literature

The method of Liapunov-Schmidt reduction has been applied in [68, 191, 192] on the semi-classical (i.e. for small parameter h) solution of the nonlinear Schrödinger equation

$$\frac{h^2}{2}\Delta U - (V - E)U + U^p = 0 \qquad (2.91)$$

in $\mathbb{R}^N$ where V is a potential function and E is a real constant to construct solutions of (2.91) close to nondegenerate critical points of V for h sufficiently small. Subsequently it has also been used in many other examples of single partial differential equations, such as [8, 9, 68, 80, 81, 191, 192, 248, 256, 257].

The approach of Liapunov-Schmidt reduction to prove the existence of multiple spikes has been applied to the Gierer-Meinhardt system in [273]. For symmetric spikes, Theorem 2.2 recovers the existence result of [226]. Thus the results of [240] have been rigorously established. We note that this approach is very flexible and can be applied to many other reaction-diffusion systems in various situations.

In [27, 43, 44] the authors showed the existence of multiple-spike ground state solutions for the Gierer-Meinhardt system on the real line using Liapunov-Schmidt reduction and dynamical systems methods (geometric singular perturbation theory), respectively. We remark that asymmetric patterns can also be obtained for the Gierer-Meinhardt system on the real line [50]. In [49] it has been shown that a general two-component, singularly perturbed system that exhibits large-amplitude pulse patterns has a leading order normal form which is given by (2.1).

For clustered spikes we have presented results from [272].

Chapter 3
The Nonlocal Eigenvalue Problem (NLEP)

For the existence of multiple spikes which were considered in the previous chapter and for the stability of multiple spikes which will be the subject of the following chapter we need to know the spectral and stability properties of certain nonlocal eigenvalue problems (NLEPs). These problems are obtained as limiting eigenvalue problems from the linearised operator of the full Gierer-Meinhardt system (2.1). Different versions of NLEPs are needed and various methods of studying them will be discussed. For the convenience of the reader, in this chapter we collect a number of results which will be referred to in other chapters. In Sect. 3.1 we will start by considering a NLEP which arises in the special case when the time relaxation parameter of the inhibitor vanishes: $\tau = 0$. In Sect. 3.2 these results will be extended to the case where $\tau > 0$ is small enough, using the method of continuation. In Sect. 3.3 the case of arbitrary finite $\tau > 0$ is considered and results concerning Hopf bifurcation will be derived. In Sect. 3.4 the case of arbitrary finite $\tau > 0$ will be studied further and an explicit characterisation of the Hopf bifurcation point will be derived, using hypergeometric functions.

3.1 A Basic Theorem for $\tau = 0$

We start with the special case which arises when the time relaxation parameter τ of the inhibitor vanishes (compare (2.1)).

Next we consider our basic NLEP for $\tau = 0$.

Theorem 3.1 *Assume that*

$$r = 2, \quad 1 < p < 1 + \frac{4}{N} \quad or \quad r = p+1, \quad 1 < p < \left(\frac{N+2}{N-2}\right)_+, \quad (3.1)$$

where $(\frac{N+2}{N-2})_+ = \frac{N+2}{N-2}$ for $N = 3, 4, \dots$ and $(\frac{N+2}{N-2})_+ = \infty$ for $N = 1, 2$. Consider the following nonlocal eigenvalue problem

J. Wei, M. Winter, *Mathematical Aspects of Pattern Formation in Biological Systems*, Applied Mathematical Sciences 189, DOI 10.1007/978-1-4471-5526-3_3, © Springer-Verlag London 2014

$$\Delta\phi - \phi + pw^{p-1}\phi - \gamma_0(p-1)\frac{\int_{\mathbb{R}} w^{r-1}\phi}{\int_{\mathbb{R}} w^r}w^p = \alpha\phi, \quad \phi \in H^1(\mathbb{R}^N). \quad (3.2)$$

(1) *If $\gamma_0 < 1$, then there is a positive eigenvalue of* (3.2).
(2) *If $\gamma_0 > 1$ and* (3.1) *holds then for any nonzero eigenvalue α of* (3.2), *we have*

$$\text{Re}(\alpha) \leq -c_0 < 0.$$

(3) *If $\gamma_0 \neq 1$ and $\alpha = 0$, then $\phi \in \text{span}\{\frac{\partial w}{\partial y_j} : j = 1, \ldots, N\}$.*

Proof We first introduce some notation and make some preparations. Set

$$L\phi := L_0\phi - \gamma_0(p-1)\frac{\int_{\mathbb{R}^N} w^{r-1}\phi}{\int_{\mathbb{R}^N} w^r}w^p, \quad \phi \in H^1(\mathbb{R}^N),$$

where $\gamma_0 = \frac{qr}{(s+1)(p-1)} > 1$ and $L_0 := \Delta - 1 + pw^{p-1}$. Note that L is not self-adjoint if $r \neq p+1$.
Let

$$X_0 := \ker(L_0) = \text{span}\left\{\frac{\partial w}{\partial y_j} : j = 1, \ldots, N\right\}.$$

Then

$$L_0 w = (p-1)w^p, \quad L_0\left(\frac{1}{p-1}w + \frac{1}{2}x \cdot \nabla w\right) = w \quad (3.3)$$

and

$$\int_{\mathbb{R}^N} (L_0^{-1}w)w\,dy = \int_{\mathbb{R}^N} w\left(\frac{1}{p-1}w + \frac{1}{2}x \cdot \nabla w\right)dy$$

$$= \left(\frac{1}{p-1} - \frac{N}{4}\right)\int_{\mathbb{R}^N} w^2\,dy, \quad (3.4)$$

$$\int_{\mathbb{R}^N} (L_0^{-1}w)w^p\,dy = \int_{\mathbb{R}^N} (L_0^{-1}w)\frac{1}{p-1}L_0w\,dy$$

$$= \frac{1}{p-1}\int_{\mathbb{R}^N} w^2\,dy. \quad (3.5)$$

We divide the rest of the proof into three cases.

Case 1: $r = 2$, $1 < p < 1 + \frac{4}{N}$.
We introduce the following self-adjoint operator:

$$L_1\phi := L_0\phi - (p-1)\frac{\int_{\mathbb{R}^N} w\phi}{\int_{\mathbb{R}^N} w^2}w^p - (p-1)\frac{\int_{\mathbb{R}^N} w^p\phi}{\int_{\mathbb{R}^N} w^2}w$$

$$+ (p-1)\frac{\int_{\mathbb{R}^N} w^{p+1}\int_{\mathbb{R}^N} w\phi}{(\int_{\mathbb{R}^N} w^2)^2}w. \tag{3.6}$$

Then we have the following result:

Lemma 3.2

(1) L_1 is self-adjoint. The kernel X_1 of L_1 satisfies

$$X_1 = \text{span}\left\{w, \frac{\partial w}{\partial y_j} : j = 1, \ldots, N\right\}.$$

(2) There exists a positive constant a_1 such that

$$L_1(\phi, \phi) := \int_{\mathbb{R}^N}\left(|\nabla\phi|^2 + \phi^2 - pw^{p-1}\phi^2\right) + \frac{2(p-1)\int_{\mathbb{R}^N} w\phi\int_{\mathbb{R}^N} w^p\phi}{\int_{\mathbb{R}^N} w^2}$$

$$- (p-1)\frac{\int_{\mathbb{R}^N} w^{p+1}}{(\int_{\mathbb{R}^N} w^2)^2}\left(\int_{\mathbb{R}^N} w\phi\right)^2$$

$$\geq a_1 d^2_{L^2(\mathbb{R}^N)}(\phi, X_1) \quad \text{for all } \phi \in H^1(\mathbb{R}^N),$$

where $d_{L^2(\mathbb{R}^N)}$ denotes the distance in the norm of $L^2(\mathbb{R}^n)$.

Proof Firstly, we compute the kernel of L_1. It is easy to see that $w \in \ker(L_1)$ and $\frac{\partial w}{\partial y_j} \in \ker(L_1)$, $j = 1, \ldots, N$. On the other hand, if $\phi \in \ker(L_1)$, then by (3.3) we get

$$L_0\phi = c_1(\phi)w + c_2(\phi)w^p$$

$$= c_1(\phi)L_0\left(\frac{1}{p-1}w + \frac{1}{2}x \cdot \nabla w\right) + c_2(\phi)L_0\left(\frac{1}{p-1}w\right),$$

where

$$c_1(\phi) = (p-1)\frac{\int_{\mathbb{R}^N} w^p\phi}{\int_{\mathbb{R}^N} w^2} - (p-1)\frac{\int_{\mathbb{R}^N} w^{p+1}\int_{\mathbb{R}^N} w\phi}{(\int_{\mathbb{R}^N} w^2)^2},$$

$$c_2(\phi) = (p-1)\frac{\int_{\mathbb{R}^N} w\phi}{\int_{\mathbb{R}^N} w^2}.$$

Hence

$$\phi - c_1(\phi)\left(\frac{1}{p-1}w + \frac{1}{2}x \cdot \nabla w\right) - c_2(\phi)\frac{1}{p-1}w \in \ker(L_0). \tag{3.7}$$

Thus

$$c_1(\phi) = (p-1)c_1(\phi)\frac{\int_{\mathbb{R}^N} w^p((1/(p-1))w + (1/2)x \cdot \nabla w)}{\int_{\mathbb{R}^N} w^2}$$

$$- (p-1)c_1(\phi)\frac{\int_{\mathbb{R}^N} w^{p+1} \int_{\mathbb{R}^N} w((1/(p-1))w + (1/2)x \cdot \nabla w)}{(\int_{\mathbb{R}^N} w^2)^2}$$

$$= c_1(\phi) - c_1(\phi)\left(\frac{1}{p-1} - \frac{N}{4}\right)\frac{\int_{\mathbb{R}^N} w^{p+1}}{\int_{\mathbb{R}^N} w^2}$$

by (3.4) and (3.5). This implies that $c_1(\phi) = 0$. By (3.7) and Lemma 13.4, this proves (1).

Secondly, we prove (2). Suppose (2) is false. Then by (1) there exists (α, ϕ) such that (i) α is real and positive, (ii) $\phi \perp w$, $\phi \perp \frac{\partial w}{\partial y_j}$, $j = 1, \ldots, N$, and (iii) $L_1\phi = \alpha\phi$.

Now we show that this is impossible. From (ii) and (iii), we have

$$(L_0 - \alpha)\phi = (p-1)\frac{\int_{\mathbb{R}^N} w^p\phi}{\int_{\mathbb{R}^N} w^2}w. \tag{3.8}$$

We first claim that $\int_{\mathbb{R}^N} w^p\phi \neq 0$. In fact, if $\int_{\mathbb{R}^N} w^p\phi = 0$, then $\alpha > 0$ is an eigenvalue of L_0. By Lemma 13.5, we get that $\alpha = \mu_1$ and ϕ has constant sign. This contradicts the fact that $\phi \perp w$. Therefore $\alpha \neq \mu_1, 0$, and hence $L_0 - \alpha$ is invertible in $X_0^\perp$. Thus (3.8) implies

$$\phi = (p-1)\frac{\int_{\mathbb{R}^N} w^p\phi}{\int_{\mathbb{R}^N} w^2}(L_0 - \alpha)^{-1}w.$$

Therefore we have

$$\int_{\mathbb{R}^N} w^p\phi = (p-1)\frac{\int_{\mathbb{R}^N} w^p\phi}{\int_{\mathbb{R}^N} w^2}\int_{\mathbb{R}^N}\left((L_0 - \alpha)^{-1}w\right)w^p$$

from which we get

$$\int_{\mathbb{R}^N} w^2 = (p-1)\int_{\mathbb{R}^N}\left((L_0 - \alpha)^{-1}w\right)w^p,$$

$$\int_{\mathbb{R}^N} w^2 = \int_{\mathbb{R}^N}\left((L_0 - \alpha)^{-1}w\right)\left((L_0 - \alpha)w + \alpha w\right),$$

and finally

$$0 = \int_{\mathbb{R}^N}\left((L_0 - \alpha)^{-1}w\right)w. \tag{3.9}$$

Let $h_1(\alpha) = \int_{\mathbb{R}^N}((L_0 - \alpha)^{-1}w)w$. Then we compute

$$h_1(0) = \int_{\mathbb{R}^N} (L_0^{-1} w) w = \int_{\mathbb{R}^N} \left(\frac{1}{p-1} w + \frac{1}{2} x \cdot \nabla w \right) w$$

$$= \left(\frac{1}{p-1} - \frac{N}{4} \right) \int_{\mathbb{R}^N} w^2 > 0$$

since $1 < p < 1 + \frac{4}{N}$. Moreover, we have

$$h_1'(\alpha) = \int_{\mathbb{R}^N} ((L_0 - \alpha)^{-2} w) w = \int_{\mathbb{R}^N} ((L_0 - \alpha)^{-1} w)^2 > 0.$$

This implies $h_1(\alpha) > 0$ for all $\alpha \in (0, \mu_1)$. Since $\lim_{\alpha \to +\infty} h_1(\alpha) = 0$, we get $h_1(\alpha) < 0$ for $\alpha \in (\mu_1, \infty)$. This contradicts (3.9) and (2) is proven. $\qquad\square$

Next we finish the proof of Theorem 3.1 in Case 1. Let $\alpha_0 = \alpha_R + i\alpha_I$ and $\phi = \phi_R + i\phi_I$. Since $\alpha_0 \neq 0$, we can choose $\phi \perp \ker(L_0)$. Then we have the following linear system for (ϕ_R, ϕ_I):

$$L_0 \phi_R - (p-1)\gamma_0 \frac{\int_{\mathbb{R}^N} w \phi_R}{\int_{\mathbb{R}^N} w^2} w^p = \alpha_R \phi_R - \alpha_I \phi_I, \qquad (3.10)$$

$$L_0 \phi_I - (p-1)\gamma_0 \frac{\int_{\mathbb{R}^N} w \phi_I}{\int_{\mathbb{R}^N} w^2} w^p = \alpha_R \phi_I + \alpha_I \phi_R. \qquad (3.11)$$

Multiplying (3.10) by ϕ_R, (3.11) by ϕ_I, integrating and adding the two resulting equations, we get

$$-\alpha_R \int_{\mathbb{R}^N} (\phi_R^2 + \phi_I^2)$$

$$= L_1(\phi_R, \phi_R) + L_1(\phi_I, \phi_I)$$

$$+ (p-1)(\gamma_0 - 2) \frac{\int_{\mathbb{R}^N} w \phi_R \int_{\mathbb{R}^N} w^p \phi_R + \int_{\mathbb{R}^N} w \phi_I \int_{\mathbb{R}^N} w^p \phi_I}{\int_{\mathbb{R}^N} w^2}$$

$$+ (p-1) \frac{\int_{\mathbb{R}^N} w^{p+1}}{(\int_{\mathbb{R}^N} w^2)^2} \left[\left(\int_{\mathbb{R}^N} w \phi_R \right)^2 + \left(\int_{\mathbb{R}^N} w \phi_I \right)^2 \right].$$

On the other hand, multiplying (3.10) by w and integrating, we have

$$(p-1) \int_{\mathbb{R}^N} w^p \phi_R - \gamma_0 (p-1) \frac{\int_{\mathbb{R}^N} w \phi_R}{\int_{\mathbb{R}^N} w^2} \int_{\mathbb{R}^N} w^{p+1}$$

$$= \alpha_R \int_{\mathbb{R}^N} w \phi_R - \alpha_I \int_{\mathbb{R}^N} w \phi_I. \qquad (3.12)$$

Multiplying (3.11) by w and integrating, we obtain

$$(p-1)\int_{\mathbb{R}^N} w^p \phi_I - \gamma_0(p-1)\frac{\int_{\mathbb{R}^N} w\phi_I}{\int_{\mathbb{R}^N} w^2}\int_{\mathbb{R}^N} w^{p+1}$$

$$= \alpha_R \int_{\mathbb{R}^N} w\phi_I + \alpha_I \int_{\mathbb{R}^N} w\phi_R. \tag{3.13}$$

Multiplying (3.12) by $\int_{\mathbb{R}^N} w\phi_R$ and (3.13) by $\int_{\mathbb{R}^N} w\phi_I$ and adding, we have

$$(p-1)\int_{\mathbb{R}^N} w\phi_R \int_{\mathbb{R}^N} w^p\phi_R + (p-1)\int_{\mathbb{R}^N} w\phi_I \int_{\mathbb{R}^N} w^p\phi_I$$

$$= \left(\alpha_R + \gamma_0(p-1)\frac{\int_{\mathbb{R}^N} w^{p+1}}{\int_{\mathbb{R}^N} w^2}\right)\left(\left(\int_{\mathbb{R}^N} w\phi_R\right)^2 + \left(\int_{\mathbb{R}^N} w\phi_I\right)^2\right).$$

Therefore we get

$$-\alpha_R \int_{\mathbb{R}^N}(\phi_R^2 + \phi_I^2)$$

$$= L_1(\phi_R, \phi_R) + L_1(\phi_I, \phi_I)$$

$$+ (p-1)(\gamma_0-2)\left(\frac{1}{p-1}\alpha_R + \gamma_0\frac{\int_{\mathbb{R}^N} w^{p+1}}{\int_{\mathbb{R}^N} w^2}\right)\frac{(\int_{\mathbb{R}^N} w\phi_R)^2 + (\int_{\mathbb{R}^N} w\phi_I)^2}{\int_{\mathbb{R}^N} w^2}$$

$$+ (p-1)\frac{\int_{\mathbb{R}^N} w^{p+1}}{(\int_{\mathbb{R}^N} w^2)^2}\left[\left(\int_{\mathbb{R}^N} w\phi_R\right)^2 + \left(\int_{\mathbb{R}^N} w\phi_I\right)^2\right].$$

Decomposing

$$\phi_R = c_R w + \phi_R^\perp, \quad \phi_R^\perp \perp X_1, \qquad \phi_I = c_I w + \phi_I^\perp, \quad \phi_I^\perp \perp X_1,$$

we get

$$\int_{\mathbb{R}^N} w\phi_R = c_R \int_{\mathbb{R}^N} w^2, \qquad \int_{\mathbb{R}^N} w\phi_I = c_I \int_{\mathbb{R}^N} w^2,$$

$$d_{L^2(\mathbb{R}^N)}^2(\phi_R, X_1) = \|\phi_R^\perp\|_{L^2}^2, \qquad d_{L^2(\mathbb{R}^N)}^2(\phi_I, X_1) = \|\phi_I^\perp\|_{L^2}^2.$$

Using a few some elementary computations, we derive

$$L_1(\phi_R, \phi_R) + L_1(\phi_I, \phi_I)$$

$$+ (\gamma_0-1)\alpha_R(c_R^2 + c_I^2)\int_{\mathbb{R}^N} w^2 + (p-1)(\gamma_0-1)^2(c_R^2 + c_I^2)\int_{\mathbb{R}^N} w^{p+1}$$

$$+ \alpha_R(\|\phi_R^\perp\|_{L^2}^2 + \|\phi_I^\perp\|_{L^2}^2) = 0.$$

By Lemma 3.2(2), we get

$$(\gamma_0-1)\alpha_R(c_R^2 + c_I^2)\int_{\mathbb{R}^N} w^2 + (p-1)(\gamma_0-1)^2(c_R^2 + c_I^2)\int_{\mathbb{R}^N} w^{p+1}$$

$$+ (\alpha_R + a_1)(\|\phi_R^\perp\|_{L^2}^2 + \|\phi_I^\perp\|_{L^2}^2) \leq 0.$$

Since $\gamma_0 > 1$, we finally have $\alpha_R < 0$. This concludes the proof of Theorem 3.1 in Case 1.

Case 2: $r = 2$, $p = 1 + \frac{4}{N}$.

We compute

$$\int_{\mathbb{R}^N} \left(L_0^{-1} w\right) w = \int_{\mathbb{R}^N} w \left(\frac{1}{p-1} w + \frac{1}{2} x \cdot \nabla w\right) = 0. \qquad (3.14)$$

We set

$$w_0 = \frac{1}{p-1} w + \frac{1}{2} x \cdot \nabla w \qquad (3.15)$$

and follow the proof of Case 1. The following lemma is similar to Lemma 3.2 and its proof is omitted.

Lemma 3.3

(1) *The kernel of L_1 is given by*

$$X_1 = \text{span}\left\{ w, w_0, \frac{\partial w}{\partial y_j}, j = 1, \ldots, N \right\}.$$

(2) *There exists a positive constant a_2 such that*

$$L_1(\phi, \phi) = \int_{\mathbb{R}^N} \left(|\nabla \phi|^2 + \phi^2 - p w^{p-1} \phi^2\right) + \frac{2(p-1) \int_{\mathbb{R}^N} w\phi \int_{\mathbb{R}^N} w^p \phi}{\int_{\mathbb{R}^N} w^2}$$

$$- (p-1) \frac{\int_{\mathbb{R}^N} w^{p+1}}{(\int_{\mathbb{R}^N} w^2)^2} \left(\int_{\mathbb{R}^N} w\phi\right)^2$$

$$\geq a_2 d_{L^2(\mathbb{R}^N)}^2 (\phi, X_1) \quad \text{for all } \phi \in H^1(\mathbb{R}^N).$$

Next we finish the proof of Theorem 3.1 in Case 2. Suppose that $\alpha_0 \neq 0$ is an eigenvalue of L. Let $\alpha_0 = \alpha_R + i\alpha_I$ and $\phi = \phi_R + i\phi_I$. Since $\alpha_0 \neq 0$, we can choose $\phi \perp \ker(L_0)$. Then, similar to Case 1, we derive the two equations (3.10) and (3.11). We decompose

$$\phi_R = c_R w + b_R w_0 + \phi_R^\perp, \quad \phi_R^\perp \perp X_1, \quad \phi_I = c_I w + b_I w_0 + \phi_I^\perp, \quad \phi_I^\perp \perp X_1$$

and obtain

$$L_1(\phi_R, \phi_R) + L_1(\phi_I, \phi_I)$$

$$+ (\gamma_0 - 1)\alpha_R \left(c_R^2 + c_I^2\right) \int_{\mathbb{R}^N} w^2 + (p-1)(\gamma_0 - 1)^2 \left(c_R^2 + c_I^2\right) \int_{\mathbb{R}^N} w^{p+1}$$

$$+ \alpha_R \left[b_R^2 \left(\int_{\mathbb{R}^N} w_0^2\right)^2 + b_I^2 \left(\int_{\mathbb{R}^N} w_0^2\right)^2 + \|\phi_R^\perp\|_{L^2}^2 + \|\phi_I^\perp\|_{L^2}^2 \right] \leq 0.$$

By Lemma 3.3(2), we have

$$(\gamma_0 - 1)\alpha_R(c_R^2 + c_I^2)\int_{\mathbb{R}^N} w^2 + (p-1)(\gamma_0 - 1)^2(c_R^2 + c_I^2)\int_{\mathbb{R}^N} w^{p+1}$$

$$+ \alpha_R\left(b_R^2\left(\int_{\mathbb{R}^N} w_0^2\right)^2 + b_I^2\left(\int_{\mathbb{R}^N} w_0^2\right)^2\right)$$

$$+ (\alpha_R + a_2)(\|\phi_R^\perp\|_{L^2}^2 + \|\phi_I^\perp\|_{L^2}^2) \le 0.$$

If $\alpha_R \ge 0$, then it follows that

$$c_R = c_I = 0, \qquad \phi_R^\perp = 0, \qquad \phi_I^\perp = 0.$$

Hence we get $\phi_R = b_R w_0$, $\phi_I = b_I w_0$ and finally

$$b_R L_0 w_0 = (b_R - b_I)w_0, \qquad b_I L_0 w_0 = (b_R + b_I)w_0.$$

This is impossible unless $b_R = b_I = 0$. This gives the desired contradiction.

Case 3: $r = p + 1$, $1 < p < (\frac{N+2}{N-2})_+$.
Let $r = p + 1$. Then L can be written as

$$L\phi = L_0\phi - \frac{qr}{s+1}\frac{\int_{\mathbb{R}^N} w^p\phi}{\int_{\mathbb{R}^N} w^{p+1}}w^p.$$

To follow the proof of Case 1, we introduce the operator

$$L_3\phi := L_0\phi - (p-1)\frac{\int_{\mathbb{R}^N} w^p\phi}{\int_{\mathbb{R}^N} w^{p+1}}w^p. \tag{3.16}$$

Then we have the following result:

Lemma 3.4

(1) *The operator L_3 is self-adjoint and its kernel is given by*

$$X_1 = \text{span}\left\{w, \frac{\partial w}{\partial y_j}, j = 1, \ldots, N\right\}.$$

(2) *There exists a positive constant a_3 such that*

$$L_3(\phi, \phi) := \int_{\mathbb{R}^N}(|\nabla\phi|^2 + \phi^2 - pw^{p-1}\phi^2) + \frac{(p-1)(\int_{\mathbb{R}^N} w^p\phi)^2}{\int_{\mathbb{R}^N} w^{p+1}}$$

$$\ge a_3 d_{L^2(\mathbb{R}^N)}^2(\phi, X_3) \quad \text{for all } \phi \in H^1(\mathbb{R}^N).$$

Proof The proof of (1) is similar to that of Lemma 3.2. We omit the details. It remains to prove (2). Suppose (2) is not true, then by (1) there exists (α, ϕ) such that (i) α is real and positive, (ii) $\phi \perp w$, $\phi \perp \frac{\partial w}{\partial y_j}$, $j = 1, \ldots, N$, and (iii) $L_3 \phi = \alpha \phi$.

Next we prove that this is impossible. From (ii) and (iii), we get

$$(L_0 - \alpha)\phi = \frac{(p-1) \int_{\mathbb{R}^N} w^p \phi}{\int_{\mathbb{R}^N} w^{p+1}} w^p. \qquad (3.17)$$

Similar to the proof of Lemma 3.2, we have that $\int_{\mathbb{R}^N} w^p \phi \neq 0$ for $\alpha \neq \mu_1, 0$, and hence $L_0 - \alpha$ is invertible in $X_0^\perp$. Thus from (3.17) we get

$$\phi = \frac{(p-1) \int_{\mathbb{R}^N} w^p \phi}{\int_{\mathbb{R}^N} w^{p+1}} (L_0 - \alpha)^{-1} w^p.$$

Finally, we have

$$\int_{\mathbb{R}^N} w^p \phi = (p-1) \frac{\int_{\mathbb{R}^N} w^p \phi}{\int_{\mathbb{R}^N} w^{p+1}} \int_{\mathbb{R}^N} \left((L_0 - \alpha)^{-1} w^p\right) w^p$$

and

$$\int_{\mathbb{R}^N} w^{p+1} = (p-1) \int_{\mathbb{R}^N} \left((L_0 - \alpha)^{-1} w^p\right) w^p. \qquad (3.18)$$

Letting

$$h_3(\alpha) = (p-1) \int_{\mathbb{R}^N} \left((L_0 - \alpha)^{-1} w^p\right) w^p - \int_{\mathbb{R}^N} w^{p+1}$$

we compute

$$h_3(0) = (p-1) \int_{\mathbb{R}^N} \left(L_0^{-1} w^p\right) w^p - \int_{\mathbb{R}^N} w^{p+1} = 0.$$

Moreover, we have

$$h_3'(\alpha) = (p-1) \int_{\mathbb{R}^N} \left((L_0 - \alpha)^{-2} w^p\right) w^p = (p-1) \int_{\mathbb{R}^N} \left((L_0 - \alpha)^{-1} w^p\right)^2 > 0.$$

Thus we get $h_3(\alpha) > 0$ for all $\alpha \in (0, \mu_1)$. On the other hand, we have $h_3(\alpha) < 0$ for $\alpha \in (\mu_1, \infty)$, which contradicts (3.18). $\qquad\qquad\qquad\qquad\qquad\qquad \square$

We finish the proof of Theorem 3.1 in Case 3. Let $\alpha_0 = \alpha_R + i\alpha_I$ and $\phi = \phi_R + i\phi_I$. Since $\alpha_0 \neq 0$, we can choose $\phi \perp \ker(L_0)$ and obtain the linear system

$$L_0 \phi_R - (p-1)\gamma_0 \frac{\int_{\mathbb{R}^N} w^p \phi_R}{\int_{\mathbb{R}^N} w^{p+1}} w^p = \alpha_R \phi_R - \alpha_I \phi_I, \qquad (3.19)$$

$$L_0 \phi_I - (p-1)\gamma_0 \frac{\int_{\mathbb{R}^N} w^p \phi_I}{\int_{\mathbb{R}^N} w^{p+1}} w^p = \alpha_R \phi_I + \alpha_I \phi_R. \qquad (3.20)$$

Multiplying (3.19) by ϕ_R, (3.20) by ϕ_I, integrating and adding the equations, we get

$$-\alpha_R \int_{\mathbb{R}^N} \left(\phi_R^2 + \phi_I^2\right) = L_3(\phi_R, \phi_R) + L_3(\phi_I, \phi_I)$$

$$+ (p-1)(\gamma_0 - 1) \frac{(\int_{\mathbb{R}^N} w^p \phi_R)^2 + (\int_{\mathbb{R}^N} w^p \phi_I)^2}{\int_{\mathbb{R}^N} w^{p+1}}.$$

By Lemma 3.4(2), we have

$$\alpha_R \int_{\mathbb{R}^N} \left(\phi_R^2 + \phi_I^2\right) + a_2 d_{L^2}^2(\phi, X_1)$$

$$+ (p-1)(\gamma_0 - 1) \frac{(\int_{\mathbb{R}^N} w^p \phi_R)^2 + (\int_{\mathbb{R}^N} w^p \phi_I)^2}{\int_{\mathbb{R}^N} w^{p+1}} \le 0.$$

Thus we have derived $\alpha_R < 0$ since $\gamma_0 > 1$ and Theorem 3.1 Part (2) in Case 3 is shown.

The proof of Part (1) is similar to the proof of Lemma 3.7 below. Part (3) is proved as in the argument immediately following the proof of Lemma 3.7, which implies that the eigenvalues will not cross through zero.   □

3.2 The Method of Continuation

In our applications to the case when $\tau > 0$, we have to deal with the situation when the coefficient γ is a function of $\tau\alpha$. Now we will extend the results from the basic case $\tau = 0$ of the previous section to the case when $\tau > 0$ is small enough by using a perturbation argument. The main point in ensuring that the perturbation argument works is showing that the eigenvalues remain uniformly bounded for τ small enough. Let $\gamma = \gamma(\tau\alpha)$ be a complex function of $\tau\alpha$. Let us suppose that

$$\gamma(0) \in \mathbb{R} \quad \text{and} \quad |\gamma(\tau\alpha)| \le C \quad \text{for } \alpha_R \ge 0, \tau \ge 0, \qquad (3.21)$$

where C is a generic constant independent of τ and α. Simple examples of $\gamma(\tau\alpha)$ satisfying (3.21) are

$$\gamma(\tau\alpha) = \frac{2}{\sqrt{1+\tau\alpha}+1} \quad \text{or} \quad \gamma(\tau\alpha) = \frac{\mu}{1+\tau\alpha} \quad (\mu > 0),$$

where $\sqrt{1+\tau\alpha}$ is the principal branch of the square root. Now we have

Theorem 3.5 *Assume that (3.1) holds and consider the nonlocal eigenvalue problem*

$$\Delta\phi - \phi + pw^{p-1}\phi - \gamma(\tau\alpha)(p-1)\frac{\int_{\mathbb{R}} w^{r-1}\phi}{\int_{\mathbb{R}} w^r} w^p = \alpha\phi, \qquad (3.22)$$

where $\gamma(\tau\alpha)$ satisfies (3.21). Then there is a small number $\tau_0 > 0$ such that for $\tau < \tau_0$:

(1) *if $\gamma(0) < 1$, then there is a positive eigenvalue of (3.2);*
(2) *if $\gamma(0) > 1$ and (3.1) holds, then for any nonzero eigenvalue α of (3.22), we have*

$$\mathrm{Re}(\alpha) \leq -c_0 < 0.$$

Proof Theorem 3.5 follows from Theorem 3.1 by a perturbation argument. To guarantee that the perturbation argument works, we have to show that if $\alpha_R \geq 0$ and $0 < \tau < 1$, then $|\alpha| \leq C$, where C is a generic constant (independent of τ). Multiplying (3.22) by the conjugate $\bar{\phi}$ of ϕ and integrating by parts, we get that

$$\int_{\mathbb{R}} \left(|\nabla\phi|^2 + |\phi|^2 - pw^{p-1}|\phi|^2 \right)$$

$$= -\alpha \int_{\mathbb{R}} |\phi|^2 - \gamma(\tau\alpha)(p-1)\frac{\int_{\mathbb{R}} w^{r-1}\phi}{\int_{\mathbb{R}} w^r} \int_{\mathbb{R}} w^p\bar{\phi}. \qquad (3.23)$$

From the imaginary part of (3.23), we obtain that

$$|\alpha_I| \leq C_1 |\gamma(\tau\alpha)|,$$

where $\alpha = \alpha_R + \sqrt{-1}\alpha_I$ and C_1 is a positive constant (independent of τ). By assumption (3.21), we have $|\gamma(\tau\alpha)| \leq C$ and so $|\alpha_I| \leq C$. Taking the real part of (3.23) and noting that

$$\text{l.h.s. of (3.23)} \geq C \int_{\mathbb{R}} |\phi|^2 \quad \text{for some } C \in \mathbb{R},$$

we obtain that $\alpha_R \leq C_2$, where C_2 is a positive constant (independent of $\tau > 0$). Therefore, $|\alpha|$ is uniformly bounded and hence the perturbation argument implies the conclusion of the theorem. $\qquad \square$

3.3 Hopf Bifurcation

Now we continue to consider the case $\tau > 0$. We relax the condition on the smallness of τ and allow τ to be any positive number. On the other hand, the function $\gamma(\tau\alpha)$ now has to be specified and the results will depend on the choice of function $\gamma(\tau\alpha)$ more explicitly than in the previous section.

In particular, we consider the following two nonlocal eigenvalue problems in the two-dimensional case:

$$L\phi := \Delta\phi - \phi + 2w\phi - \gamma\frac{\int_{\mathbb{R}^2} w\phi}{\int_{\mathbb{R}^2} w^2}w^2 = \lambda_0\phi, \quad \phi \in H^1(\mathbb{R}^2), \qquad (3.24)$$

where either (a) $\gamma = \frac{\mu}{1+\tau\lambda_0}$ with $\mu > 0$, $\tau \geq 0$, or (b) $\gamma = \frac{2(K+\eta_0(1+\tau\lambda_0))}{(K+\eta_0)(1+\tau\lambda_0)}$ with $\eta_0 > 0$, $\tau \geq 0$.

Case (a) will be studied in Theorem 3.6 and Case (b) will be considered in Theorem 3.9.

First we consider Case (a):

Theorem 3.6 *Let* $\gamma = \frac{\mu}{1+\tau\lambda_0}$, *where* $\mu > 0$, $\tau \geq 0$ *and let the operator L be defined by (3.24).*

(1) *Suppose that* $\mu > 1$. *Then there exists a unique* $\tau = \tau_1 > 0$ *such that for* $\tau < \tau_1$, *(3.24) admits a positive eigenvalue, and for* $\tau > \tau_1$, *all nonzero eigenvalues of problem (3.24) satisfy* $\mathrm{Re}(\lambda) \leq -c < 0$. *At* $\tau = \tau_1$, *L has a Hopf bifurcation.*
(2) *Suppose that* $\mu < 1$. *Then L admits a positive eigenvalue* λ_0.

Proof Theorem 3.6 will be proved by the following two lemmas.

Lemma 3.7 *If* $\mu < 1$, *then L has a positive eigenvalue* λ_0.

Proof We conclude that ϕ is a radially symmetric function: $\phi \in H_r^2(\mathbb{R}^2) = \{u \in H^1(\mathbb{R}^2)|u = u(|y|)\}$. Let L_0 be defined as in (2.40). Then, by Lemma 13.5, it follows that L_0 is invertible in $H_r^2(\mathbb{R}^2)$. We denote its inverse by L_0^{-1}. By Lemma 13.5, the operator L_0 has a unique positive eigenvalue μ_1. Using $\int_{\mathbb{R}^2} w\Phi_0 > 0$, we conclude that $\lambda_0 \neq \mu_1$. Now $\lambda_0 > 0$ is an eigenvalue of (3.24) if and only if it is a solution of the algebraic equation

$$\int_{\mathbb{R}^2} w^2 = \frac{\mu}{1+\tau\lambda_0} \int_{\mathbb{R}^2} [((L_0 - \lambda_0)^{-1}w^2)w]. \tag{3.25}$$

The algebraic equation (3.25) can be simplified to

$$\rho(\lambda_0) := ((\mu - 1) - \tau\lambda_0)\int_{\mathbb{R}^2} w^2 + \mu\lambda_0 \int_{\mathbb{R}^2}[((L_0 - \lambda_0)^{-1}w)w] = 0, \tag{3.26}$$

where $\rho(0) = (\mu - 1)\int_{\mathbb{R}^2} w^2 < 0$. Further, using $\lambda_0 \to \mu_1$, $\lambda_0 < \mu_1$, we get

$$\int_{\mathbb{R}^2} ((L_0 - \lambda_0)^{-1}w)w \to +\infty$$

and thus $\rho(\lambda_0) \to +\infty$. By continuity, there is a $\lambda_0 \in (0, \mu_1)$ such that $\rho(\lambda_0) = 0$, and λ_0 is an eigenvalue of L. $\qquad\square$

Next we study the case $\mu > 1$. It suffices to restrict our attention to radially symmetric functions. By Theorem 1.4 of [249], for $\tau = 0$ (and by perturbation, for τ small), all eigenvalues are located on the left half of the complex plane. By [37], for τ large, there are unstable eigenvalues.

It is easy to see that the eigenvalues will not cross through zero: If $\lambda_0 = 0$, then we get

$$L_0\phi - \mu \frac{\int_{\mathbb{R}^2} w\phi}{\int_{\mathbb{R}^2} w^2} w^2 = 0$$

which implies that

$$L_0\left(\phi - \mu \frac{\int_{\mathbb{R}^2} w\phi}{\int_{\mathbb{R}^2} w^2} w\right) = 0$$

and by Lemma 13.4 we get

$$\phi - \mu \frac{\int_{\mathbb{R}^2} w\phi}{\int_{\mathbb{R}^2} w^2} w \in X_0.$$

This is impossible since ϕ is a radially symmetric function and $\phi \neq cw$ for all $c \in R$.

Hence there is a point τ_1 at which L has a Hopf bifurcation, i.e., L has a purely imaginary eigenvalue $\alpha = \sqrt{-1}\alpha_I$. To conclude the proof of Theorem 3.6(1), it suffices to show that τ_1 is unique.

Lemma 3.8 *Let $\mu > 1$. Then there exists a unique $\tau_1 > 0$ such that L has a Hopf bifurcation.*

Proof Let $\lambda_0 = \sqrt{-1}\alpha_I$ be an eigenvalue of L. Without loss of generality, we may assume that $\alpha_I > 0$. (Note that then $-\sqrt{-1}\alpha_I$ is also an eigenvalue of L.) Letting $\phi_0 = (L_0 - \sqrt{-1}\alpha_I)^{-1}w^2$, (3.24) becomes

$$\frac{\int_{\mathbb{R}^2} w\phi_0}{\int_{\mathbb{R}^2} w^2} = \frac{1 + \tau\sqrt{-1}\alpha_I}{\mu}. \tag{3.27}$$

Decomposing $\phi_0 = \phi_0^R + \sqrt{-1}\phi_0^I$, from (3.27) we derive the linear system

$$\frac{\int_{\mathbb{R}^2} w\phi_0^R}{\int_{\mathbb{R}^2} w^2} = \frac{1}{\mu}, \tag{3.28}$$

$$\frac{\int_{\mathbb{R}^2} w\phi_0^I}{\int_{\mathbb{R}^2} w^2} = \frac{\tau\alpha_I}{\mu}. \tag{3.29}$$

Note that only (3.29) depends on τ, whereas (3.28) is independent of τ.

Next we compute $\int_{\mathbb{R}^2} w\phi_0^R$. Using the fact that (ϕ_0^R, ϕ_0^I) satisfies

$$L_0\phi_0^R = w^2 - \alpha_I\phi_0^I, \qquad L_0\phi_0^I = \alpha_I\phi_0^R,$$

we get $\phi_0^R = \alpha_I^{-1}L_0\phi_0^I$ and

$$\phi_0^I = \alpha_I(L_0^2 + \alpha_I^2)^{-1}w^2, \qquad \phi_0^R = L_0(L_0^2 + \alpha_I^2)^{-1}w^2. \tag{3.30}$$

Substituting (3.30) into (3.28) and (3.29), we have

$$\frac{\int_{\mathbb{R}^2}[wL_0(L_0^2+\alpha_I^2)^{-1}w^2]}{\int_{\mathbb{R}^2}w^2} = \frac{1}{\mu}, \tag{3.31}$$

$$\frac{\int_{\mathbb{R}^2}[w(L_0^2+\alpha_I^2)^{-1}w^2]}{\int_{\mathbb{R}^2}w^2} = \frac{\tau}{\mu}. \tag{3.32}$$

Setting

$$h(\alpha_I) = \frac{\int_{\mathbb{R}^2}wL_0(L_0^2+\alpha_I^2)^{-1}w^2}{\int_{\mathbb{R}^2}w^2} = h(\alpha_I) = \frac{\int_{\mathbb{R}^2}w^2(L_0^2+\alpha_I^2)^{-1}w^2}{\int_{\mathbb{R}^2}w^2},$$

we compute $h'(\alpha_I) = -2\alpha_I\frac{\int_{\mathbb{R}^2}w^2(L_0^2+\alpha_I^2)^{-2}w^2}{\int_{\mathbb{R}^2}w^2} < 0$. Since

$$h(0) = \frac{\int_{\mathbb{R}^2}w(L_0^{-1}w^2)}{\int_{\mathbb{R}^2}w^2} = 1,$$

$h(\alpha_I) \to 0$ as $\alpha_I \to \infty$ and $\mu > 1$, there exists a unique $\alpha_I > 0$ such that (3.31) holds. Substituting α_I into (3.32), we get a unique $\tau = \tau_1 > 0$ and the proof of Lemma 3.8 is finished.   □

   Theorem 3.6 follows from Lemmas 3.7 and 3.8.   □
   Finally we study Case (b) by considering the NLEP

$$\Delta\phi - \phi + 2w\phi - \frac{2(K+\eta_0(1+\tau\lambda_0))}{(K+\eta_0)(1+\tau\lambda_0)}\frac{\int_{\mathbb{R}^2}w\phi}{\int_{\mathbb{R}^2}w^2}w^2 = \lambda_0\phi, \quad \phi \in H^1(\mathbb{R}^2), \tag{3.33}$$

where $0 < \eta_0 < +\infty$ and $0 \le \tau < +\infty$.
   We have the following result:

Theorem 3.9

(1) If $\eta_0 < K$, then for τ small enough problem (3.33) is stable and for τ large enough it is unstable.
(2) If $\eta_0 > K$, then there exist $0 < \tau_2 \le \tau_3$ such that problem (3.33) is stable for $\tau < \tau_2$ or $\tau > \tau_3$.

Proof Setting

$$f(\tau\lambda) = \frac{2(K+\eta_0(1+\tau\lambda))}{(K+\eta_0)(1+\tau\lambda)}, \tag{3.34}$$

we note that

$$\lim_{\tau\lambda\to+\infty}f(\tau\lambda) = \frac{2\eta_0}{K+\eta_0} =: f_\infty.$$

If $\eta_0 < K$, then by Theorem 3.6(2), the problem (3.33) with $\mu = f_\infty$ possesses a positive eigenvalue α_1. Using a regular perturbation argument, this implies that for τ large enough problem (3.33) has an eigenvalue near $\alpha_1 > 0$. We conclude that for τ large enough problem (3.33) is unstable.

Next we show that problem (3.33) does not possess any nonzero eigenvalues with nonnegative real part, provided that either τ is small or $\eta_0 > K$ and τ is large. (It is immediately seen that $f(\tau\lambda) \to 2$ as $\tau\lambda \to 0$ and $f(\tau\lambda) \to \frac{2\eta_0}{\eta_0+K} > 1$ as $\tau\lambda \to +\infty$ if $\eta_0 > K$ and thus Theorem 3.6 should apply. However, we do not have control on $\tau\lambda$. Here we provide a rigorous proof.)

We apply the following inequality (see Lemma 3.2): For any (real-valued function) $\phi \in H_r^1(\mathbb{R}^2)$, we have

$$\int_{\mathbb{R}^2} \left(|\nabla\phi|^2 + \phi^2 - 2w\phi^2 \right)$$
$$+ 2\frac{\int_{\mathbb{R}^2} w\phi \int_{\mathbb{R}^2} w^2\phi}{\int_{\mathbb{R}^2} w^2} - \frac{\int_{\mathbb{R}^2} w^3}{(\int_{\mathbb{R}^2} w^2)^2}\left(\int_{\mathbb{R}^2} w\phi \right)^2 \geq 0, \qquad (3.35)$$

where equality holds if and only if ϕ is a multiple of w.

Let $\lambda_0 = \lambda_R + \sqrt{-1}\lambda_I$, $\phi = \phi_R + \sqrt{-1}\phi_I$ satisfy (3.33). Then we get

$$L_0\phi - f(\tau\lambda_0)\frac{\int_{\mathbb{R}^2} w\phi}{\int_{\mathbb{R}^2} w^2}w^2 = \lambda_0\phi. \qquad (3.36)$$

Multiplying (3.36) by the complex conjugate $\bar{\phi}$ of the function ϕ and integrating over $\mathbb{R}^2$, we have

$$\int_{\mathbb{R}^2} \left(|\nabla\phi|^2 + |\phi|^2 - 2w|\phi|^2 \right) = -\lambda_0 \int_{\mathbb{R}^2} |\phi|^2 - f(\tau\lambda_0)\frac{\int_{\mathbb{R}^2} w\phi}{\int_{\mathbb{R}^2} w^2} \int_{\mathbb{R}^2} w^2\bar{\phi}. \quad (3.37)$$

Multiplying (3.36) by w and integrating over $\mathbb{R}^2$, we obtain

$$\int_{\mathbb{R}^2} w^2\phi = \left(\lambda_0 + f(\tau\lambda_0)\frac{\int_{\mathbb{R}^2} w^3}{\int_{\mathbb{R}^2} w^2} \right) \int_{\mathbb{R}^2} w\phi. \qquad (3.38)$$

Taking the complex conjugate of (3.38) gives

$$\int_{\mathbb{R}^2} w^2\bar{\phi} = \left(\bar{\lambda}_0 + f(\tau\bar{\lambda}_0)\frac{\int_{\mathbb{R}^2} w^3}{\int_{\mathbb{R}^2} w^2} \right) \int_{\mathbb{R}^2} w\bar{\phi}. \qquad (3.39)$$

Substituting (3.39) into (3.37), it follows that

$$\int_{\mathbb{R}^2} \left(|\nabla\phi|^2 + |\phi|^2 - 2w|\phi|^2 \right)$$
$$= -\lambda_0 \int_{\mathbb{R}^2} |\phi|^2 - f(\tau\lambda_0)\left(\bar{\lambda}_0 + f(\tau\bar{\lambda}_0)\frac{\int_{\mathbb{R}^2} w^3}{\int_{\mathbb{R}^2} w^2} \right)\frac{|\int_{\mathbb{R}^2} w\phi|^2}{\int_{\mathbb{R}^2} w^2}. \qquad (3.40)$$

Next we consider the real part of (3.40). Applying the inequality (3.35) and using (3.39), we get

$$-\lambda_R \geq \mathrm{Re}\left(f(\tau\lambda_0)\left(\bar{\lambda}_0 + f(\tau\bar{\lambda}_0)\frac{\int_{\mathbb{R}^2} w^3}{\int_{\mathbb{R}^2} w^2} \right) \right)$$
$$- 2\,\mathrm{Re}\left(\bar{\lambda}_0 + f(\tau\bar{\lambda}_0)\frac{\int_{\mathbb{R}^2} w^3}{\int_{\mathbb{R}^2} w^2} \right) + \frac{\int_{\mathbb{R}^2} w^3}{\int_{\mathbb{R}^2} w^2},$$

where $\lambda_0 = \lambda_R + \sqrt{-1}\lambda_I$ with $\lambda_R, \lambda_I \in R$.

Assuming that $\lambda_R \geq 0$, we get

$$\frac{\int_{\mathbb{R}^2} w^3}{\int_{\mathbb{R}^2} w^2}\left| f(\tau\lambda_0) - 1 \right|^2 + \mathrm{Re}\big(\bar{\lambda}_0\big(f(\tau\lambda_0) - 1\big)\big) \leq 0. \qquad (3.41)$$

By the Pohozaev identity for (6.9) (multiplying (6.9) by $y \cdot \nabla w(y)$ and integrating by parts), we have

$$\int_{\mathbb{R}^2} w^3 = \frac{3}{2}\int_{\mathbb{R}^2} w^2. \qquad (3.42)$$

Substituting (3.42) and the expression (3.34) for $f(\tau\lambda)$ into (3.41), we get

$$\frac{3}{2}\big|\eta_0 + K + (\eta_0 - K)\tau\lambda\big|^2$$
$$+ \mathrm{Re}\big[(\eta_0 + K)(1 + \tau\bar{\lambda}_0)\big((\eta_0 + K)\bar{\lambda}_0 + (\eta_0 - K)\tau|\lambda_0|^2\big)\big] \leq 0.$$

This is equivalent to

$$\frac{3}{2}(1 + \mu_0\tau\lambda_R)^2 + \lambda_R + \big(\mu_0\tau + \tau + \mu_0\tau^2|\lambda_0|^2\big)\lambda_R$$
$$+ \left(\frac{3}{2}\mu_0^2\tau^2 + \mu_0\tau - \tau\right)\lambda_I^2 \leq 0, \qquad (3.43)$$

where $\mu_0 := \frac{\eta_0 - K}{\eta_0 + K}$.

If $\eta_0 > K$ (i.e., $\mu_0 > 0$) and τ is large, then

$$\frac{3}{2}\mu_0^2\tau^2 + \mu_0\tau - \tau \geq 0. \qquad (3.44)$$

Thus (3.43) does not hold for $\lambda_R \geq 0$.

To consider the case when τ is small, we next derive an upper bound for λ_I.

By (3.37), we get

$$\lambda_I \int_{\mathbb{R}^2} |\phi|^2 = \mathrm{Im}\left(-f(\tau\lambda_0)\frac{\int_{\mathbb{R}^2} w\phi}{\int_{\mathbb{R}^2} w^2}\int_{\mathbb{R}^2} w^2\bar{\phi} \right).$$

Thus we have

$$|\lambda_I| \leq |f(\tau\lambda_0)|\sqrt{\frac{\int_{\mathbb{R}^2} w^4}{\int_{\mathbb{R}^2} w^2}} \leq C \tag{3.45}$$

where C is independent of λ_0.

Substituting (3.45) into (3.43), we conclude that (3.43) does not hold for $\lambda_R \geq 0$, if τ is small. □

Remark 3.10 The proof of Theorem 3.9 allows us to obtain explicit values for τ_2 and τ_3. (In fact, first from (3.44) we obtain a value for τ_3. Then from (3.43) and (3.45) we get a value for τ_2.)

3.4   The Method of Hypergeometric Functions

In this section, we use the method of hypergeometric functions for eigenvalue problems on the real line to determine analytically for which $\tau > 0$ Hopf bifurcation occurs and the eigenvalue problem becomes unstable. We begin by studying two eigenvalue problems in $\mathbb{R}$. The first is a local eigenvalue problem with complex coefficients

$$\begin{cases} \Delta\phi - \phi + w\phi + \sigma w\phi = \lambda\phi, \\ \sigma = \sigma_R + i\sigma_I = |\sigma|e^{i\theta}, \quad |\sigma| > 0, \theta \in (-\pi, \pi], \phi \in H^1(\mathbb{R}), \end{cases} \tag{3.46}$$

where w is defined by (2.3).

The second is a scalar nonlocal eigenvalue problem (NLEP):

$$\Delta\phi - \phi + 2w\phi - \frac{2(1-\eta)}{\eta\sqrt{1+\tau\lambda}+1-\eta} \frac{\int_R w\phi}{\int_R w^2} w^2 = \lambda\phi, \quad \phi \in H^1(\mathbb{R}), \tag{3.47}$$

where

$$0 < \eta < 1, \qquad \tau \geq 0, \qquad \lambda \in \mathbb{C}, \qquad \lambda = \lambda_R + i\lambda_I, \qquad \lambda_R \geq 0$$

and we take the principal branch of $\sqrt{1+\tau\lambda}$.

To study (3.46) and (3.47), we first collect some important properties associated with the function w.

Lemma 3.11

(1) *The eigenvalue problem*

$$\begin{cases} \Delta\phi - \phi + \mu w\phi = 0, \\ \phi \in H^1(\mathbb{R}) \end{cases} \tag{3.48}$$

has the following eigenvalues

$$\mu_1 = 1, \qquad v_1 = \text{span}\{w\},$$

$$\mu_2 = 2, \qquad v_2 = \text{Ker}(L_0),$$

$$\mu_n = \frac{(1+n)(2+n)}{6} > 2, \quad \text{for } n \geq 3.$$

(2) *If $\mu_R > 0$, then the eigenvalue problem*

$$\begin{cases} \Delta\phi - \phi + w\phi + \mu_R w\phi = \lambda\phi, \\ \mu_R > 0, \qquad \phi \in H^1(\mathbb{R}) \end{cases}$$

has a positive (principal) eigenvalue λ_1 such that

$$-\lambda_1 = \inf_{\phi \in H^1(\mathbb{R})\setminus\{0\}} \frac{\int_{\mathbb{R}} (\phi')^2 + \phi^2 - (1+\mu_R)w\phi^2}{\int_{\mathbb{R}} \phi^2} < 0.$$

Moreover, in case $\mu_R = 1$, there is only one positive eigenvalue (which is the principal one).

(3) *Let ϕ (complex-valued) satisfy the following eigenvalue problem*

$$\begin{cases} \Delta\phi - \phi + w\phi + \sigma w\phi = \lambda\phi \\ \text{Re}(\sigma) \leq 0, \qquad \phi \in H^1(\mathbb{R}), \qquad \lambda \neq 0. \end{cases}$$

Then we have

$$\text{Re}(\lambda) \leq -c_0 < 0.$$

Proof For Part (1), the fact that $\mu_1 = 1$ and $\mu_2 = 2$ is proved in Lemma 4.1 of [245]. In fact, in this case, $\lambda = 0$, $\gamma = 1$ and the eigenvalues are given by

$$a = 2\gamma - \alpha = -(n-1), \quad n = 1, 2, \ldots$$

where $\alpha^2 + \alpha - 6\mu = 0$. Thus $\mu_n = \frac{\alpha^2 + \alpha}{6}$, $\alpha = n + 1$.

Part (2) follows by the variational characterisation of eigenvalues:

$$-\lambda_1 = \inf_{\phi \in H^1(\mathbb{R}), \phi \neq 0} \frac{\int_{\mathbb{R}} (\phi')^2 + \phi^2 - (1+\mu_R)w\phi^2}{\int_{\mathbb{R}} \phi^2} < 0$$

since by the last inequality for $\phi = w$ we have

$$-\lambda_1 \leq -\mu_R \frac{\int_{\mathbb{R}} w^3}{\int_{\mathbb{R}} w^2} < 0.$$

For $\mu_R = 1$ there exists only one positive eigenvalue (which is the principal one). See Lemma 13.5.

To prove Part 3(3), we note that

$$\sigma = \sigma_R + i\sigma_I, \qquad \phi = \phi_R + i\phi_I, \qquad \lambda = \lambda_R + i\lambda_I$$

and write the eigenvalue problem for real and imaginary parts separately:

$$\Delta\phi_R - \phi_R + (1+\sigma_R)w\phi_R - \sigma_I w\phi_I = \lambda_R\phi_R - \lambda_I\phi_I, \qquad (3.49)$$

$$\Delta\phi_R - \phi_I + (1+\sigma_R)w\phi_I + \sigma_I w\phi_R = \lambda_R\phi_I + \lambda_I\phi_R. \qquad (3.50)$$

Multiplying (3.49) by ϕ_R, (3.50) by ϕ_I, integrating over $\mathbb{R}$, and adding up, we get

$$\int_{\mathbb{R}}\left[-(\phi_R')^2 - \phi_R^2 + (1+\sigma_R)w\phi_R^2\right] + \int_{\mathbb{R}}\left[-|\phi_I'|^2 - \phi_I^2 + (1+\sigma_R)w\phi_I^2\right]$$

$$= \lambda_R\int_{\mathbb{R}}\phi_R^2 + \phi_I^2.$$

Since in the last equation l.h.s. ≤ 0 we also have r.h.s. ≤ 0. Therefore we get $\lambda_R \leq 0$. Now we assume that $\lambda_R = 0$. Then by Part (2) we get $\phi_R = c_1 w$, $\phi_I = c_2 w$ (with $c_1, c_2 \in \mathbb{R}$) and $\sigma_R = 0$. But this implies $\lambda_I = 0$, $\sigma_I = 0$ and we have $\lambda = 0$, contrary to what we assumed. Therefore λ_R cannot vanish and we get that $\mathrm{Re}\,\lambda \leq -c_0 < 0$.

$\square$

Now we are ready to study the first eigenvalue problem (3.46). By symmetry, we may assume that $\theta \in [0, \frac{\pi}{2}]$. We consider θ as a parameter. By Lemma 3.11(2) and a perturbation argument, for $|\theta|$ near 0, there exists an unstable eigenvalue λ of problem (3.46), i.e. $\lambda = \lambda_R + i\lambda_I$ where $\lambda_R > 0$. On the other hand, by Lemma 3.11(3), for $|\theta| \geq \frac{\pi}{2}$, problem (3.46) has only stable eigenvalues, i.e. $\lambda = \lambda_R + i\lambda_I$, where $\lambda_R < 0$. Varying θ, there must be a point $\theta^h \in (0, \frac{\pi}{2})$ such that for $\theta = \theta^h$, problem (3.46) has a Hopf bifurcation, i.e. there is an eigenvalue $\lambda = \sqrt{-1}\lambda_I$. Next we compute θ^h. It turns out that we can obtain the exact value for θ^h which is given in the following result:

Lemma 3.12 *Let ϕ (complex-valued) satisfy the eigenvalue problem (3.46) with $\sigma = \sigma_R + \sqrt{-1}\sigma_I$, $\sigma_R > 0$. Setting*

$$f(\sigma) := (12\sigma_R + 5)^2(3\sigma_R^2 + 2\sigma_R) - 3\sigma_I^2, \qquad (3.51)$$

we have the following three cases:

(1) *If $f(\sigma) < 0$, problem (3.46) is stable.*
(2) *If $f(\sigma) > 0$, problem (3.46) is unstable.*
(3) *If $f(\sigma) = 0$, there exists an eigenvalue λ with $\lambda = \sqrt{-1}\lambda_I$.*

Proof To determine a Hopf bifurcation in problem (3.46), for

$$\sigma = \sigma_R + i\sigma_I$$

we have to solve

$$\Delta\phi - \phi + (1+\sigma)w\phi = \lambda\phi \tag{3.52}$$

with

$$\lambda = \sqrt{-1}\lambda_I$$

(i.e. the real part λ_R of λ vanishes).

Letting

$$\gamma = \sqrt{1+\lambda}, \qquad \mu = 1+\sigma, \qquad \phi = w^\gamma F,$$

we have

$$F'' + 2\gamma\frac{w'}{w}F' + \left(\mu - \left(\gamma + \frac{2}{3}\gamma(\gamma-1)\right)\right)w^{p-1}F = 0. \tag{3.53}$$

Introducing the variable

$$z = \frac{1}{2}\left(1 - \frac{w'}{w}\right), \tag{3.54}$$

we get

$$\frac{w'}{w} = 1 - 2z, \qquad w = 6z(1-z), \qquad \frac{dz}{dx} = z(1-z).$$

Thus we have the following equation for F as a function of z:

$$z(1-z)F'' + \left(c - (a+b+1)z\right)F' - abF = 0, \tag{3.55}$$

where

$$a+b+1 = 2+4\gamma, \qquad ab = 2\left(2\gamma(\gamma-1) - 3(\mu-\gamma)\right), \qquad c = 1+2\gamma. \tag{3.56}$$

The solutions to (3.55) are standard hypergeometric functions. There are two such solutions which can be stated as follows:

$$F(a, b; c; z), \qquad z^{1-c}F(a-c+1, b-c+1; 2-c; z).$$

By our construction $F(a, b; c; z)$ is regular at $z = 0$ and at $z = 1$ it has a singularity which satisfies

$$\lim_{z\to 1}(1-z)^{-(c-a-b)}F(a, b; c; z) = \frac{\Gamma(c)\Gamma(a+b-c)}{\Gamma(a)\Gamma(b)},$$

where $c - a - b = -2\gamma < 0$. Note that since $\gamma = \sqrt{1+\sqrt{-1}\lambda_I}$, the real part of γ is positive. Thus we can only have a solution that is regular at both $z = 0$ and $z = 1$ if $\Gamma(x)$ has a pole at a or b, respectively. Written differently, we need that $a, b = 0, -1, -2, \ldots$.

From (3.56), we get that

$$a = 2\gamma - \alpha \quad \text{or} \quad b = 2\gamma - \alpha,$$

where α satisfies

$$\alpha^2 + \alpha - 6\mu = 0. \tag{3.57}$$

By symmetry, we may assume that $a = 2\gamma - \alpha = -l$, $l \geq 0$ and $\alpha = \alpha_R + \sqrt{-1}\alpha_I$. Thus we have to solve the system

$$\begin{cases} \alpha_R^2 + \alpha_R - \alpha_I^2 - 6(1 + \sigma_R) = 0, \\ 2\gamma = \alpha - l, \quad l = 0, 1, 2, \ldots. \end{cases} \tag{3.58}$$

Since we consider the principal branch of $\gamma = \sqrt{1 + \sqrt{-1}\lambda_I}$, we get that

$$\alpha_R > l.$$

Moreover, we have

$$4 = (\alpha_R - l)^2 - \alpha_I^2$$

which implies that

$$\alpha_R \geq l + 2. \tag{3.59}$$

On the other hand, we have

$$4 = (\alpha_R - l)^2 - \alpha_I^2 = \alpha_R^2 - \alpha_I^2 - 2l\alpha_R + l^2$$
$$= -(2l + 1)\alpha_R + l^2 + 6(1 + \sigma_R).$$

Thus we get

$$\alpha_R = \frac{1}{2l + 1}(l^2 + 2 + 6\sigma_R).$$

By (3.59), we obtain

$$\frac{1}{2l + 1}(l^2 + 2 + 6\sigma_R) \geq l + 2.$$

This is impossible unless $l = 0$ or $l = 1$. For $l = 1$, we derive that the eigenvalue $\lambda = 0$ with corresponding eigenfunction w' given by Lemma 13.4. This clearly does not correspond to Hopf bifurcation.

In conclusion, to get Hopf bifurcation we must have $a = 0$ or $b = 0$. This gives

$$\alpha_R = 2 + 6\sigma_R, \alpha_I = \frac{6}{(2\alpha_R + 1)}\sigma_I. \tag{3.60}$$

Substitution into (3.58) implies that (σ_R, σ_I) is a zero of the polynomial f defined by (3.51).

In summary, Hopf bifurcation can occur only at a point (σ_R^h, σ_I^h) which satisfies $f(\sigma) = 0$.

Note that the set $\{f(\sigma) = 0\}$ defines a monotone curve within the set $\{(\sigma_R, \sigma_I)|$ $\sigma_R > 0, \sigma_I > 0\}$. Since $f(0, \sigma_I) < 0$ for $\sigma_I > 0$ and $f(\sigma_R, 0) > 0$ for $\sigma_R > 0$, we conclude that for $f(\sigma) < 0$ problem (3.46) is stable, while for $f(\sigma) > 0$ problem (3.46) is unstable. □

We next study the scalar NLEP (3.47). Using a perturbation argument we get the following result:

Lemma 3.13 *Consider the nonlocal eigenvalue problem* (3.47).

(1) *Suppose that $0 \leq \tau < \tau_0$ where τ_0 is sufficiently small and $0 < \eta < \frac{1}{2}$. Let $\lambda_0 \neq 0$ be an eigenvalue of* (3.47). *Then we have* $\mathrm{Re}(\lambda_0) \leq -c_1$ *for some $c_1 > 0$.*
(2) *Suppose that $\tau > 0$ and $\frac{1}{2} < \eta < 1$, then problem* (3.47) *admits a real eigenvalue λ_0 with $\lambda_0 \geq c_2 > 0$ for some $c_2 > 0$.*

Proof

(1) For $\tau = 0$, we get

$$\frac{2(1 - \eta)}{\eta\sqrt{1 + \tau\lambda} + 1 - \eta} = 2(1 - \eta) > 1$$

if $0 < \eta < \frac{1}{2}$. By Theorem 3.5 the result follows.
(2) Assume that $\frac{1}{2} < \eta < 1$. We now show that (3.47) possesses a positive eigenvalue for all $\tau > 0$.

By Lemma 3.2(2), L_0 admits only one positive eigenvalue λ_1. Considering the function

$$h(\alpha) = \int_{\mathbb{R}} ((L_0 - \alpha)^{-1} w) w, \quad 0 < \alpha < \lambda_1, \tag{3.61}$$

we get

$$h'(\alpha) = \int_{\mathbb{R}} ((L_0 - \alpha)^{-2} w) w = \int_{\mathbb{R}} [(L_0 - \alpha)^{-1} w]^2 > 0,$$

and finally

$$\lim_{\alpha \to \lambda_1} h(\alpha) = +\infty.$$

Next, using the function

$$\rho(\lambda) = \frac{\eta\sqrt{1 + \tau\lambda} + 1 - \eta}{2(1 - \eta)} - 1 - \left(\int_{\mathbb{R}} w^2\right)^{-1} \lambda h(\lambda), \tag{3.62}$$

we compute

$$\rho(0) = \frac{1}{2(1 - \eta)} - 1 > 0$$

since $\frac{1}{2} < \eta < 1$. On the other hand, we have

$$\lim_{\lambda \to \lambda_1 -} \rho(\lambda) = -\infty.$$

Thus there is a $\lambda_0 \in (0, \lambda_1)$ such that $\rho(\lambda_0) = 0$. It is easy to see that $\lambda_0 > 0$ is an eigenvalue of (3.47) and Part (2) of Lemma 3.13 follows. $\qquad\square$

In the general case when τ is large and $0 < \eta < \frac{1}{2}$, we can use hypergeometric functions and generalised hypergeometric functions to reduce problem (3.47) to a computable one. We recall that by Lemma 13.5 for $\tau = 0$ all eigenvalues are stable. Thus by varying τ, we either obtain stability or Hopf bifurcation. Thus it remains to compute when Hopf bifurcation happens.

Let $a_1, a_2, \ldots, a_A$ and $b_1, b_2, \ldots, b_B$ be two sequences of numbers. Then the series

$$
1 + \frac{a_1 a_2 \cdots a_A}{b_1 b_2 \cdots b_B} \frac{z}{1!} + \frac{(a_1 + 1)(a_2 + 1) \cdots (a_A + 1)}{(b_1 + 1)(b_2 + 1) \cdots (b_B + 1)} \frac{z^2}{2!} + \cdots
$$

$$
\equiv {}_A F_B \left\{ \begin{array}{c} a_1, a_2, \ldots, a_A; \\ z \\ b_1, b_2, \ldots, b_B; \end{array} \right\} \tag{3.63}
$$

is called a generalised Gauss function or generalised hypergeometric function.

Then we have the following lemma, whose proof is based on hypergeometric functions:

Lemma 3.14   *Let* $\lambda = \sqrt{-1}\lambda_I$ *be an eigenvalue of problem (3.47). Then* λ *is a solution of the algebraic equation*

$$
\frac{\eta\sqrt{1 + \tau\lambda} + 1 - \eta}{2(1 - \eta)} - 1
$$

$$
= {}_{-4}F_3 \left\{ \begin{array}{c} 1, 3, -\frac{1}{2}, 2; \\ 1 \\ 2 + \gamma, 2 - \gamma, \frac{5}{2}; \end{array} \right\}
$$

$$
+ \frac{2\lambda}{3} b_1 \frac{\Gamma(1 + \gamma)\Gamma(5/2)}{\Gamma(\gamma + (3/2))} {}_3 F_2 \left\{ \begin{array}{c} 2 + \gamma, -\frac{3}{2} + \gamma, 1 + \gamma; \\ 1 \\ 1 + 2\gamma, \frac{3}{2} + \gamma; \end{array} \right\}, \tag{3.64}
$$

where $\gamma = \sqrt{1 + \lambda}$ *and* b_1 *is given by (3.74) below.*

Proof   We will proceed by reducing problem (3.47) to (3.64).

Let $_A F_B$ be given by (3.63). Then $_A F_B$ satisfies the following integral property, whose proof can be found in [217]:

$$
_{A+1} F_{B+1} \left\{ \begin{array}{c} a_1, a_2, \ldots, a_A, c; \\ z \\ b_1, b_2, \ldots, b_B, d; \end{array} \right\}
$$

$$
= \frac{\Gamma(d)}{\Gamma(c)\Gamma(d-c)} \int_0^1 t^{c-1}(1-t)^{d-c-1} {}_A F_B \left\{ \begin{array}{c} a_1, a_2, \ldots, a_A; \\ tz \\ b_1, b_2, \ldots, b_B; \end{array} \right\} dt. \quad (3.65)
$$

Let w be the unique solution of (2.3). Integrating (2.3) it follows that

$$
(w')^2 = w^2 - \frac{2}{3} w^3. \tag{3.66}
$$

Let us first consider the problem

$$
\Delta \phi_0 - \phi_0 + 2w\phi_0 = w^2 + \lambda \phi_0, \quad \phi_0 \in H^1(\mathbb{R}). \tag{3.67}
$$

Since w is an even function, we may assume that ϕ_0 is also even. Note that ϕ_0 is unique. We denote the variable by t and set

$$
\gamma = \sqrt{1 + \lambda},
$$

where we take the principal branch of $\sqrt{1 + \lambda}$. Further, let

$$
f(\lambda) = \frac{2(1 - \eta)}{\eta \sqrt{1 + \tau \lambda} + 1 - \eta}.
$$

Then the NLEP (3.47) becomes

$$
\frac{1}{f(\lambda)} = \frac{\int_{\mathbb{R}} w\phi_0}{\int_{\mathbb{R}} w^2} = \frac{\int_0^{+\infty} w\phi_0 dt}{\int_0^{+\infty} w^2 dt}. \tag{3.68}
$$

Using the transformation

$$
\phi_0 = w^\gamma G,
$$

then by some simple computations G satisfies

$$
\frac{d^2 G}{dt^2} + 2\gamma \frac{w'}{w} \frac{dG}{dt} + \left(2 - \frac{\gamma}{3}(1 + 2\gamma) \right) wG = w^{1-\gamma}. \tag{3.69}
$$

Next we perform the following change of variables

$$
z = \frac{2}{3} w. \tag{3.70}
$$

Note that $w(0) = \frac{3}{2}$ and so z is a homeomorphism from $[0, +\infty]$ to $[0, 1]$.

After some elementary computations, using (3.66), we can write the equation for $G(z)$ as

$$z(1-z)G'' + \left(c - (a+b+1)z\right)G' - abG = \left(\frac{3}{2}\right)^{2-\gamma} z^{1-\gamma}, \qquad (3.71)$$

where

$$a = 2+\gamma, \qquad b = \gamma - \frac{3}{2}, \qquad c = 1+2\gamma. \qquad (3.72)$$

To determine a particular solution of (3.71), we use as an ansatz a power series:

$$G_p(z) = z^s \sum_{k=0}^{+\infty} c_k z^k.$$

Substitution into (3.71) gives

$$\sum_{k=0}^{+\infty} c_k z^{s+k-1}(s+k)(s+k-1+c) - \sum_{k=1}^{+\infty} c_k z^{s+k}(s+k+a)(s+k+b)$$

$$= \left(\frac{3}{2}\right)^{2-\gamma} z^{1-\gamma}.$$

This implies

$$s - 1 = 1 - \gamma, \qquad c_0 s(s-1+c) = \left(\frac{3}{2}\right)^{2-\gamma},$$

$$c_k(s+k)(s+k-1+c) = c_{k-1}(s+k-1+a)(s+k-1+b).$$

Regrouping the coefficients, we get

$$G_p(z) = \left(\frac{3}{2}\right)^{2-\gamma} (4-\gamma^2)^{-1} z^{2-\gamma} {}_3F_2 \left\{ \begin{array}{c} 1, \frac{1}{2}, 4; \\ \\ 3-\gamma, 3+\gamma; \end{array} z \right\}. \qquad (3.73)$$

A homogeneous solution to (3.71) is given by

$$G_h = a_1 {}_2F_1 \left\{ \begin{array}{c} a, b; \\ \\ c; \end{array} z \right\} + a_2 z^{1-c} {}_2F_1 \left\{ \begin{array}{c} b-c-1, a-c+1; \\ \\ 2-c; \end{array} z \right\},$$

where a_1 and a_2 are arbitrary real constants. Combining homogeneous and inhomogeneous solutions, we have

$$\phi_0 = b_1 z^\gamma \, _2F_1 \left\{ \begin{array}{c} 2+\gamma, \gamma - 3/2; \\ \\ 1+2\gamma; \end{array} \; z \right\} + b_2 z^{-\gamma} \, _2F_1 \left\{ \begin{array}{c} -\gamma - 3/2, 1-\gamma; \\ \\ 1-2\gamma; \end{array} \; z \right\}$$

$$+ \left(\frac{3}{2}\right)^2 (4-\gamma^2)^{-1} z^2 \, _3F_2 \left\{ \begin{array}{c} 1, \frac{1}{2}, 4; \\ \\ 3-\gamma, 3+\gamma; \end{array} \; z \right\}$$

where the real constants b_1 and b_2 are yet to be determined. Since $\phi_0(0)$ is bounded, we have $b_2 = 0$. Next we choose b_1 such that $\phi_0'(0) = 0$. Note that

$$\frac{d\phi}{dy} = \frac{d\phi}{dz} z \sqrt{1-z}.$$

Letting

$$f(z) = \, _3F_2 \left\{ \begin{array}{c} 1, \frac{1}{2}, 4; \\ \\ 3-\gamma, 3+\gamma; \end{array} \; z \right\}$$

we compute

$$f(z) = \sum_{k=0}^{\infty} c_k z^k,$$

where

$$c_0 = 1, \qquad c_k = \frac{(k+3)(k-1/2)}{(k+2-\gamma)(k+2+\gamma)} c_{k-1}, \qquad k = 1, 2, \ldots.$$

Using that

$$\frac{(k+3)(k-1/2)}{(k+2-\gamma)(k+2+\gamma)} = 1 - \frac{3}{2k} + O\left(\frac{1}{k^2}\right) \quad \text{as } k \to \infty,$$

we derive

$$c_k = \prod_{j=1}^{k} \left(1 - \frac{3}{2j} + O\left(\frac{1}{j^2}\right)\right) = \hat{c}_1 \exp\left(\sum_{j=2}^{k} \ln\left(1 - \frac{3}{2j} + O\left(\frac{1}{j^2}\right)\right)\right)$$

$$= \hat{c}_2 \exp\left(-\frac{3}{2}\sum_{j=2}^{k} \frac{1}{j}\right)\left(1 + O\left(\frac{1}{k}\right)\right) = \frac{\hat{c}}{k^{3/2}}\left(1 + O\left(\frac{1}{k}\right)\right) \quad \text{as } k \to \infty,$$

where $\hat{c}_1, \hat{c}_2$ are suitable positive constants and

$$\hat{c} = \lim_{K \to \infty} K^{3/2} \prod_{k=1}^{K} \frac{(k+3)(k-1/2)}{(k+2-\gamma)(k+2+\gamma)}.$$

In the limit $z \to 1$, the infinite sum for $f(z)$ satisfies

$$f(z) = \hat{c} \sum_{n=1}^{\infty} \frac{z^n}{n^{3/2}} + \sum_{n=1}^{\infty} O\left(\frac{1}{n^{5/2}}\right) z^n.$$

Taking derivatives, we get

$$f'(z) = \hat{c} \sum_{n=1}^{\infty} \frac{z^{n-1}}{n^{1/2}} + O(1) \quad \text{as } z \to 1.$$

This implies that $f'(z) \to \infty$ as $z \to 1$. We will now show that the limit $\lim_{z \to 1} f'(z)\sqrt{1-z}$ is finite.

Consider

$$u(h) = \sum_{n=1}^{\infty} \frac{(1-h)^{n-1}}{\sqrt{n}} \sqrt{h}.$$

In the limit $h \to 0$, we compute

$$u(h) = \sum_{n=1}^{\infty} \frac{\exp(-nh)}{\sqrt{n}} \sqrt{h}\left(1 + O\left(\frac{1}{n}\right)\right) = \int_0^{\infty} \frac{\exp(-th)}{\sqrt{t}} \sqrt{h}\,dt + o(1)$$

$$= \int_0^{\infty} 2\exp\left(-s^2\right)ds + o(1) = \sqrt{\pi} + o(1) \quad \text{as } h \to 0.$$

Thus

$$\lim_{z \to 1} f'(z)\sqrt{1-z} = \hat{c}\sqrt{\pi}.$$

Let

$$g(z) = {}_2F_1\left\{ \begin{matrix} 2+\gamma, \gamma - 3/2; \\ \\ 1+2\gamma; \end{matrix} \; z \right\}.$$

Similar to the above, we derive that

$$\lim_{z \to 1} g'(z)\sqrt{1-z} = \hat{d}\sqrt{\pi},$$

where

$$\hat{d} = \lim_{K \to \infty} K^{3/2} \prod_{k=1}^{K} \frac{(k+1+\gamma)(k-5/2+\gamma)}{(k+2\gamma)k}.$$

Taken together, this gives

$$\phi_0'(0) = b_1\hat{d}\sqrt{\pi} + \frac{9}{4}\frac{1}{1-\gamma^2}\hat{c}\sqrt{\pi} = 0.$$

Thus we have

$$b_1 = -\frac{9}{4}\frac{1}{1-\gamma^2}\prod_{k=1}^{\infty}\frac{(k+3)(k-1/2)(k+2\gamma)k}{(k+2-\gamma)(k+2+\gamma)(k+1+\gamma)(k-5/2+\gamma)}$$

$$= \prod_{k=0}^{\infty}\frac{(k+4)(k+1/2)(k+1+2\gamma)(k+1)}{(k+3-\gamma)(k+3+\gamma)(k+2+\gamma)(k-3/2+\gamma)}.$$

Using the identity

$$\prod_{k=0}^{\infty}\frac{(k+a-b)(k+b+c)}{(k+a+d)(k+c-d)} = \frac{\Gamma(a+d)\Gamma(c-d)}{\Gamma(a-b)\Gamma(b+c)},$$

we can simplify

$$\prod_{k=0}^{\infty}\frac{(k+4)(k+1+2\gamma)}{(k+3+\gamma)(k+2+\gamma)} = \frac{\Gamma(3+\gamma)\Gamma(2+\gamma)}{\Gamma(4)\Gamma(1+2\gamma)}$$

and

$$\prod_{k=0}^{\infty}\frac{(k+1/2)(k+1)}{(k+3-\gamma)(k-3/2+\gamma)} = \frac{\Gamma(3-\gamma)\Gamma(-3/2+\gamma)}{\Gamma(1/2)\Gamma(1)}.$$

This gives

$$b_1 = \frac{9}{4}\frac{1}{4-\gamma^2}\frac{\Gamma(3+\gamma)\Gamma(2+\gamma)}{\Gamma(4)\Gamma(1+2\gamma)}\frac{\Gamma(3-\gamma)\Gamma(-3/2+\gamma)}{\Gamma(1/2)\Gamma(1)}.$$

Using the identities

$$\Gamma(1-z)\Gamma(z) = \frac{\pi}{\sin(\pi z)}, \qquad \Gamma(2z) = \Gamma(z)\Gamma\left(z+\frac{1}{2}\right)2^{2z-1}\pi^{-1/2},$$

$$\Gamma\left(\frac{1}{2}\right) = \sqrt{\pi},$$

this implies

$$b_1 = \frac{9}{4}\frac{(1-\gamma)^2\gamma(\gamma-1)}{6(-3/2+\gamma)(-1/2+\gamma)2^{2\gamma}}\frac{\pi}{\sin(\pi(\gamma-1))}. \tag{3.74}$$

Now we can compute

$$\int_0^{+\infty} w\phi_0 dt = \frac{3}{2}\int_0^1 w^{1+\gamma}G(z)\frac{dz}{-w'}$$

$$= \frac{3}{2}\int_0^1 w^{1+\gamma}G_p(z)\frac{dz}{-w'} + \int_0^1 w^{\gamma}G_h(z)\frac{dz}{-w'}.$$

For the first integral, we get by (3.65)

$$
\frac{3}{2}\int_0^1 w^{1+\gamma}G_p(z)\frac{dz}{-w'}
$$

$$
=\left(\frac{3}{2}\right)^{1+\gamma}\int_0^1 z^\gamma(1-z)^{-1/2}G(z)dz
$$

$$
=\left(\frac{3}{2}\right)^3(4-\gamma^2)^{-1}\int_0^1 z^2(1-z)^{-1/2}{}_3F_2\left\{\begin{matrix}1,\frac{1}{2},4;\\[4pt]3-\gamma,3+\gamma;\end{matrix}\ z\right\}dz
$$

$$
=\int_0^{+\infty}w\phi_0 dt
$$

$$
=\left(\frac{3}{2}\right)^3(4-\gamma^2)^{-1}\frac{\Gamma(3)\Gamma(1/2)}{\Gamma(7/2)}{}_4F_3\left\{\begin{matrix}1,\frac{1}{2},4,3;\\[4pt]3-\gamma,3+\gamma,\frac{7}{2};\end{matrix}\ 1\right\}.\qquad(3.75)
$$

For the second integral, we have

$$
\int_0^1 w^\gamma G_h(z)\frac{dz}{-w'}
$$

$$
=\frac{3}{2}b_1\frac{\Gamma(1+\gamma)\Gamma(1/2)}{\Gamma(3/2+\gamma)}{}_3F_2\left\{\begin{matrix}2+\gamma,-3/2+\gamma,1+\gamma;\\[4pt]1+2\gamma,3/2+\gamma;\end{matrix}\ 1\right\}
$$

$$
=\left(\frac{3}{2}\right)^3\frac{(1-\gamma)^2\gamma(\gamma-1)}{6(-3/2+\gamma)(-1/2+\gamma)2^{2\gamma}}\frac{\pi}{\sin(\pi(\gamma-1))}
$$

$$
\times\frac{\Gamma(1+\gamma)\Gamma(1/2)}{\Gamma(3/2+\gamma)}{}_3F_2\left\{\begin{matrix}2+\gamma,-3/2+\gamma,1+\gamma;\\[4pt]1+2\gamma,3/2+\gamma;\end{matrix}\ 1\right\}.\qquad(3.76)
$$

On the other hand, it is easy to compute that

$$
\int_0^{+\infty}w^2 dt=\left(\frac{3}{2}\right)^2\int_0^1 z^2(1-z)^{-1/2}dz=\left(\frac{3}{2}\right)^2\frac{\Gamma(2)\Gamma(1/2)}{\Gamma(2+1/2)}.\qquad(3.77)
$$

By (3.75), (3.76), (3.77) and (3.68), we obtain (3.64).   □

By Lemma 3.14, problem (3.47) can be solved using Mathematica. We will not produce any numerical results here.

In the next chapter we will use the results of this chapter and apply them to the study of the stability problem.

3.5 Notes on the Literature

The proof of Theorem 3.1 follows [249]. Extensions to general parameters are given in [264]. The main idea is to introduce a quadratic form which is positive definite and then use a continuation argument. See [37, 178, 179, 252, 281] for related studies on nonlocal eigenvalue problems with arbitrary powers.

The method of continuation and Hopf bifurcation follow [261]. Theorem 3.6 is true in $\mathbb{R}^N$, $N \leq 4$. Here we have proved the uniqueness of Hopf bifurcation. In the one-dimensional case, the existence of Hopf bifurcation for increasing τ has been proved in [243] and the uniqueness of Hopf bifurcation has been shown for the shadow system [242]. The existence of a Hopf bifurcation has also been studied in [37, 178, 179, 281]. Some explicit results for the one-dimensional case with $p = 2r - 3$, $r > 2$ have been derived in [167]. It turns out that in the two-dimensional case a result similar to Lemma 3.12 holds, but the exact threshold for θ^h cannot be established explicitly as in [260].

For hypergeometric functions we use the approach of [47, 117, 118, 260, 263, 278]. More background on hypergeometric functions can be found in [217].

Chapter 4
Stability of Spikes for the Gierer-Meinhardt System in One Dimension

In this chapter we study the stability of multi-spike steady states in one space dimension. There are two types of solutions: symmetric and asymmetric multiple spikes. We will show that symmetric spikes in one dimension may be stable or unstable. Asymmetric spikes in one dimension are always unstable, however they may be metastable.

4.1 Symmetric Multiple Spikes: Stability

Now we study the stability of the N-spike solutions constructed in Theorem 2.2. The stability result is stated as follows:

Theorem 4.1 *Let* (A_ϵ, H_ϵ) *be the multi-spike steady states constructed in Theorem 2.2. Assume that* $\epsilon \ll 1$.
 Let $\mathcal{B}$ *and* $\mathcal{M}$ *be the matrices introduced in (2.21) and (2.28), respectively.*

(1) (*Stability*) *If*

$$2 \min_{\sigma \in \sigma(\mathcal{B})} \sigma > 1 \tag{4.1}$$

and

$$\sigma(\mathcal{M}) \subseteq \{\sigma \,|\, \mathrm{Re}(\sigma) > 0\}, \tag{4.2}$$

there exists a $\tau_0 > 0$ *such that* (A_ϵ, H_ϵ) *is linearly stable for* $0 \leq \tau < \tau_0$.
(2) (*Instability*) *If*

$$2 \min_{\sigma \in \sigma(\mathcal{B})} \sigma < 1, \tag{4.3}$$

there exists $\tau_0 > 0$ *such that* (A_ϵ, H_ϵ) *is linearly unstable for* $0 \leq \tau < \tau_0$.

J. Wei, M. Winter, *Mathematical Aspects of Pattern Formation in Biological Systems*,
Applied Mathematical Sciences 189, DOI 10.1007/978-1-4471-5526-3_4,
© Springer-Verlag London 2014

(3) *(Instability) If there exists*

$$\sigma \in \sigma(\mathcal{M}), \quad \mathrm{Re}(\sigma) < 0, \tag{4.4}$$

then (A_ϵ, H_ϵ) is linearly unstable for all $\tau > 0$.

Remark 4.2 By Theorems 2.2 and 4.1, the existence and stability of two-spike solutions are completely determined by the two matrices $\mathcal{B}$ and $\mathcal{M}$. They are related to the asymptotic behaviour of large eigenvalues which tend to a nonzero limit and small eigenvalues which tend to zero as $\epsilon \to 0$, respectively. In Sect. 4.1.3 we will compute the eigenvalues of the matrices $\mathcal{B}$ and $\mathcal{M}$ in the symmetric two-spike case. Then the stability of symmetric two-spike solutions is completely characterised and the following result is established rigorously.

Theorem 4.3 *Let (A_ϵ, H_ϵ) be the symmetric two-spike solutions given in Theorem 2.2.*

(a) *(Stability) Assume that $0 \le \tau < \tau_0$ for some small τ_0 and*

$$D < D_2 := \frac{1}{4(\log(1+\sqrt{2}))^2}, \tag{4.5}$$

then the symmetric two-spike solution is linearly stable.
(b) *(Instability) If*

$$D > D_2, \tag{4.6}$$

where D_2 is given by (4.5), then the two-spike solution is linearly unstable for all $\tau \ge 0$.

Remark 4.4 For the case of N spikes in an interval of length L the stability condition given in (4.5) generalises to the following (see [273]):

$$D < D_N := \frac{L}{2N^2(\log(1+\sqrt{2}))^2}, \quad N = 2, 3, \ldots. \tag{4.7}$$

We first study the eigenvalue problem for the vectorial linear operator L given in (2.39):

$$L\Phi := \Phi'' - \Phi + 2w\Phi - 2\mathcal{B}\frac{\int_{\mathbb{R}} w\Phi}{\int_{\mathbb{R}} w^r}w^2 = \alpha\Phi, \quad \Phi \in \left(H^1(\mathbb{R})\right)^N.$$

We have the following result:

Lemma 4.5 *Assume that all the eigenvalues of $\mathcal{B}$ are real. Then the following holds:*

(1) *If $2\min_{\sigma \in \sigma(\mathcal{B})} > 1$, then for any nonzero eigenvalue α of (2.39) we have $\alpha \le -c_0 < 0$.*

(2) *If there exists a $\sigma \in \sigma(\mathcal{B})$ such that $2\sigma < 1$, then there exists a positive eigen-value of (2.39).*

Proof Let (Φ, α) satisfy (2.39), where $\alpha = \alpha_R + i\alpha_I$ and $\alpha_R > 0$. Similarly to Lemma 2.5, we diagonalise (2.39) and obtain

$$\Phi'' - \Phi + 2w\Phi - 2\left(\int_{\mathbb{R}} w^2\right)^{-1}\left(\int_{\mathbb{R}} wJ\Phi\right)w^2 = \alpha\Phi. \tag{4.8}$$

Then the l-th equation of system (4.8) reads

$$\phi_l'' - \phi_l + 2w\phi_l - 2\sigma_l\left(\int_{\mathbb{R}} w^2\right)^{-1}\left(\int_{\mathbb{R}} w\phi_l\right)w^2 = \alpha\phi_l.$$

(i) Since $2\sigma_l > 1$, by Theorem 3.1(2) we have

$$\alpha \leq -c_0 < 0$$

and Lemma 4.5(1) follows.

(ii) Since $2\sigma_l < 1$ for some $\sigma_l \in \sigma(\mathcal{B})$, the equation corresponding to σ_l becomes

$$\phi_l'' - \phi_l + 2w\phi_l - 2\sigma_l\frac{\int_{\mathbb{R}} w\phi_l}{\int_{\mathbb{R}} w^2} = \alpha\phi_l.$$

Then by Theorem 3.1(1) there exists an eigenvalue $\alpha_0 > 0$ and an eigenfunction ϕ_0 such that

$$L_0\phi_0 - 2\sigma_l\frac{\int_{\mathbb{R}} w\phi_0}{\int_{\mathbb{R}} w^2} = \alpha_0\phi_0.$$

Let us take $\phi_l = \phi_0$ and $\phi_2 = 0$. Then (Φ, α) satisfy (2.39) and Lemma 4.5(2) is proved.   □

4.1.1 Large Eigenvalues

In this subsection, we study the large eigenvalues with $\lambda_\epsilon \to \lambda_0 \neq 0$ as $\epsilon \to 0$.

We need to analyse the following eigenvalue problem for $(\phi_\epsilon, \psi_\epsilon) \in (H_N^2(\Omega))^2$:

$$\tilde{L}_\epsilon\phi_\epsilon = \epsilon^2\phi_\epsilon'' - \phi_\epsilon + \frac{2A_\epsilon\phi_\epsilon}{T[A_\epsilon]} - \frac{A_\epsilon^2}{(T[A_\epsilon])^2}\psi_\epsilon = \lambda_\epsilon\phi_\epsilon, \tag{4.9}$$

where ψ_ϵ satisfies

$$D\psi_\epsilon'' - \psi_\epsilon + 2\xi_\epsilon A_\epsilon\phi_\epsilon = \tau\lambda_\epsilon\psi_\epsilon. \tag{4.10}$$

Here λ_ϵ is some complex number and $A_\epsilon = w_{\epsilon,t^\epsilon} + \phi_{\epsilon,t^\epsilon}$ is the exact solution with $\mathbf{t}^\epsilon$ determined in Sect. 2.3.4.

We first consider the case $\tau = 0$. At the end, we shall explain the necessary modifications if we assume that $\tau > 0$, where τ is small enough. By (4.10) we have

$$\psi_\epsilon = T'[A_\epsilon](\phi_\epsilon). \tag{4.11}$$

Since we are concerned only with eigenvalues satisfying $\text{Re}(\lambda_\epsilon) \geq c_1$, by following a similar argument as in Theorem 3.5(2), we conclude that $|\lambda_\epsilon| \leq c$ for some constant $c > 0$ (independent of $\epsilon > 0$).

Recall the definition of $\phi_{\epsilon,j}$ given in (2.75).

From (4.9) and the facts that $\text{Re}(\lambda_\epsilon) \geq c_1$ and w_{ϵ,t^ϵ} has exponential decay, we get that

$$\phi_\epsilon = \sum_{j=1}^{2} \phi_{\epsilon,j} + \text{e.s.t.}$$

Next we extend $\phi_{\epsilon,j}$ to a function defined on $\mathbb{R}^1$ such that

$$\|\phi_{\epsilon,j}\|_{H^2(\mathbb{R}^1)} \leq C\|\phi_{\epsilon,j}\|_{H^2(\Omega_\epsilon)}, \quad j = 1, 2.$$

Without loss of generality, we have $\|\phi_\epsilon\|_\epsilon = \|\phi_\epsilon\|_{H^2(\Omega_\epsilon)} = 1$. Then we get $\|\phi_{\epsilon,j}\|_\epsilon \leq C$ and, taking a subsequence of ϵ, we may also assume that $\phi_{\epsilon,j} \to \phi_j$ as $\epsilon \to 0$ in $H^2(\mathbb{R})$ for $j = 1, 2$.

Taking the limit $\epsilon \to 0$ and using $\lambda_\epsilon \to \lambda_0$, we have

$$L\Phi = \Delta\Phi - \Phi + 2w\Phi - 2B\frac{\int_\mathbb{R} wB\Phi}{\int_\mathbb{R} w^2}w^2 = \lambda_0\Phi, \tag{4.12}$$

where

$$\Phi = \begin{pmatrix} \phi_1 \\ \phi_2 \end{pmatrix} \in \left(H^2(\mathbb{R})\right)^2.$$

Now we can rigorously establish the following result:

Theorem 4.6 *Let λ_ϵ be an eigenvalue of (4.9) and (4.10) such that $\text{Re}(\lambda_\epsilon) \geq c$ for some $c > 0$.*

(1) *Suppose that (for suitable sequences $\epsilon_n \to 0$) we have $\lambda_{\epsilon_n} \to \lambda_0 \neq 0$. Then λ_0 is an eigenvalue of the problem (NLEP) given in (4.12).*
(2) *Let $\lambda_0 \neq 0$ with $\text{Re}(\lambda_0) > 0$ be an eigenvalue of the problem (NLEP) given in (4.12). Then for ϵ small enough, there is an eigenvalue λ_ϵ of (4.9) and (4.10) with $\lambda_\epsilon \to \lambda_0$ as $\epsilon \to 0$.*

Proof Theorem 4.6(1) is shown by standard asymptotic analysis.

To prove Theorem 4.6(2), we use the argument given in Sect. 2 of [37], where the following eigenvalue problem has been investigated:

$$\begin{cases} \epsilon^2 \Delta h - h + p u_\epsilon^{p-1} h - \frac{qr}{s+1+\tau\lambda_\epsilon} \frac{\int_\Omega u_\epsilon^{r-1} h}{\int_\Omega u_\epsilon^r} u_\epsilon^p = \lambda_\epsilon h & \text{in } \Omega, \\ h = 0 \quad \text{on } \partial\Omega, \end{cases} \tag{4.13}$$

where u_ϵ is a solution of the partial differential equation

$$\begin{cases} \epsilon^2 \Delta u_\epsilon - u_\epsilon + u_\epsilon^p = 0 & \text{in } \Omega, \\ u_\epsilon > 0 & \text{in } \Omega, \\ u_\epsilon = 0 & \text{on } \partial\Omega. \end{cases}$$

Further, we have $1 < p < \frac{n+2}{n-2}$ if $n \geq 3$ and $1 < p < +\infty$ if $n = 1, 2$, $\frac{qr}{(s+1)(p-1)} > 1$ and $\Omega \subset \mathbb{R}^N$ is a smooth bounded domain. If u_ϵ is a single interior peak solution, then it has been shown in [249] that the limiting eigenvalue problem is given by the NLEP

$$\Delta\phi - \phi + p w^{p-1}\phi - \frac{qr}{s+1+\tau\lambda_0} \frac{\int_{\mathbb{R}^N} w^{r-1}\phi}{\int_{\mathbb{R}^N} w^r} w^p = \lambda_0\phi, \tag{4.14}$$

where w is the corresponding ground state solution in $\mathbb{R}^N$:

$$\Delta w - w + w^p = 0, \quad w > 0 \text{ in } \mathbb{R}^N, w = w(|y|) \in H^2(\mathbb{R}^N).$$

Dancer in [37] proved that if $\lambda_0 \neq 0$, $\text{Re}(\lambda_0) > 0$ is an unstable eigenvalue of (4.14), then there exists an eigenvalue λ_ϵ of (4.13) such that $\lambda_\epsilon \to \lambda_0$. We next explain the key step in his argument. Let $\lambda_0 \neq 0$ be an eigenvalue of problem (4.12) with $\text{Re}(\lambda_0) > 0$. Then we write (4.9) as follows:

$$\phi_\epsilon = R_\epsilon(\lambda_\epsilon)\left[\frac{2A_\epsilon\phi_\epsilon}{H_\epsilon} - \frac{A_\epsilon^2}{H_\epsilon^2}\psi_\epsilon\right], \tag{4.15}$$

where $R_\epsilon(\lambda_\epsilon)$ is the inverse of $-\Delta + (1+\lambda_\epsilon)$ in $H^2(\mathbb{R})$ (which exists if $\text{Re}(\lambda_\epsilon) > -1$ or $\text{Im}(\lambda_\epsilon) \neq 0$), and $\psi_\epsilon = T_\epsilon'[A_\epsilon](\phi_\epsilon)$ is given in (4.11). The crucial point is that $R_\epsilon(\lambda_\epsilon)$ is a compact operator if ϵ is small enough. The rest of the argument follows in the same way as in [37]. □

We now study the stability of the eigenvalue problem (4.9), (4.10) for large eigenvalues explicitly and derive the conditions (4.1) and (4.3) of Theorem 4.1.

Suppose that we have

$$2 \min_{\sigma \in \sigma(\mathcal{B})} \sigma < 1. \tag{4.16}$$

Then by Theorem 3.1(1), the problem (4.12) has a positive eigenvalue. Thus, by Theorem 4.6, the problem (4.9), (4.10) possesses an eigenvalue with $\text{Re}(\lambda_\epsilon) > c_0$ for some positive number c_0. Therefore (A_ϵ, H_ϵ) is unstable.

On the other hand, let us suppose that

$$2 \min_{\sigma \in \sigma(\mathcal{B})} \sigma > 1. \tag{4.17}$$

Then by Theorem 3.1(2), for any nonzero eigenvalue λ_0 of L we have

$$\text{Re}(\lambda_0) \leq c_0 < 0 \quad \text{for some } c_0 > 0.$$

Thus by Theorem 4.6, for ϵ small enough all nonzero large eigenvalues of (4.9), (4.10) have strictly negative real part. We conclude that in this case for all eigenvalues λ_ϵ of (4.9), (4.10), with $|\lambda_\epsilon| \geq c > 0$, we have $\text{Re}(\lambda_\epsilon) \leq -c < 0$ for ϵ small enough.

Finally, we remark that when $\tau \neq 0$ and τ is small, the matrix $\mathcal{B}$ is replaced by the new matrix $\mathcal{B}_{\tau\lambda_\epsilon}$ which depends on $\tau\lambda_\epsilon$. To derive it, we replace the Green's function G_D by the solution of

$$D\Delta G - (1 + \tau\lambda_\epsilon)G + \delta_z = 0, \quad G'(\pm 1, z) = 0. \tag{4.18}$$

It is easy to check that the new matrix has eigenvalues satisfying (3.21). Using the method of continuation and Theorem 3.5, the conclusion follows.

We have finished the study of large eigenvalues. In the next subsection, we shall investigate the small eigenvalues.

4.1.2 Small Eigenvalues

In this subsection, we study the small eigenvalues for (4.9) and (4.10) which satisfy $\lambda_\epsilon \to 0$ as $\epsilon \to 0$.

Let the exact solution be denoted by

$$\bar{w}_\epsilon = w_{\epsilon,t^\epsilon} + \phi_{\epsilon,t^\epsilon}, \qquad \bar{H}_\epsilon = T[w_{\epsilon,t^\epsilon} + \phi_{\epsilon,t^\epsilon}], \tag{4.19}$$

where $\mathbf{t}^\epsilon = (t_1^\epsilon, t_2^\epsilon)$.

After rescaling, the eigenvalue problem (4.9), (4.10) becomes the following problem for $(\phi_\epsilon, \psi_\epsilon) \in (H_N^2(\Omega))^2$:

$$\epsilon^2 \Delta\phi_\epsilon - \phi_\epsilon + \frac{2\bar{w}_\epsilon}{\bar{H}_\epsilon}\phi_\epsilon - \frac{\bar{w}_\epsilon^2}{\bar{H}_\epsilon^2}\psi_\epsilon = \lambda_\epsilon\phi_\epsilon, \tag{4.20}$$

$$D\Delta\psi_\epsilon - \psi_\epsilon + \xi_\epsilon 2\bar{A}_\epsilon\phi_\epsilon = \lambda_\epsilon\tau\psi_\epsilon, \tag{4.21}$$

where ξ_ϵ is given by (2.18).

For simplicity, we assume that $\tau = 0$. Due to the estimate $\tau\lambda_\epsilon \ll 1$ the results in this subsection are also valid for any finite $\tau > 0$, since they can easily be extended to the case $\tau > 0$ using the Green's function (4.18).

We will prove that the small eigenvalues are of the order $O(\epsilon^2)$. In contrast to the single interior peak case we need to expand the eigenfunction up to the order $O(\epsilon)$ term. (Such a higher-order expansion is also required in the study of boundary spikes for the shadow system, see [8] and [253].)

We define

$$\tilde{w}_{\epsilon,j}(x) = \chi\left(\frac{x - t_j^\epsilon}{r_0}\right)\bar{w}_\epsilon(x), \quad j = 1, 2, \tag{4.22}$$

where $\chi(x)$ and r_0 are given in (2.51) and (2.50). Similarly as in Sect. 2.3.3, we introduce

$$\mathcal{K}_{\epsilon,t^\epsilon}^{new} := \text{span}\{\tilde{w}_{\epsilon,j}', j = 1, 2\} \subset H_N^2(\Omega_\epsilon),$$

$$\mathcal{C}_{\epsilon,t^\epsilon}^{new} := \text{span}\{\tilde{w}_{\epsilon,j}', j = 1, 2\} \subset L^2(\Omega_\epsilon).$$

Note that in contrast to Sect. 2.3.3, here we linearise the equations around the exact solutions which are now at our disposal. Then it follows that

$$\bar{w}_\epsilon(x) = \sum_{j=1}^{2} \tilde{w}_{\epsilon,j}(x) + \text{e.s.t.} \tag{4.23}$$

Note that $\tilde{w}_{\epsilon,j}(x) \sim \hat{\xi}_j w(\frac{x - t_j^\epsilon}{\epsilon})$ in $H_{loc}^2(-1, 1)$ and $\tilde{w}_{\epsilon,j}$ satisfies

$$\epsilon^2(\tilde{w}_{\epsilon,j})'' - \tilde{w}_{\epsilon,j} + \frac{(\tilde{w}_{\epsilon,j})^2}{\bar{H}_\epsilon} + \text{e.s.t.} = 0.$$

Therefore $\tilde{w}_{\epsilon,j}' := \frac{d\tilde{w}_{\epsilon,j}}{dx}$ solves

$$\epsilon^2 \Delta \tilde{w}_{\epsilon,j}' - \tilde{w}_{\epsilon,j}' + \frac{2\tilde{w}_{\epsilon,j}}{\bar{H}_\epsilon}\tilde{w}_{\epsilon,j}' - \frac{\tilde{w}_{\epsilon,j}^2}{(\bar{H}_\epsilon)^2}\bar{H}_\epsilon' + \text{e.s.t.} = 0. \tag{4.24}$$

Next we decompose

$$\phi_\epsilon = \epsilon \sum_{j=1}^{2} a_j^\epsilon \tilde{w}_{\epsilon,j}' + \phi_\epsilon^\perp, \tag{4.25}$$

where a_j^ϵ are complex numbers and

$$\phi_\epsilon^\perp \perp \mathcal{K}_{\epsilon,t^\epsilon}^{new}.$$

Suppose that $\|\phi_\epsilon\|_{H^2(\Omega_\epsilon)} = 1$. Then we have $|a_j^\epsilon| \leq C$.
Similarly, we decompose

$$\psi_\epsilon = \epsilon \sum_{j=1}^{2} a_j^\epsilon \psi_{\epsilon,j} + \psi_\epsilon^\perp, \tag{4.26}$$

where $\psi_{\epsilon,j} \in H_N^2(\Omega)$ satisfies

$$D(\psi_{\epsilon,j})'' - \psi_{\epsilon,j} + 2\xi_\epsilon \bar{w}_\epsilon \tilde{w}_{\epsilon,j}' = 0 \tag{4.27}$$

and $\psi_\epsilon^\perp \in H_N^2(\Omega)$ solves

$$D(\psi_\epsilon^\perp)'' - \psi_\epsilon^\perp + 2\xi_\epsilon \bar{w}_\epsilon \phi_\epsilon^\perp = 0. \tag{4.28}$$

Substituting the decompositions (4.25) and (4.26) of ϕ_ϵ and ψ_ϵ, respectively, into (4.20), we get

$$\epsilon \sum_{j=1}^2 a_j^\epsilon \left(\frac{(\tilde{w}_{\epsilon,j})^2}{\bar{H}_\epsilon^2} \bar{H}_\epsilon' - \frac{(\bar{w}_\epsilon)^2}{\bar{H}_\epsilon^2} \psi_{\epsilon,j} \right)$$

$$+ \epsilon^2 (\phi_\epsilon^\perp)'' - \phi_\epsilon^\perp + \frac{2\bar{w}_\epsilon}{\bar{H}_\epsilon} \phi_\epsilon^\perp - \frac{\bar{w}_\epsilon^2}{\bar{H}_\epsilon^2} \psi_\epsilon^\perp - \lambda_\epsilon \phi_\epsilon^\perp + \text{e.s.t.}$$

$$= \lambda_\epsilon \left(\epsilon \sum_{j=1}^2 a_j^\epsilon \tilde{w}_{\epsilon,j}' \right). \tag{4.29}$$

In the previous problem, we first compute

$$I_4 := \epsilon \sum_{j=1}^2 a_j^\epsilon \left(\frac{(\tilde{w}_{\epsilon,j})^2}{\bar{H}_\epsilon^2} \bar{H}_\epsilon' - \frac{(\bar{w}_\epsilon)^2}{\bar{H}_\epsilon^2} \psi_{\epsilon,j} \right)$$

$$= \epsilon \sum_{j=1}^2 a_j^\epsilon \left(\frac{(\tilde{w}_{\epsilon,j})^2}{\bar{H}_\epsilon^2} (\bar{H}_\epsilon' - \psi_{\epsilon,j}) \right) - \epsilon \sum_{j=1}^2 a_j^\epsilon \sum_{k \neq j} \frac{(\tilde{w}_{\epsilon,k})^2}{\bar{H}_\epsilon^2} \psi_{\epsilon,j} + \text{e.s.t.}$$

$$= \epsilon \sum_{j=1}^2 a_j^\epsilon \frac{(\tilde{w}_{\epsilon,j})^2}{\bar{H}_\epsilon^2} [-\psi_{\epsilon,j} + \bar{H}_\epsilon'] - \epsilon \sum_{j=1}^N \sum_{k \neq j} a_k^\epsilon \frac{\tilde{w}_{\epsilon,j}^2}{\bar{H}_\epsilon^2} \psi_{\epsilon,k}$$

and write I_4 as follows

$$I_4 = -\epsilon \sum_{j=1}^2 \sum_{k=1}^N a_k^\epsilon \frac{\tilde{w}_{\epsilon,j}^2}{\bar{H}_\epsilon^2} \left(\psi_{\epsilon,k} - \bar{H}_\epsilon' \delta_{jk} \right) + \text{e.s.t.} \tag{4.30}$$

Next we introduce

$$\tilde{L}_\epsilon \phi_\epsilon^\perp := \epsilon^2 (\phi_\epsilon^\perp)'' - \phi_\epsilon^\perp + \frac{2\bar{w}_\epsilon}{\bar{H}_\epsilon} \phi_\epsilon^\perp - \frac{\bar{w}_\epsilon^2}{\bar{H}_\epsilon^2} \psi_\epsilon^\perp \tag{4.31}$$

and

$$\mathbf{a}_\epsilon := \left(a_1^\epsilon, a_2^\epsilon\right)^T.$$ (4.32)

Multiplying both sides of (4.29) by $\tilde{w}'_{\epsilon,l}$ and integrating over $(-1, 1)$, we have

$$\text{r.h.s.} = \epsilon\lambda_\epsilon \sum_{j=1}^{2} a_j^\epsilon \int_{-1}^{1} \tilde{w}'_{\epsilon,j}\tilde{w}'_{\epsilon,l}$$

$$= \lambda_\epsilon a_l^\epsilon \hat{\xi}_l^2 \int_{\mathbb{R}} \left(w'(y)\right)^2 dy\left(1 + O(\epsilon)\right)$$ (4.33)

and

$$\text{l.h.s.} = \left(-\epsilon \sum_{j=1}^{2}\sum_{k=1}^{2} a_k^\epsilon \int_{-1}^{1} \frac{\tilde{w}^2_{\epsilon,j}}{\bar{H}_\epsilon^2}\left(\psi_{\epsilon,k} - \bar{H}'_\epsilon\delta_{jk}\right)\tilde{w}'_{\epsilon,l}\right.$$

$$\left. + \int_{-1}^{1} \frac{\tilde{w}^2_{\epsilon,l}}{\bar{H}_\epsilon^2}\left(\bar{H}'_\epsilon\phi_\epsilon^\perp\right) - \int_{-1}^{1} \frac{\tilde{w}^2_{\epsilon,l}}{\bar{H}_\epsilon^2}\left(\psi_\epsilon^\perp\tilde{w}'_{\epsilon,l}\right)\right)\left(1 + o(1)\right)$$

$$=: (J_{1,l} + J_{2,l} + J_{3,l})\left(1 + o(1)\right).$$

To simplify the notation, we introduce the vectors

$$\mathbf{J}_i = (J_{i,1}, J_{i,2})^T, \quad i = 1, 2, 3$$ (4.34)

and the matrix

$$\mathcal{P}(\mathbf{t}) := \left(I - 2\mathcal{G}_D(\mathbf{t})\mathcal{H}(\mathbf{t})\right)^{-1}.$$ (4.35)

Then we can state the following key lemma.

Lemma 4.7 *We have*

$$\mathbf{J}_1 = c_1\epsilon^2\mathcal{H}(\mathbf{t}^\epsilon)\left[\nabla^2\mathcal{G}_D(\mathbf{t}^\epsilon) - \mathcal{Q}\right]\left(\mathcal{H}(\mathbf{t}^\epsilon)\right)^2\mathbf{a}_\epsilon + o(\epsilon^2),$$ (4.36)

$$\mathbf{J}_2 = o\left(\epsilon^2\right),$$ (4.37)

and

$$\mathbf{J}_3 = c_1\epsilon^2 2\mathcal{H}(\mathbf{t}^\epsilon)\nabla G_D(\mathbf{t}^\epsilon)\mathcal{H}(\mathbf{t}^\epsilon)\mathcal{P}(\mathbf{t}^\epsilon)\left(\nabla\mathcal{G}_D(\mathbf{t}^\epsilon)\right)^T\left(\mathcal{H}(\mathbf{t}^\epsilon)\right)^2\mathbf{a}_\epsilon + o(\epsilon^2),$$ (4.38)

where c_1 is given by (2.85) and $\mathcal{P}(\mathbf{t}^\epsilon)$ is defined by (4.35). Recall that $\mathcal{G}_D(\mathbf{t}^\epsilon)$ and its partial derivatives are introduced in (2.20) and (2.23), respectively, $\mathcal{H}(\mathbf{t}^\epsilon)$ is defined in (2.31) and $\mathbf{a}_\epsilon$ is given in (4.32).

Let us now show that from Lemma 4.7 we can derive Theorem 4.1.

Combining the estimates for J_1, J_2, J_3, we get

$$l.h.s. = J_1 + J_2 + J_3$$
$$= c_1\epsilon^2 \mathcal{H}(\mathbf{t}^\epsilon)[\nabla^2 \mathcal{G}_D(\mathbf{t}^\epsilon) - \mathcal{Q} + 2\nabla\mathcal{G}_D(\mathbf{t}^\epsilon)\mathcal{H}(\mathbf{t}^\epsilon)\mathcal{P}(\mathbf{t}^\epsilon)(\nabla\mathcal{G}_D(\mathbf{t}^\epsilon))^T]$$
$$\times (\mathcal{H}(\mathbf{t}^\epsilon))^2 \mathbf{a}_\epsilon + o(\epsilon^2)$$
$$= c_1\epsilon^2 (\mathcal{H}(\mathbf{t}^\epsilon))^2 \mathcal{M}(\mathbf{t}^\epsilon)\mathbf{a}_\epsilon + o(\epsilon^2).$$

Equating with r.h.s., we get

$$c_1\epsilon^2 (\mathcal{H}(\mathbf{t}^\epsilon))^2 \mathcal{M}(\mathbf{t}^\epsilon)\mathbf{a}_\epsilon + o(\epsilon^2)$$
$$= \lambda_\epsilon (\mathcal{H}(\mathbf{t}^\epsilon))^2 \mathbf{a}_\epsilon \int_{\mathbb{R}} (w'(y))^2 dy (1 + O(\epsilon)). \tag{4.39}$$

Using the expansion $\mathcal{M}(\mathbf{t}^\epsilon) = \mathcal{M}(\mathbf{t}^0) + O(|t^\epsilon - t^0|)$, from (4.39) we can estimate the small eigenvalues of (4.20) in leading order by

$$\lambda_\epsilon \sim \epsilon^2 c_2 \sigma (\mathcal{M}(\mathbf{t}^0)),$$

where $c_2 = \frac{c_1}{\int_{\mathbb{R}} (w')^2} < 0$. This implies the following two cases: If all the eigenvalues of $\mathcal{M}(\mathbf{t}^0)$ are positive, then the small eigenvalues have negative real part for ϵ small enough. On the other hand, if $\mathcal{M}(\mathbf{t}^0)$ has a negative eigenvalue, then we can construct eigenfunctions and corresponding eigenvalues to make the eigenvalue problem unstable by embedding the finite-dimensional problem into infinite-dimensional Sobolev space and using nondegeneracy properties.

This proves Theorem 4.1.

Finally, we prove Lemma 4.7 through a series of lemmas. We first study the asymptotic behaviour of $\psi_{\epsilon,j}$ near t_l^ϵ.

Lemma 4.8 *We have*

$$(\psi_{\epsilon,k} - \bar{H}'_\epsilon \delta_{kl})(t_l^\epsilon) = -(\mathcal{H}(\mathbf{t}^\epsilon))^2 \nabla\mathcal{G}_D(\mathbf{t}^\epsilon) + O(\epsilon). \tag{4.40}$$

Proof For $l \neq k$, we compute

$$(\psi_{\epsilon,k} - \bar{H}'_\epsilon \delta_{kl})(t_l^\epsilon) = \psi_{\epsilon,k}(t_l^\epsilon)$$
$$= 2\xi_\epsilon \int_{-1}^{1} G_D(t_l^\epsilon, z)\bar{w}_\epsilon \tilde{w}'_{\epsilon,k} dz$$
$$= -\nabla_{t_k^\epsilon} G_D(t_k^\epsilon, t_l^\epsilon)\hat{\xi}_k^2 + O(\epsilon). \tag{4.41}$$

Next we consider $\psi_{\epsilon,l} - \bar{H}'_\epsilon$ for $|t_l^\epsilon - x| = O(\epsilon^{2/3})$:

$$\bar{H}_\epsilon(x) = \xi_\epsilon \int_{-1}^{1} G_D(x, z)\tilde{w}_\epsilon^2$$

$$= \xi_\epsilon \int_{-\infty}^{+\infty} K_D(|z|)\tilde{w}_{\epsilon,l}^2(x + z)dz - \xi_\epsilon \int_{-1}^{1} H_D(x, z)\tilde{w}_{\epsilon,l}^2 dz$$

$$+ \xi_\epsilon \sum_{k\neq l} \int_{-1}^{1} G_D(x, z)\tilde{w}_{\epsilon,k}^2.$$

Hence, we have

$$\bar{H}_\epsilon'(x) = \xi_\epsilon \int_{-\infty}^{+\infty} K_D(|z|)2\tilde{w}_{\epsilon,l}(x + z)\tilde{w}_{\epsilon,l}'(x + z)dz - \xi_\epsilon \int_{-1}^{1} \nabla_x H_D(x, z)\tilde{w}_{\epsilon,l}^2 dz$$

$$+ \xi_\epsilon \sum_{k\neq l} \int_{-1}^{1} \nabla_x G_D(x, z)\tilde{w}_{\epsilon,k}^2 dz.$$

Thus we have

$$\bar{H}_\epsilon'(x) - \psi_{\epsilon,l}(x) = -\xi_\epsilon \int_{-1}^{1} \nabla_x H_D(x, z)\tilde{w}_{\epsilon,l}^2 dz + \xi_\epsilon \sum_{k\neq l} \int_{-1}^{1} \nabla_x G_D(x, z)\tilde{w}_{\epsilon,k}^2 dz$$

$$- \left(-\xi_\epsilon \int_{-1}^{1} H_D(x, z)2\tilde{w}_{\epsilon,l}\tilde{w}_{\epsilon,l}' dz\right).$$

Using (H3), we get

$$\bar{H}_\epsilon'(t_l^\epsilon) - \psi_{\epsilon,l}(t_l^\epsilon) = -\xi_\epsilon \int_{-1}^{1} \nabla_{t_l^\epsilon} H(t_l^\epsilon, z)\tilde{w}_{\epsilon,l}^2 dz + \xi_\epsilon \int_{-1}^{1} \nabla_{t_l^\epsilon} G(t_l^\epsilon, z)\tilde{w}_{\epsilon,3-l}^2 dz$$

$$- \nabla_{t_l^\epsilon} H_D(t_l^\epsilon, t_l^\epsilon)\hat{\xi}_l^2 + O(\epsilon)$$

$$= -\nabla_{t_l^\epsilon} H_D(t_l^\epsilon, t_l^\epsilon)\hat{\xi}_l^2 + O(\epsilon). \tag{4.42}$$

Combining the relations (4.41) and (4.42), statement (4.40) follows.   □

As in Lemma 4.8, we derive

Lemma 4.9 *We have*

$$(\psi_{\epsilon,k} - \bar{H}_\epsilon' \delta_{lk})(t_l^\epsilon + \epsilon y) - (\psi_{\epsilon,k} - \bar{H}_\epsilon' \delta_{lk})(t_l^\epsilon)$$

$$= -\epsilon y[\nabla_{t_l^\epsilon} \nabla_{t_k^\epsilon} G_D(t_l^\epsilon, t_k^\epsilon) - q_{lk}\delta_{lk}]\hat{\xi}_k^2 + O(\epsilon^2 y^2), \tag{4.43}$$

where q_{lk} is defined in (2.34).

Proof The computations and the result both resemble those of (2.35) and are therefore omitted.   □

Next we consider the asymptotic expansion of $\phi_\epsilon^\perp$. We introduce

$$\phi_{\epsilon,j}^1 = -\sum_{l=1}^2 (\nabla_{t_j^\epsilon} \hat{\xi}_l \tilde{w}_{\epsilon,l}), \qquad \phi_\epsilon^1 := \epsilon \sum_{j=1}^2 a_j^\epsilon \phi_{\epsilon,j}^1. \qquad (4.44)$$

Lemma 4.10 *For ϵ small enough, we have*

$$\|\phi_\epsilon^\perp - \phi_\epsilon^1\|_{H^2(-1/\epsilon,1/\epsilon)} = O(\epsilon^2). \qquad (4.45)$$

Proof We begin by deriving a relation between $\psi_\epsilon^\perp$ and $\phi_\epsilon^\perp$. As in the proof of Proposition 2.7, the linearised operator around the exact solution $\tilde{L}_\epsilon$ defined in (4.9), (4.10) is invertible from $(\mathcal{K}_\epsilon^{new})^\perp$ to $(\mathcal{C}_\epsilon^{new})^\perp$. Using Lemma 4.8, we get

$$\|\phi_\epsilon^\perp\|_{H^2(-1/\epsilon,1/\epsilon)} = O(\epsilon). \qquad (4.46)$$

Let us decompose

$$\tilde{\phi}_{\epsilon,j} = \frac{\phi_\epsilon^\perp}{\epsilon} \chi\left(\frac{x - t_j^\epsilon}{r_0}\right). \qquad (4.47)$$

Then

$$\phi_\epsilon^\perp = \epsilon \sum_{j=1}^2 \tilde{\phi}_{\epsilon,j} + \text{e.s.t.}$$

Suppose that

$$\tilde{\phi}_{\epsilon,j} \to \phi_j \quad \text{in } H^2. \qquad (4.48)$$

Let

$$\Phi_0 = (\phi_1, \phi_2)^T.$$

Then, as in the proof of Lemma 4.8, we compute

$$\psi_\epsilon^\perp(t_j^\epsilon) = 2\epsilon \sum_{k=1}^2 \xi_\epsilon \int_{-1}^1 G_D(t_j^\epsilon, z) \bar{w}_\epsilon \tilde{\phi}_{\epsilon,k} dz + \text{e.s.t.}$$

$$= -2\epsilon \sum_{k=1}^2 G_D(t_j^\epsilon, t_k^\epsilon) \hat{\xi}_k \frac{\int_{\mathbb{R}} w \phi_k}{\int_{\mathbb{R}} w^2} + O(\epsilon^2). \qquad (4.49)$$

Thus we have

$$(\psi_\epsilon^\perp(t_1^\epsilon), \psi_\epsilon^\perp(t_2^\epsilon))^T = -2\epsilon \mathcal{G}_D(\mathbf{t}^\epsilon) \mathcal{H}(\mathbf{t}^\epsilon) \frac{\int_{\mathbb{R}} w \Phi_0}{\int_{\mathbb{R}} w^2} + O(\epsilon^2). \qquad (4.50)$$

Substituting (4.50) into (4.29) and taking the limit $\epsilon \to 0$, we get

$$(\Phi_0)'' - \Phi_0 + 2w\Phi_0 - 2\mathcal{G}_D(\mathbf{t}^0)\mathcal{H}(\mathbf{t}^0)\frac{\int_{\mathbb{R}} w\Phi_0}{\int_{\mathbb{R}} w^2}w^2 + (\nabla\mathcal{G}_D(\mathbf{t}^0))^T (\mathcal{H}(\mathbf{t}^0))^2 \mathbf{a}^0 w^2 = 0,$$

where

$$\mathbf{a}^0 = \lim_{\epsilon \to 0} \mathbf{a}^\epsilon.$$

This implies

$$\begin{aligned}
\Phi_0 &= -(I - 2\mathcal{G}_D(\mathbf{t}^0)\mathcal{H}(\mathbf{t}^0))^{-1}(\nabla\mathcal{G}_D(\mathbf{t}^0))^T (\mathcal{H}(\mathbf{t}^0))^2 \mathbf{a}^0 w \\
&= -\mathcal{P}(\mathbf{t}^0)(\nabla\mathcal{G}_D(\mathbf{t}^0))^T (\mathcal{H}(\mathbf{t}^0))^2 \mathbf{a}^0 w \\
&= -\nabla\xi(\mathbf{t}^0)\mathbf{a}^0 w
\end{aligned} \tag{4.51}$$

by (2.33).

Next we compare Φ_0 with ϕ_ϵ^1. By definition, we have

$$\begin{aligned}
\phi_\epsilon^1 &= -\epsilon \sum_{j=1}^{2} a_j^\epsilon \sum_{l=1}^{2} ((\nabla_{t_j^\epsilon}\hat{\xi}_l(\mathbf{t}^\epsilon))\tilde{w}_{\epsilon,l}) \\
&= -\epsilon \sum_{l=1}^{2} \left[\sum_{j=1}^{2}(\nabla_{t_j^\epsilon}\hat{\xi}_l)a_j^\epsilon\right]\tilde{w}_{\epsilon,l}.
\end{aligned} \tag{4.52}$$

On the other hand, we compute

$$\begin{aligned}
\phi_\epsilon^\perp &= \epsilon \sum_{j=1}^{2} \tilde{\phi}_{\epsilon,j} + O(\epsilon^2) \\
&= \epsilon \sum_{j=1}^{2} \phi_j\left(\frac{x - t_j^\epsilon}{\epsilon}\right) + O(\epsilon^2).
\end{aligned} \tag{4.53}$$

Combining (4.52) and (4.53) with (4.51) we get (4.45) and the result follows. $\square$

Now Lemma 4.10 implies that

$$\begin{aligned}
&(\psi_\epsilon^\perp(t_1^\epsilon), \psi_\epsilon^\perp(t_2^\epsilon))^T \\
&= 2\epsilon\mathcal{G}_D(\mathbf{t}^\epsilon)\mathcal{H}(\mathbf{t}^\epsilon)\mathcal{P}(\mathbf{t}^\epsilon)(\nabla\mathcal{G}_D(\mathbf{t}^\epsilon))^T (\mathcal{H}(\mathbf{t}^\epsilon))^2 \mathbf{a}^\epsilon + O(\epsilon^2)
\end{aligned} \tag{4.54}$$

and

$$\psi_\epsilon^\perp\left(t_j^\epsilon + \epsilon y\right) - \psi_\epsilon^\perp\left(t_j^\epsilon\right)$$

$$= -2\epsilon^2 y \sum_{k=1}^{2} \nabla_{t_j^\epsilon} G_D\left(t_j^\epsilon, t_k^\epsilon\right)\hat{\xi}_k\left(\mathbf{t}^\epsilon\right)\frac{\int_{\mathbb{R}} w\phi_k}{\int_{\mathbb{R}} w^2} + O\left(\epsilon^3 y^2\right). \qquad (4.55)$$

Finally, we prove the key lemma, Lemma 4.7.

Proof of Lemma 4.7 To compute J_1 using Lemma 4.9 and the estimate $\bar{H}_\epsilon' = o(1)$, we begin with

$$J_{1,l} = -\epsilon \sum_{k=1}^{2} a_k^\epsilon \int_{-1}^{1} \frac{\tilde{w}_{\epsilon,l}^2}{\bar{H}_\epsilon^2}\left(\psi_{\epsilon,k} - \bar{H}_\epsilon'\delta_{lk}\right)\tilde{w}_{\epsilon,l}'dx + \text{e.s.t.}$$

$$= -\epsilon \sum_{k=1}^{2} a_k^\epsilon \int_{-1}^{1} \frac{\tilde{w}_{\epsilon,l}^2}{\bar{H}_\epsilon^2}\left(\left[\psi_{\epsilon,k}(x) - \bar{H}_\epsilon'(x)\delta_{lk}\right] - \left[\psi_{\epsilon,k}\left(t_l^\epsilon\right) - \bar{H}_\epsilon'\left(t_l^\epsilon\right)\delta_{lk}\right]\right)\tilde{w}_{\epsilon,l}'dx$$

$$+ o\left(\epsilon^2\right)$$

$$= \epsilon^2 \int_{\mathbb{R}}\left(yw^2w'(y)\right)dy \sum_{k=1}^{2}\left[\hat{\xi}_l\left(\mathbf{t}^\epsilon\right)\nabla_{t_j^\epsilon}\nabla_{t_k^\epsilon}G_D\left(t_j^\epsilon, t_k^\epsilon\right)\hat{\xi}_k^2\left(\mathbf{t}^\epsilon\right) - q_{lk}\delta_{lk}\right]a_k^\epsilon + o\left(\epsilon^2\right),$$

where q_{lk} has been defined in (2.34). This implies (4.36).

Next we show (4.37) by using Lemma 4.10 and the following estimates:

$$\bar{H}_\epsilon\left(t_j^\epsilon\right) = \hat{\xi}_j\left(\mathbf{t}^\epsilon\right) + O\left(\epsilon^2\right),$$

$$\bar{H}_\epsilon'\left(t_j^\epsilon + \epsilon y\right) - \bar{H}_\epsilon'\left(t_j^\epsilon\right) = \epsilon \times \text{odd function in } y + O\left(\epsilon^2\right).$$

Therefore we derive $\mathbf{J}_2 = o(\epsilon^2)$ and (4.37) is proved.

Finally we prove (4.38). First we compute

$$J_{3,l} = -\int_{-1}^{1}\frac{\tilde{w}_{\epsilon,l}^2}{\bar{H}_\epsilon^2}\left(\psi_\epsilon^\perp w_{\epsilon,l}'\right)dx$$

$$= -\int_{-1}^{1}\frac{\tilde{w}_{\epsilon,l}^2}{\bar{H}_\epsilon^2}\left(\psi_\epsilon^\perp\left(t_l^\epsilon\right)w_{\epsilon,l}'\right)dx - \int_{-1}^{1}\frac{\tilde{w}_{\epsilon,l}^2}{\bar{H}_\epsilon^2}\left(\psi_\epsilon^\perp(x) - \psi_\epsilon^\perp\left(t_l^\epsilon\right)\right)w_{\epsilon,l}'dx$$

$$= -\int_{-1}^{1}\frac{\tilde{w}_{\epsilon,l}^2}{\bar{H}_\epsilon^2}\left(\psi_\epsilon^\perp(x) - \psi_\epsilon^\perp\left(t_l^\epsilon\right)\right)w_{\epsilon,l}'dx + o\left(\epsilon^2\right)$$

using (4.54). Combining this result with (4.51) and (4.55), we get (4.38).   □

4.1.3   The Spectrum of the Matrices $\mathcal{B}$ and $\mathcal{M}$

In this subsection, we sketch the computation of the eigenvalues of $\mathcal{B}$ and $\mathcal{M}$ for the case of symmetric two-spike solutions. Then Theorem 4.3 follows from Theorem 4.1.

We begin by considering the matrices $\mathcal{G}_D$, $\nabla \mathcal{G}_D$ and $\nabla^2 \mathcal{G}_D$ which were introduced in (2.20) and (2.23).

Recalling from (2.4) that

$$t_1^0 = -\frac{1}{2}, \qquad t_2^0 = \frac{1}{2}, \qquad \theta = \frac{1}{\sqrt{D}},$$

we compute

$$\mathcal{G}_D = \frac{\theta}{\sinh(2\theta)}(a_{ij}), \qquad \nabla \mathcal{G}_D = \frac{\theta^2}{\sinh(2\theta)}(b_{ij}), \qquad \nabla^2 \mathcal{G}_D = \frac{\theta^3}{\sinh(2\theta)}(c_{ij}),$$

where

$$a_{ij} = \begin{cases} \cosh(\theta(1 + t_i^0))\cosh(\theta(1 - t_j^0)) & \text{for } i \leq j; \\ \cosh(\theta(1 - t_i^0))\cosh(\theta(1 + t_j^0)) & \text{for } i > j, \end{cases} \tag{4.56}$$

$$b_{ij} = \begin{cases} \sinh(\theta(1 + t_i^0))\cosh(\theta(1 - t_j^0)) & \text{for } i < j; \\ \frac{1}{2}\sinh(2\theta t_i^0) & \text{for } i = j; \\ -\sinh(\theta(1 - t_i^0))\cosh(\theta(1 + t_j^0)) & \text{for } i > j, \end{cases} \tag{4.57}$$

and

$$c_{ij} = \begin{cases} -\sinh(\theta(1 + t_i^0))\sinh(\theta(1 - t_j^0)) & \text{for } i < j; \\ -\sinh(\theta(1 + t_i^0))\sinh(\theta(1 - t_j^0)) + \frac{1}{2}\sinh(2\theta) & \text{for } i = j; \\ -\sinh(\theta(1 - t_i^0))\sinh(\theta(1 + t_j^0)) & \text{for } i > j. \end{cases} \tag{4.58}$$

In the symmetric two-spike case, we have $\hat{\xi}_1^0 = \hat{\xi}_2^0 = \hat{\xi}_0$. Hence, we get

$$\mathcal{H} = \hat{\xi}_0 I.$$

We can compute $\hat{\xi}_0$ explicitly as follows:

$$\hat{\xi}_0 = \frac{2}{\theta}\tanh\frac{\theta}{2}. \tag{4.59}$$

Thus

$$\mathcal{Q} = \frac{\theta^3}{2}\left(1 - \coth\frac{\theta}{2}\right)I. \tag{4.60}$$

Firstly, we consider

$$\mathcal{G}_D = \frac{\theta \cosh(\theta/2)}{\sinh(2\theta)} \begin{pmatrix} d_1 & f_1 \\ f_1 & d_1 \end{pmatrix}, \tag{4.61}$$

where

$$d_1 = \cosh\frac{3\theta}{2}, \qquad f_1 = \cosh\frac{\theta}{2}.$$

Using the fact that $\mathcal{G}_D$ is a symmetric matrix, we determine its eigenvalues and eigenvectors as follows:

$$\lambda_1 = \frac{\theta}{2}\coth\frac{\theta}{2}, \qquad \lambda_2 = \frac{\theta}{2}\tanh\theta,$$

$$\mathbf{v}_1^t = \frac{1}{\sqrt{2}}(1, 1), \qquad \mathbf{v}_2^t = \frac{1}{\sqrt{2}}(1, -1). \tag{4.62}$$

In summary, if we choose

$$\mathcal{P}_1 = (\mathbf{v}_1, \mathbf{v}_2),$$

then we have

$$\mathcal{P}_1^{-1}\mathcal{G}_D\mathcal{P}_1 = \begin{pmatrix} \lambda_1 & 0 \\ 0 & \lambda_2 \end{pmatrix}. \tag{4.63}$$

Secondly, we study

$$\nabla^2 \mathcal{G}_D = -\frac{\theta^3 \sinh(\theta/2)}{\sinh(2\theta)} \begin{pmatrix} f_2 & d_2 \\ d_2 & f_2 \end{pmatrix}, \tag{4.64}$$

where

$$f_2 = \sinh(3\theta/2), \qquad d_2 = \sinh(\theta/2).$$

Since $\nabla^2 \mathcal{G}_D$ is a symmetric matrix, we compute its eigenvalues and eigenvectors as follows:

$$\mu_1 = -\frac{\theta^3}{2}\tanh\theta, \qquad \mu_2 = -\frac{\theta^3}{2}\tanh\frac{\theta}{2},$$

$$\mathbf{v}_1^t = \frac{1}{\sqrt{2}}(1, 1), \qquad \mathbf{v}_2^t = \frac{1}{\sqrt{2}}(1, -1). \tag{4.65}$$

Thus, for

$$\mathcal{P}_1 = (\mathbf{v}_1, \mathbf{v}_2)$$

we have

$$\mathcal{P}_1^{-1}\nabla^2\mathcal{G}_D\mathcal{P}_1 = \begin{pmatrix} \mu_1 & 0 \\ 0 & \mu_2 \end{pmatrix}. \tag{4.66}$$

Thirdly, we diagonalise

$$\nabla \mathcal{G}_D = -\frac{\theta^2}{4\cosh\theta}\begin{pmatrix} -1 & 1 \\ -1 & 1 \end{pmatrix}. \tag{4.67}$$

Thus $\nabla \mathcal{G}_D$ has eigenvector

$$\mathbf{v}_1^t = \frac{1}{\sqrt{2}}(1, 1)$$

and Jordan chain vector

$$\mathbf{v}_2^t = \frac{1}{\sqrt{2}}(1, -1)$$

both with eigenvalue 0.

Hence, for

$$\mathcal{P}_1 = (\mathbf{v}_1, \mathbf{v}_2)$$

we have

$$\mathcal{P}_1^{-1}\nabla \mathcal{G}_D \mathcal{P}_1 = \frac{\theta^2}{2\cosh\theta}\begin{pmatrix} 0 & 1 \\ 0 & 0 \end{pmatrix}. \tag{4.68}$$

Next by (4.63), (4.66) and (4.68), the eigenvalues of $\mathcal{M}$ are as follows:

$$m_j = \hat{\xi}_0\left[\mu_j - q_0 + 2\hat{\xi}_0(1 - 2\hat{\xi}_0\lambda_j)^{-1}\frac{\theta^4}{2\cosh^2\theta}\delta_{1j}\right], \quad j = 1, 2,$$

where μ_j is given in (4.65) and

$$q_0 = \frac{\theta^3}{2}\left(1 - \coth\frac{\theta}{2}\right).$$

For stability, we need

$$2\hat{\xi}_0 \min_{j=1,2}\lambda_j > 1 \tag{4.69}$$

and

$$\min_{j=1,2}m_j > 0. \tag{4.70}$$

The first condition (4.69) comes from the large eigenvalues and it gives us the following criterion for $\tau = 0$:

$$1 < 2\hat{\xi}_0 \min_{j=1,2}\lambda_j = 2\tanh\frac{\theta}{2}\tanh\theta = 2\left(1 - \frac{1}{\cosh\theta}\right)$$

which implies

$$D < D_2^1 = \frac{1}{\theta_{2,1}^2}, \quad \theta_{2,1} = \log(2 + \sqrt{3}). \tag{4.71}$$

The second condition (4.70) originates from the small eigenvalues and is given by

$$0 < \frac{1}{\xi} \min_{j=1,2} m_j = -\tanh\theta + \coth\frac{\theta}{2} + \frac{2\tanh(\theta/2)}{(1 - 2\tanh(\theta/2)\tanh\theta)\cosh^2\theta}.$$

Using the identity

$$1 - 2\tanh\frac{\theta}{2}\tanh\theta = \frac{1 - 2\sinh^2(\theta/2)}{1 + 2\sinh^2(\theta/2)},$$

we derive the polynomial inequality

$$(2\alpha + 1)(\alpha - 1) > 0,$$

where

$$\alpha = \sinh^2\frac{\theta}{2}.$$

This leads to the critical threshold

$$D < D_2^2 = \frac{1}{\theta_{2,2}^2}, \qquad \theta_{2,2} = 2\log(1 + \sqrt{2}). \tag{4.72}$$

It is easy to see that $D_2^1 > D_2^2$. Thus the symmetric two-spike solutions are stable for $D < D_2 \equiv D_2^2$ and unstable for $D > D_2$. We remark that the estimates for small eigenvalues do not depend on $\tau \geq 0$.

Theorem 4.3 is now proved.

4.2 Notes on the Literature

For stability we have to assume that $0 \leq \tau < \tau_0$ for some $\tau_0 > 0$ which we do not know explicitly. Stability for the case of large τ has been investigated in [242]. We remark that stability for the case of large τ for the shadow system has been studied in [37]. The rigorous relation between the eigenvalue problem for small parameter ϵ and the limiting case $\epsilon = 0$ uses an idea of Dancer [37].

For an N-spike solution with general N the matrices $\mathcal{B}$ and $\mathcal{M}$ including their eigenvalues have been computed in [99, 273].

The eigenvalues of $\mathcal{B}$ and its eigenvectors are

$$\beta_j = e_1 + 2f_1 \cos\left(\frac{\pi(j-1)}{N}\right), \qquad j = 1, \ldots, N,$$

where

$$e_1 = 2\coth\frac{2\theta}{N}, \qquad f_1 = -\operatorname{csch}\frac{2\theta}{N},$$

and

$$v_{l,j} = \sqrt{\frac{2}{N}} \cos\left(\frac{\pi(j-1)}{N}\left(l - \frac{1}{2}\right)\right), \quad j = 1, \ldots, N,$$

respectively. The eigenvalues of $\mathcal{M}$ are given by

$$m_j = \mu_j + 2\hat{\xi}_0 v_j^2 (1 - 2\hat{\xi}_0 \lambda_j)^{-1}, \quad j = 1, \ldots, N,$$

where

$$\mu_j = \theta^3 - \theta^3\left(e_1 + 2f_1 \cos\left(\frac{\pi(j-1)}{N}\right)\right)^{-1}, \quad j = 2, \ldots, N,$$

$$\mu_1 = \theta^3 - \theta^3(e_1 - 2f_1)^{-1},$$

(4.73)

and

$$v_j = \operatorname{csch}\left(\frac{2\theta}{N}\right)\sin\left(\frac{\pi(j-1)}{N}\right)\lambda_j, \quad j = 1, \ldots, N, \tag{4.74}$$

$$\lambda_j = \theta\left(e_1 + 2f_1 \cos\left(\frac{\pi(j-1)}{N}\right)\right)^{-1}, \quad j = 1, \ldots, N. \tag{4.75}$$

The stability of asymmetric N-spike solutions is studied in [240], using matched asymptotic expansions. The matrices $\mathcal{B}$ and $\mathcal{M}$ change and they have to be computed again. This will alter the stability thresholds. A rigorous proof of the results of [99] is contained in [273].

It turns out that for multi-pulses on the real line [46, 47, 49] with increasing inhibitor diffusivity D the instability first occurs because of large eigenvalues, in contrast to the case of a finite interval, where the small eigenvalues cause the instability. A dynamics approach which covers the case of general $\tau \geq 0$ but is restricted to the whole real line or to periodic boundary conditions is contained in [49].

In [212] absolute instabilities of standing pulses in reaction-diffusions coming from the homogeneous background state are analysed, in particular, the impact of pitchfork, Turing and oscillatory bifurcations. The methods employed are the Evans functions approach and blow-up arguments. A survey is given in [211].

Chapter 5
Existence of Spikes for the Shadow Gierer-Meinhardt System

5.1 The Shadow Gierer-Meinhardt System

We consider functions $(A(x), H(x)) = (\xi u(x), \xi)$, where $A(x) > 0$ and $\xi > 0$, which are positive solutions of (1.1), (1.2), (1.3). By a simple scaling argument, it follows that the function $u(x)$ satisfies the boundary value problem

$$\begin{cases} \epsilon^2 \Delta u - u + u^p = 0 & \text{in } \Omega, \\ u > 0 & \text{in } \Omega \\ \frac{\partial u}{\partial \nu} = 0 & \text{on } \partial \Omega. \end{cases} \tag{5.1}$$

This system is called the shadow system of (1.3), following the terminology of Nishiura [181]. In this chapter, we assume that

$$1 < p < \frac{N+2}{N-2} \quad \text{if } N \geq 3, \qquad 1 < p \quad \text{if } N = 2. \tag{5.2}$$

We recall that $\Delta = \sum_{i=1}^{N} \frac{\partial^2}{\partial x_i^2}$ is the Laplace operator, Ω is a bounded smooth domain in $\mathbb{R}^N$, $\epsilon > 0$ is a small positive constant, and $\nu(x)$ denotes the normal derivative at $x \in \partial \Omega$.

Our method will be Liapunov-Schmidt reduction in combination with a variational structure given by an energy functional. We call it the Variational Liapunov-Schmidt Method (or Localised Energy Method). We will lean on the presentation in Sect. 2.3.3 rather than giving a full proof, but the modifications and novelties will be highlighted.

Our aim is to construct a family of boundary peak steady states for (5.1).

Let $\Lambda \subset \partial \Omega$ be an open set such that

$$\min_{P \in \partial \Lambda} H(P) > \min_{P \in \Lambda} H(P). \tag{5.3}$$

Now we state the main result of this section.

J. Wei, M. Winter, *Mathematical Aspects of Pattern Formation in Biological Systems*, Applied Mathematical Sciences 189, DOI 10.1007/978-1-4471-5526-3_5, © Springer-Verlag London 2014

Theorem 5.1 *Assume that conditions* (5.2) *and* (5.3) *hold. Then, for ε small enough, the shadow system* (5.1) *possesses a solution* $u_\varepsilon > 0$ *which has exactly one local maximum point* $Q^\varepsilon \in \Lambda$. *Further, we have* $H(Q^\varepsilon) \to \min_{P \in \Lambda} H(P)$ *and*

$$u_\varepsilon(x) \le a \exp\left(-\frac{b|x - Q^\varepsilon|}{\epsilon}\right) \tag{5.4}$$

for certain positive constants a, b.

Remark 5.2 Throughout this section, we will use the notation $f(u) = u^p$. The reason for this choice is that the result can be readily extended to more general nonlinearities which satisfy certain smoothness, growth and nondegeneracy conditions. For more details, we refer to [36, 81].

Before outlining the main ideas of the proof, we first need to introduce some notation and definitions.

The equation

$$\begin{cases} \Delta w - w + u^p = 0 & \text{in } \mathbb{R}^N, \\ w > 0, \quad w(0) = \max_{z \in \mathbb{R}^N} w(z), \\ w \to 0 & \text{as } |z| \to \infty \end{cases} \tag{5.5}$$

has a least energy solution $w(y)$ which is nondegenerate. This follows by combining results of [71] and [124–126]. See Sect. 13.2 for a complete proof.

By [71] function w is radially symmetric, decreasing and it satisfies

$$\lim_{|y| \to \infty} w(y)e^{|y|}|y|^{(N-1)/2} = c_0 > 0.$$

Associated with problem (5.1) is the energy functional

$$J_\epsilon(u) = \frac{1}{2} \int_\Omega \left(\epsilon^2 |\nabla u|^2 + u^2\right) - \int_\Omega F(u)$$

where $F(u) = \int_0^u f(s)ds = \frac{u^{p+1}}{p+1}$ and $u \in H^2(\Omega)$.

For any smooth bounded domain U, we next introduce the following projection P_U from $H^2(U)$ into

$$H_N^2(U) = \left\{v \in H^2(U) : \frac{\partial v}{\partial \nu} = 0 \text{ on } \partial U\right\}.$$

For given function $w \in H^2(U)$, let $P_U w$ be the unique solution $u \in H_N^2(U)$ of the semilinear partial differential equation

$$\Delta u - u + f(w) = 0 \quad \text{in } U. \tag{5.6}$$

Using this projection, we will be able to construct solutions in the space $H_N^2(U)$. Choosing all the approximate solutions in this space we only have to solve the partial differential equation and the boundary condition is always satisfied.

For $P \in \overline{\Lambda}$, we use the notation

$$\Omega_\epsilon = \{y : \epsilon y \in \Omega\}, \qquad \Omega_{\epsilon, P} = \{y : \epsilon y + P \in \Omega\}$$

and set

$$Pw(y) = P_{\Omega_{\epsilon, P}} w(y), \quad y \in \Omega_{\epsilon, P}.$$

Our solution will be decomposed as

$$u = Pw + \Phi_{\epsilon, P} \in H_N^2(\Omega_{\epsilon, P})$$

and we use the approximate kernel and cokernel defined by

$$\mathcal{K}_{\epsilon, P} = \mathrm{span} \left\{ \frac{\partial Pw}{\partial \tau_{P_j}} : j = 1, \ldots, N - 1 \right\} \subset H_N^2(\Omega_\epsilon),$$

$$\mathcal{C}_{\epsilon, P} = \mathrm{span} \left\{ \frac{\partial Pw}{\partial \tau_{P_j}} : j = 1, \ldots, N - 1 \right\} \subset L^2(\Omega_\epsilon),$$

where τ_{P_j} are the $(N - 1)$ tangential derivatives at P (by a suitable choice of coordinates, we may assume that the inward normal derivative at P is e_N).

Using the Liapunov-Schmidt reduction method, we find a solution in two steps. First, we first solve the equation for $\Phi_{\epsilon, P} \in \mathcal{K}_{\epsilon, P}^\perp$ up to $\mathcal{C}_{\epsilon, P}^\perp$. Secondly, we show that $\Phi_{\epsilon, P}$ is C^1 in P and define a finite-dimensional function M_ϵ by

$$M_\epsilon : \overline{\Lambda} \to \mathbb{R}, \quad M_\epsilon(P) = J_\epsilon(Pw + \Phi_{\epsilon, P}). \tag{5.7}$$

Then we maximise $M_\epsilon(P)$ over $\overline{\Lambda}$ noting that condition (5.3) guarantees that $M_\epsilon(P)$ attains its maximum in the interior of Λ. We conclude by showing that the resulting solution has the properties stated in Theorem 5.1.

5.2 The Existence Proof

Now we give the complete existence proof in detail.

5.2.1 Technical Analysis

In this subsection, we provide some results which will be needed later.

We define $\mathbb{R}_+^N = \{(x', x_N) : x_N > 0\}$. Let w be the unique solution of (5.5) and set

$$I(w) = \frac{1}{2} \int_{\mathbb{R}^N} \left(|\nabla w|^2 + w^2 \right) - \int_{\mathbb{R}^N} F(w).$$

For $P \in \overline{\Lambda}$, we define a diffeomorphism straightening the boundary in a neighbourhood of P. After a rotation of the coordinate system, we may assume that the inward normal to $\partial \Omega$ at P is pointing in the direction of the positive x_N-axis. Defining $x' = (x_1, \ldots, x_{N-1})$ and

$$B'(R_0) = \{x' \in \mathbb{R}^{N-1} : |x'| < R_0\}, \qquad B(P, R_0) = \{x \in \mathbb{R}^N : |x - P| < R_0\},$$

we further introduce

$$\Omega_0 = \Omega \cap B(P, R_0) = \{(x', x_N) \in B(P, R_0) : x_N - P_N > \rho(x' - P')\}.$$

Since $\partial \Omega$ is smooth, there is a constant $R_0 > 0$ such that $\partial \Omega \cap \overline{\Omega_0}$ is given by the graph of a smooth function $\rho_P : B'(R_0) \to \mathbb{R}$ where $\rho_P(0) = 0$, $\nabla \rho_P(0) = 0$.

For simplicity of notation, we omit the subscript P in ρ_P if this does not cause confusion. The mean curvature of $\partial \Omega$ at P is given by $H(P) = \frac{1}{N-1} \sum_{i=1}^{N-1} \rho_{ii}(0)$ where

$$\rho_i = \frac{\partial \rho}{\partial x_i}, \quad i = 1, \ldots, N-1.$$

Further, higher-order derivatives are defined accordingly. Using Taylor expansion, we get

$$\rho(x' - P') = \frac{1}{2} \sum_{i,j=1}^{N-1} \rho_{ij}(0)(x_i - P_i)(x_j - P_j)$$

$$+ \frac{1}{6} \sum_{i,j,k=1}^{N-1} \rho_{ijk}(0)(x_i - P_i)(x_j - P_j)(x_k - P_k) + O(|x' - P'|^4).$$

Setting $h_{\epsilon,P}(x) = w(\frac{x-P}{\epsilon}) - P_{\Omega_{\epsilon,P}} w(\frac{x-P}{\epsilon})$, $x \in \Omega$, we compute for $v = h_{\epsilon,P}$:

$$\begin{cases} \epsilon^2 \Delta v - v = 0 & \text{in } \Omega, \\ \frac{\partial v}{\partial \nu}(x) = \frac{\partial}{\partial \nu} w(\frac{x-P}{\epsilon}) & \text{on } \partial \Omega. \end{cases} \tag{5.8}$$

We let

$$\|v\|_{\epsilon}^2 = \epsilon^{-N} \int_{\Omega} [\epsilon^2 |\nabla v|^2 + v^2] dx$$

and, for $x \in \Omega_0$, we set

$$\begin{cases} \epsilon y' = x' - P', \\ \epsilon y_N = x_N - P_N - \rho(x' - P'). \end{cases} \tag{5.9}$$

Further, for $x \in \Omega_0$, we introduce the transformation T by

$$\begin{cases} T_i(x') = x_i, & i = 1, \ldots, N-1, \\ T_N(x') = x_N - P_N - \rho(x' - P'). \end{cases} \tag{5.10}$$

Then we have

$$y = \frac{1}{\epsilon} T(x).$$

Let v_1 be the unique solution of the problem

$$
\begin{cases}
\Delta v - v = 0 & \text{in } \mathbb{R}^N_+, \\
\frac{\partial v}{\partial y_N} = -\frac{w'}{|y|} \frac{1}{2} \sum_{i,j=1}^{N-1} \rho_{ij}(0) y_i y_j & \text{on } \partial \mathbb{R}^N_+
\end{cases}
\tag{5.11}
$$

where w' is the radial derivative of w, i.e. $w' = w_r(r)$, and $r = |\frac{x-P}{\epsilon}|$. Then v_1 is an even function in $y' = (y_1, \ldots, y_{N-1})$ and $|v_1| \le Ce^{-\mu|y|}$ for some $0 < \mu < 1$.

Let χ be the smooth cutoff function introduced in (2.51).

Setting

$$h_{\epsilon,P}(x) = \epsilon v_1(y) \chi \left(\frac{2|x-P|}{R_0} \right) + \epsilon^2 \Psi_{\epsilon,P}(x), \quad x \in \Omega,$$

we have

Proposition 5.3

$$\|\Psi_{\epsilon,P}\|_\epsilon \le C.$$

Proof Proposition 5.3 was proved in [256] using Taylor expansion and a rigorous estimate for the remainder based on elliptic estimates. □

Similarly, we know from [256] that

Proposition 5.4

$$\left[\frac{\partial w}{\partial \tau_{P_j}} - \frac{\partial P_{\Omega_{\epsilon,P}} w}{\partial \tau_{P_j}} \right] \left(\frac{x-P}{\epsilon} \right) = w_1(y) \chi \left(\frac{2|x-P|}{R_0} \right) + \epsilon w_2^\epsilon(x), \quad x \in \Omega,$$

where $\epsilon y = T(x)$ and w_1 satisfies

$$
\begin{cases}
\Delta v - v = 0 & \text{in } \mathbb{R}^N_+, \\
\frac{\partial v}{\partial y_N} = -\frac{1}{2} \left(\frac{w''}{|y|^2} - \frac{w'}{|y|^3} \right) \sum_{k,l=1}^{N-1} \rho_{kl}(0) y_k y_l y_j \\
\qquad - \frac{w'}{|y|} \sum_{k=1}^{N-1} \rho_{jk}(0) y_k & \text{on } \partial \mathbb{R}^N_+
\end{cases}
\tag{5.12}
$$

and

$$\|w_2^\epsilon\|_\epsilon \le C.$$

Note that w_1 is an odd function in y' and $|w_1| \le Ce^{-\mu|y|}$ for some $0 < \mu < 1$.

Finally, letting

$$L_0 = \Delta - 1 + f'(w),$$

we have

Lemma 5.5

$$\text{Ker}(L_0) \cap H^2_N(\mathbb{R}^N_+) = \text{span}\left\{ \frac{\partial w}{\partial y_1}, \ldots, \frac{\partial w}{\partial y_{N-1}} \right\},$$

where $H^2_N(\mathbb{R}^N_+) = \{u \in H^2(\mathbb{R}^N_+), \frac{\partial u}{\partial y_N} = 0 \text{ on } \partial\mathbb{R}^N_+\}$.

Proof This result follows easily from Lemma 13.4.   □

Next we are going to prove three technical lemmas. We begin by stating several relations of integrals over $\mathbb{R}^{N-1}$ associated with w.

Letting

$$\gamma_1 = \frac{1}{N+1} \int_{\mathbb{R}^{N-1}} |\nabla w|^2 |y'|^2 dy', \tag{5.13}$$

we have

Lemma 5.6

$$\frac{N-3}{2}\gamma_1 = \int_{\mathbb{R}^{N-1}} F\big(w(|y'|)\big)|y'|^2 dy' - \frac{1}{2}\int_{\mathbb{R}^{N-1}} |w|^2 |y'|^2 dy', \tag{5.14}$$

$$(N+1)\gamma_1 = \frac{N-1}{2}\int_{\mathbb{R}^{N-1}} |w|^2 dy' - \int_{\mathbb{R}^{N-1}} |w|^2 |y'|^2 dy'$$

$$+ \int_{\mathbb{R}^{N-1}} f(w)w|y'|^2 dy'. \tag{5.15}$$

Proof Let $y = (y', y_N)$. Note that the operators Δ and ∇ below will be considered with variables $y \in \mathbb{R}^N$, whereas integrations will be taken with respect to $y' = (y', 0) \in \mathbb{R}^{N-1}$. Instead of $|y'|$ we will sometimes use the notation r.

We compute

$$\int_{\mathbb{R}^{N-1}} |y'|^2 \Delta w(\nabla w \cdot y) dy'$$

$$= \omega_{N-2} \int_0^\infty \left(w''(r) + \frac{N-1}{r} w'(r) \right) w'(r) r^{N+1} dr$$

$$= \frac{N-3}{2}\omega_{N-2} \int_0^\infty r^N \big(w'(r)\big)^2 dr$$

$$= \frac{(N+1)(N-3)}{2}\gamma_1 \tag{5.16}$$

and

$$\int_{\mathbb{R}^{N-1}} |y'|^2 w(\nabla w \cdot y) dy'$$

$$= \omega_{N-2} \int_0^\infty w'(r)w(r) r^{N+1} dr$$

$$= -\frac{N+1}{2}\omega_{N-2}\int_0^\infty r^N w^2 dr$$

$$= -\frac{N+1}{2}\int_{\mathbb{R}^{N-1}} w^2 |y'|^2 dy'. \tag{5.17}$$

Further, we have

$$\int_{\mathbb{R}^{N-1}} |y'|^2 f(w)(\nabla w \cdot y) dy'$$

$$= \omega_{N-2}\int_0^\infty f(w)w'(r)r^{N+1}dr$$

$$= -(N+1)\omega_{N-2}\int_0^\infty r^N F(w)dr$$

$$= -(N+1)\int_{\mathbb{R}^{N-1}} F(w)|y'|^2 dy'. \tag{5.18}$$

Multiplying (5.5) by $|y'|^2(\nabla w \cdot y)$ and integrating with respect to y' in $\mathbb{R}^{N-1}$, we derive (5.14).

Using the identity

$$\int_{\mathbb{R}^{N-1}} |y'|^2 w \Delta w dy' = \omega_{N-2}\int_0^\infty w(r)\left(w''(r) + \frac{N-1}{r}w'(r)\right)r^N dr$$

$$= -\omega_{N-2}\int_0^\infty r^N (w'(r))^2 dr + \frac{N-1}{2}\omega_{N-2}\int_0^\infty r^{N-2} w^2 dr$$

$$= -(N+1)\gamma_1 + \frac{N-1}{2}\int_{\mathbb{R}^{N-1}} w^2 dy',$$

multiplying (5.5) by $|y'|^2 w$ and integrating over $\mathbb{R}^{N-1}$, (5.15) follows.

Lemma 5.6 is proven. $\qquad\square$

Lemma 5.7 *Let $G(t)$ be a function in $C^{1+\sigma}([0,\infty))$ with $G(0) = G'(0) = 0$. Then we have*

$$\int_{\Omega_{\epsilon,P}} G(w(y))dy = \int_{\mathbb{R}_+^N} G(w(y))dy$$

$$- \epsilon H(P)\frac{1}{2}\int_{\mathbb{R}^{N-1}} G(w(y',0))|y'|^2 dy' + o(\epsilon).$$

Proof Since w decays exponentially with respect to y at infinity, we compute

$$\int_{\Omega_{\epsilon,P}} G(w(y))dy$$

$$= \int_{(\Omega_0)_{\epsilon,P}} G(w(y))dy + o(\epsilon)$$

$$= \int_{B^+(R_0/\epsilon)} G(w(y))dy - \int_{B^+(R_0/\epsilon)\backslash(\Omega_0)_{\epsilon,P}} G(w(y))dy + o(\epsilon)$$

$$= \int_{\mathbb{R}^N_+} G(w(y))dy - \int_{|y'|\leq R_0/\epsilon} \int_0^{(1/\epsilon)\rho(\epsilon y')} G(w(y', y_N))dy_N dy' + o(\epsilon)$$

$$= \int_{\mathbb{R}^N_+} G(w(y))dy - \int_{|y'|\leq R_0/\epsilon} \int_0^{(1/\epsilon)\rho(\epsilon y')} G(w(y', 0))dy_N dy'$$

$$+ \int_{|y'|\leq R_0/\epsilon} \int_0^{(1/\epsilon)\rho(\epsilon y')} (G(w(y', y_N)) - G(w(y', 0)))dy_N dy' + o(\epsilon)$$

$$= \int_{\mathbb{R}^N_+} G(w(y))dy - \int_{|y'|\leq R_0/\epsilon} G(w(y', 0))\frac{\rho(\epsilon y')}{\epsilon}dy'$$

$$+ \int_{|y'|\leq R_0/\epsilon} O\left(|w(y', 0)|^\sigma \left(\frac{\rho(\epsilon y')}{\epsilon}\right)^2\right)dy' + o(\epsilon)$$

$$= \int_{\mathbb{R}^N_+} G(w(y))dy - \frac{1}{2}\epsilon \int_{|y'|\leq R_0/\epsilon} G(w(y', 0)) \sum_{i,j=1}^{N-1} \rho_{ij}(0)y_i y_j dy'$$

$$+ \int_{|y'|\leq R_0/\epsilon} O(\epsilon^2|y'|^3)dy' + o(\epsilon)$$

$$= \int_{\mathbb{R}^N_+} G(w(y))dy - \epsilon H(P)\frac{1}{2} \int_{|y'|\leq R_0/\epsilon} G(w(y', 0))|y'|^2 dy' + o(\epsilon)$$

$$= \int_{\mathbb{R}^N_+} G(w(y))dy - \epsilon H(P)\frac{1}{2} \int_{\mathbb{R}^{N-1}} G(w(y', 0))|y'|^2 dy' + o(\epsilon), \quad (5.19)$$

where

$$B^+\left(\frac{R_0}{\epsilon}\right) = B\left(\frac{R_0}{\epsilon}\right) \cap \mathbb{R}^N_+$$

and

$$(\Omega_0)_{\epsilon,P} = \{y|\epsilon y + P \in \Omega_0\}.$$

Hence Lemma 5.7 follows.   □

Finally, we have

Lemma 5.8

$$\int_{\mathbb{R}^N_+} f(w)(P_{\Omega_{\epsilon,P}}w - w)dy = \epsilon H(P)\frac{N-1}{4} \int_{\mathbb{R}^{N-1}} |w|^2 dy' + o(\epsilon).$$

Proof Using (5.5), (5.11) and the exponential decay of w and v_1, we compute

$$\int_{\mathbb{R}^N_+} f(w(y))v_1(y)dy = \int_{\mathbb{R}^N_+} (w - \Delta w)v_1(y)dy$$

$$= \int_{\mathbb{R}^N_+} w(v_1 - \Delta v_1)dy + \int_{\mathbb{R}^{N-1}} \left(v_1 \frac{\partial w}{\partial y_N} - w\frac{\partial v_1}{\partial y_N}\right)dy'$$

$$= \frac{1}{2}\int_{\mathbb{R}^{N-1}} w(r)w'(r)r^{-1} \sum_{i,j=1}^{N-1} \rho_{ij}(0)y_iy_jdy'$$

$$= \frac{1}{2}\int_{\mathbb{R}^{N-1}} w(r)w'(r)r^{-1} \sum_{i=1}^{N-1} \rho_{ii}(0)|y_i|^2dy'$$

$$= \frac{1}{2}H(P)\omega_{N-2}\int_0^\infty w(r)w'(r)r^{N-1}dr$$

$$= -\frac{N-1}{4}H(P)\int_{\mathbb{R}^{N-1}} w^2dy'. \tag{5.20}$$

By Proposition 5.3, Lemma 5.8 follows.   □

The next lemma is the key result in this section.

Lemma 5.9 *For any $P \in \overline{\Lambda}$ and ϵ small enough, we have*

$$J_\epsilon(Pw) = \epsilon^N\left[\frac{1}{2}I(w) - \epsilon\gamma_1 H(P) + o(\epsilon)\right], \tag{5.21}$$

where γ_1 has been defined in (5.13).

Proof Using Proposition 5.3, Lemma 5.7 and Lemma 5.8, we get

$$\epsilon^2\int_\Omega \left|\nabla P_{\Omega_\epsilon,P}w\left(\frac{x-P}{\epsilon}\right)\right|^2 dx + \int_\Omega \left|P_{\Omega_\epsilon,P}w\left(\frac{x-P}{\epsilon}\right)\right|^2 dx$$

$$= \epsilon^N\int_{\Omega_{\epsilon,P}} f(w)P_{\Omega_\epsilon,P}wdy$$

$$= \epsilon^N\left(\int_{\Omega_{\epsilon,P}} f(w)wdy + \int_{\Omega_{\epsilon,P}} f(w)(P_{\Omega_\epsilon,P}w - w)\right)dy$$

$$= \epsilon^N\left(\int_{\mathbb{R}^N_+} f(w)wdy - \epsilon H(P)\frac{1}{2}\int_{\mathbb{R}^{N-1}} f(w)w|y'|^2dy'\right.$$

$$\left. + \epsilon H(P)\frac{N-1}{4}\int_{\mathbb{R}^{N-1}} |w|^2dy' + o(\epsilon)\right). \tag{5.22}$$

Similarly, we derive

$$\int_{\Omega} F\left(P_{\Omega_{\epsilon,P}} w\left(\frac{x-P}{\epsilon}\right)\right) dx$$

$$= \epsilon^N \left(\int_{\Omega_{\epsilon,P}} F(w)dy + \int_{\Omega_{\epsilon,P}} (F(P_{\Omega_{\epsilon,P}} w) - F(w))dy\right)$$

$$= \epsilon^N \left(\int_{\mathbb{R}_+^N} F(w)dy - \epsilon H(P)\frac{1}{2}\int_{\mathbb{R}^{N-1}} f(w)w|y'|^2 dy'\right.$$

$$\left. + \epsilon H(P)\frac{N-1}{4}\int_{\mathbb{R}^{N-1}} |w|^2 dy' + o(\epsilon)\right). \tag{5.23}$$

Then we have

$$J_\epsilon(P_{\Omega_{\epsilon,P}} w) = \epsilon^N \left(\frac{1}{2}I(w) - \gamma_1 H(P) + o(\epsilon)\right).$$

The proof is finished.   □

5.2.2 The Liapunov-Schmidt Reduction Method

In this subsection, we reduce the infinite-dimensional problem (5.1) to finite dimensions, using the Liapunov-Schmidt method.

For $u \in H_N^2(\Omega_{\epsilon,P})$, we define the nonlinear operator

$$S_\epsilon(u) = \Delta u - u + f(u).$$

Then (5.1) is equivalent to the problem

$$S_\epsilon(u) = 0, \quad u \in H_N^2(\Omega_{\epsilon,P}).$$

For fixed $P \in \overline{\Lambda}$, we consider the linearised operator

$$\tilde{L}_\epsilon : u \mapsto \Delta u - u + f'(Pw)u, \quad H_N^2(\Omega_{\epsilon,P}) \to L^2(\Omega_{\epsilon,P}).$$

Using integration by parts, it follows easily that the cokernel of $\tilde{L}_\epsilon$ coincides with its kernel. Choosing approximate cokernel and kernel as

$$\mathcal{C}_{\epsilon,P} = \mathcal{K}_{\epsilon,P} = \text{span}\left\{\frac{\partial Pw}{\partial \tau_{P_j}} : j = 1, \ldots, N-1\right\},$$

where $\mathcal{K}_{\epsilon,P} \subset H_N^2(\Omega_{\epsilon,P})$ and $\mathcal{C}_{\epsilon,P} \subset L^2(\Omega_{\epsilon,P})$ and letting $\pi_{\epsilon,P}$ denote the projection from $L^2(\Omega_{\epsilon,P})$ onto $\mathcal{C}_{\epsilon,P}^{\perp}$, then, following the approach in Sect. 2.3.3, we can show that the equation

$$\pi_{\epsilon,P} \circ S_\epsilon (Pw + \Phi_{\epsilon,P}) = 0 \tag{5.24}$$

has a unique solution $\Phi_{\epsilon,P} \in \mathcal{K}_{\epsilon,P}^\perp$ for ϵ small enough and $P \in \overline{\Lambda}$.

As a preparation, the following two propositions are needed which guarantee the invertibility of the corresponding linearised operator.

Proposition 5.10 Let $L_{\epsilon,P} = \pi_{\epsilon,P} \circ \tilde{L}_\epsilon$. There are positive constants $\overline{\epsilon}, \lambda$ such that for all $\epsilon \in (0, \overline{\epsilon})$ and $P \in \overline{\Lambda}$, we have

$$\|L_{\epsilon,P} \Phi\|_{L^2(\Omega_{\epsilon,P})} \geq \lambda \|\Phi\|_{H^2(\Omega_{\epsilon,P})} \quad \text{for all } \Phi \in \mathcal{K}_{\epsilon,P}^\perp. \tag{5.25}$$

Proposition 5.11 For any $\epsilon \in (0, \tilde{\epsilon})$ and $P \in \overline{\Lambda}$ the mapping

$$L_{\epsilon,P} = \pi_{\epsilon,P} \circ \tilde{L}_\epsilon : \mathcal{K}_{\epsilon,P}^\perp \to \mathcal{C}_{\epsilon,P}^\perp$$

is surjective.

The proofs of Propositions 5.10 and 5.11 are similar to those in Sect. 2.3.3 and are therefore omitted.

We are now in a position to solve the equation. We will be able to proceed as in Sect. 2.3.3, after proving the following lemma:

Lemma 5.12 For ϵ sufficiently small, we have

$$\left\| S_\epsilon (Pw) \right\|_{L^2(\Omega_{\epsilon,P})} \leq C\epsilon. \tag{5.26}$$

Proof Using Proposition 5.3 and the facts that Pw, w and v_1 decay exponentially at infinity, we get

$$\left\| S_\epsilon (Pw) \right\|_{L^2(\Omega_{\epsilon,P})} = O(\epsilon). \qquad \square$$

Using the contraction mapping principle, we derive the following result.

Lemma 5.13 There exists an $\overline{\epsilon} > 0$ such that for every pair (ϵ, P) with $0 < \epsilon < \overline{\epsilon}$ and $P \in \overline{\Lambda}$, there is a unique $\Phi_{\epsilon,P} \in \mathcal{K}_{\epsilon,P}^\perp$ satisfying $S_\epsilon (Pw + \Phi_{\epsilon,P}) \in \mathcal{C}_{\epsilon,P}$ and

$$\|\Phi_{\epsilon,P}\|_{H^2(\Omega_{\epsilon,P})} \leq C\epsilon. \tag{5.27}$$

The next lemma is our main estimate.

Lemma 5.14 Let $\Phi_{\epsilon,P}$ be defined by Lemma 5.13. Then we have

$$J_\epsilon (Pw + \Phi_{\epsilon,P}) = \epsilon^N \left[\frac{1}{2} I(w) - \gamma_1 \epsilon H(P) + o(\epsilon) \right],$$

where γ_1 is defined in (5.13).

Proof For any $P \in \overline{\Lambda}$, we have

$$\epsilon^{-N} J_\epsilon(Pw + \Phi_{\epsilon,P}) = \epsilon^{-N} J_\epsilon(Pw) + g_{\epsilon,P}(\Phi_{\epsilon,P}) + O\big(\|\Phi_{\epsilon,P}\|^2_{H^2(\Omega_{\epsilon,P})}\big),$$

where

$$
\begin{aligned}
g_{\epsilon,P}(\Phi_{\epsilon,P}) &= \int_{\Omega_{\epsilon,P}} (\nabla Pw \nabla \Phi_{\epsilon,P} + Pw\Phi_{\epsilon,P})dy - \int_{\Omega_{\epsilon,P}} f(Pw)\Phi_{\epsilon,P}dy \\
&= \int_{\Omega_{\epsilon,P}} \big[f(w) - f(Pw)\big]\Phi_{\epsilon,P}dy \\
&\leq \|f(w) - f(Pw)\|_{L^2(\Omega_{\epsilon,P})}\|\Phi_{\epsilon,P}\|_{L^2(\Omega_{\epsilon,P})} \\
&= O(\epsilon^2)
\end{aligned}
$$

by Proposition 5.3 and Lemma 5.13.

Now Lemma 5.14 follows from Lemmas 5.9 and 5.13.   □

Finally, we show that the solution $\Phi_{\epsilon,P}$ is smooth with respect to $P \in \overline{\Lambda}$.

Lemma 5.15 *Let $\Phi_{\epsilon,P}$ be defined by Lemma 5.13. Then $\Phi_{\epsilon,P} \in C^1$ for $P \in \overline{\Lambda}$.*

Proof Recall that $\Phi_{\epsilon,P}$ is a solution of the equation

$$\pi_{\epsilon,P} \circ S_\epsilon(Pw + \Phi_{\epsilon,P}) = 0 \tag{5.28}$$

such that

$$\Phi_{\epsilon,P} \in \mathcal{K}_{\epsilon,P}^\perp. \tag{5.29}$$

We differentiate equation (5.28) twice with respect to $P \in \overline{\Lambda}$ and conclude that the functions Pw_i and $\frac{\partial^2 Pw}{\partial \tau_{P_j} \partial \tau_{P_k}}$ are C^1 for $P \in \overline{\Lambda}$. This implies that the projection $\pi_{\epsilon,P}$ is C^1 for $P \in \overline{\Lambda}$. Applying $\partial/\partial \tau_{P_j}$, we compute

$$\pi_{\epsilon,P} \circ DS_\epsilon(Pw + \Phi_{\epsilon,P})\left(\frac{\partial Pw}{\partial \tau_{P_j}} + \frac{\partial \Phi_{\epsilon,P}}{\partial \tau_{P_j}}\right)$$

$$+ \frac{\partial \pi_{\epsilon,P}}{\partial \tau_{P_j}} \circ S_\epsilon(Pw + \Phi_{\epsilon,P}) = 0, \tag{5.30}$$

where

$$DS_\epsilon(Pw + \Phi_{\epsilon,P}) = \Delta - 1 + f'(Pw + \Phi_{\epsilon,P}).$$

Decomposing $\frac{\partial \Phi_{\epsilon,P}}{\partial \tau_{P_j}}$ into two parts, we get

$$\frac{\partial \Phi_{\epsilon,P}}{\partial \tau_{P_j}} = \left(\frac{\partial \Phi_{\epsilon,P}}{\partial \tau_{P_j}}\right)_1 + \left(\frac{\partial \Phi_{\epsilon,P}}{\partial \tau_{P_j}}\right)_2,$$

where $(\frac{\partial \Phi_{\epsilon,P}}{\partial \tau_{P_j}})_1 \in \mathcal{K}_{\epsilon,P}$ and $(\frac{\partial \Phi_{\epsilon,P}}{\partial \tau_{P_j}})_2 \in \mathcal{K}_{\epsilon,P}^{\perp}$.

Now it easily follows that $(\frac{\partial \Phi_{\epsilon,P}}{\partial \tau_{P_j}})_1$ is continuous for $P \in \overline{\Lambda}$ since

$$\int_{\Omega_{\epsilon,P}} \Phi_{\epsilon,P} \frac{\partial Pw}{\partial \tau_{P_l}} dy = 0, \quad l = 1, \ldots, N-1,$$

and

$$\int_{\Omega_{\epsilon,P}} \frac{\partial \Phi_{\epsilon,P}}{\partial \tau_{P_j}} \frac{\partial Pw}{\partial \tau_{P_l}} dy + \int_{\Omega_{\epsilon,P}} \Phi_{\epsilon,P} \frac{\partial^2 Pw}{\partial \tau_{P_j}} \partial \tau_{P_l} dy = 0, \quad l, j = 1, \ldots, N-1.$$

Next we rewrite equation (5.30) as follows:

$$\pi_{\epsilon,P} \circ DS_{\epsilon}(Pw + \Phi_{\epsilon,P})\left(\frac{\partial \Phi_{\epsilon,P}}{\partial \tau_{P_j}}\right)_2$$

$$+ \pi_{\epsilon,P} \circ DS_{\epsilon}(Pw + \Phi_{\epsilon,P})\left(\frac{\partial Pw}{\partial \tau_{P_j}} + \left(\frac{\partial \Phi_{\epsilon,P}}{\partial \tau_{P_j}}\right)_1\right)$$

$$+ \frac{\partial \pi_{\epsilon,P}}{\partial \tau_{P_j}} \circ S_{\epsilon}(Pw + \Phi_{\epsilon,P}) = 0. \tag{5.31}$$

Similarly to Propositions 5.10 and 5.11, it can now be shown that the operator

$$\pi_{\epsilon,P} \circ DS_{\epsilon}(Pw + \Phi_{\epsilon,P})$$

is invertible from $\mathcal{K}_{\epsilon,P}^{\perp}$ to $\mathcal{C}_{\epsilon,P}^{\perp}$. Thus we can take the inverse of $\pi_{\epsilon,P} \circ DS_{\epsilon}(Pw + \Phi_{\epsilon,P})$ in the above equation, where the inverse is continuous for $P \in \overline{\Lambda}$.

Since the functions $\frac{\partial Pw}{\partial \tau_{P_j}}, (\frac{\partial \Phi_{\epsilon,P}}{\partial \tau_{P_j}})_1 \in \mathcal{K}_{\epsilon,P}$ are continuous with respect to $P \in \overline{\Lambda}$ and the same property holds for $\frac{\partial \pi_{\epsilon,P}}{\partial \tau_{P_j}}$, it follows that $(\frac{\partial \Phi_{\epsilon,P}}{\partial \tau_{P_j}})_2$ is also continuous for $P \in \overline{\Lambda}$. This results in the C^1 dependence of $\Phi_{\epsilon,P}$ for $P \in \overline{\Lambda}$ and concludes the proof.   □

5.2.3   The Reduced Problem: A Finite-Dimensional Maximisation Problem

In this subsection, we study a finite-dimensional maximisation problem.

For $P \in \overline{\Lambda}$, let $\Phi_{\epsilon,P}$ be the unique solution given by Lemma 5.13. We define a new functional by

$$M_{\epsilon} : \overline{\Lambda} \to \mathbb{R}, \quad M_{\epsilon}(P) = J_{\epsilon}(Pw + \Phi_{\epsilon,P}). \tag{5.32}$$

We shall prove

Proposition 5.16 *For ϵ small enough, the maximisation problem*

$$\max\{M_\epsilon(P) : P \in \overline{\Lambda}\} \tag{5.33}$$

has a solution $P^\epsilon \in \Lambda$.

Proof Since $J_\epsilon(P_{\Omega_\epsilon,P} w + \Phi_{\epsilon,P})$ is continuous for $P \in \overline{\Lambda}$, the maximisation problem has a solution $P^\epsilon \in \overline{\Lambda}$. Let $M_\epsilon(P^\epsilon)$ be the corresponding maximum. Next we show that $P^\epsilon \in \Lambda$.

For any $P \in \overline{\Lambda}$, we have by Lemma 5.14

$$M_\epsilon(P) = \epsilon^N \left[\frac{K}{2} I(w) - \epsilon \gamma H(P) + o(\epsilon) \right].$$

Since $M_\epsilon(P^\epsilon)$ is the maximum, we have

$$\gamma_1 H\left(P^\epsilon\right) \leq \gamma_1 H(P) + o(1)$$

for any $P \in \overline{\Lambda}$.

Choosing P such that $H(P) \to \min_{P \in \Lambda} H(P)$, this implies that

$$\gamma_1 H(P) \leq \gamma_1 \min_{P \in \Lambda} H(P) + \delta$$

for any $\delta > 0$.

By definition of Λ, if $P \in \partial \Lambda$, then $H(P) = \min_{P \in \overline{\Lambda}} H(P) + \eta_0$ for some $\eta_0 > 0$. This yields a contradiction if δ is small enough.

Thus we conclude that $P^\epsilon \in \Lambda$ and Proposition 5.16 follows. $\qquad\square$

5.2.4 The Completion of the Existence Proof

In this subsection, we complete the proof of Theorem 5.1.

By Lemmas 5.13 and 5.15, there is an $\epsilon_0 > 0$ such that for all $0 < \epsilon < \epsilon_0$ we have a C^1 map which, to any $P \in \overline{\Lambda}$, associates $\Phi_{\epsilon,P} \in \mathcal{K}_{\epsilon,P}^\perp$ via the equation

$$S_\epsilon(Pw + \Phi_{\epsilon,P}) = \sum_{l=1,\dots,N-1} \alpha_l \frac{\partial Pw}{\partial \tau_{P_l}} \tag{5.34}$$

for some constants $\alpha_l \in \mathbb{R}$, $l = 1, \dots, N-1$.

By Proposition 5.16, at some $P^\epsilon \in \Lambda$, the maximum of the maximisation problem given in (5.33) is attained. Let $\Phi_\epsilon = \Phi_{\epsilon,P^\epsilon}$ and $u_\epsilon = P_{\Omega_\epsilon,P^\epsilon} w + \Phi_{\epsilon,P^\epsilon}$. Then we have

$$\frac{\partial}{\partial \tau_{P_j}}\bigg|_{P=P^\epsilon} M_\epsilon\left(P^\epsilon\right) = 0, \quad j = 1, \dots, N-1.$$

This implies

$$\int_{\Omega_{\epsilon,P^\epsilon}} \left[\nabla u_\epsilon \nabla \frac{\partial (Pw + \Phi_{\epsilon,P})}{\partial \tau_{P_j}} \Bigg|_{P=P^\epsilon} + u_\epsilon \frac{\partial (Pw + \Phi_{\epsilon,P})}{\partial \tau_{P_j}} \Bigg|_{P=P^\epsilon} \right.$$
$$\left. - f(u_\epsilon) \frac{\partial (Pw + \Phi_{\epsilon,P})}{\partial \tau_{P_j}} \Bigg|_{P=P^\epsilon} \right] = 0$$

for $j = 1, \ldots, N - 1$. Therefore, we get

$$\sum_{l=1,\ldots,N-1} \alpha_l \int_{\Omega_{\epsilon,P}} \frac{\partial Pw}{\partial \tau_{P_l}} \frac{\partial (Pw + \Phi_{\epsilon,P})}{\partial \tau_{P_j}} \Bigg|_{P=P^\epsilon} = 0. \qquad (5.35)$$

Since $\Phi_{\epsilon,P} \in \mathcal{K}_{\epsilon,P}^\perp$, we have that

$$\left| \int_{\Omega_{\epsilon,P}} \frac{\partial Pw}{\partial \tau_{P_l}} \frac{\partial \Phi_{\epsilon,P}}{\partial \tau_{P_j}} \right| = \left| - \int_{\Omega_{\epsilon,P}} \frac{\partial^2 Pw}{\partial \tau_{P_l} \partial \tau_{P_j}} \Phi_{\epsilon,P} \right|$$
$$\leq \left\| \frac{\partial^2 Pw}{\partial \tau_{P_l} \partial \tau_{P_j}} \right\|_{L^2} \|\Phi_{\epsilon,P}\|_{L^2} = O\left(\frac{1}{\epsilon}\right) \quad \text{by (5.27).}$$

Note that

$$\int_{\Omega_{\epsilon,P}} \frac{\partial Pw}{\partial \tau_{P_l}} \frac{\partial Pw}{\partial \tau_{P_j}} = \frac{1}{\epsilon^2} \delta_{lj} (A + o(1)),$$

where

$$A = \int_{\mathbb{R}_+^N} \left(\frac{\partial w}{\partial y_1} \right)^2 dy' > 0.$$

Hence (5.35) is a linear system of homogeneous equations with the unknown α_l whose matrix is diagonally dominant and therefore nonsingular. Therefore we have $\alpha_l \equiv 0, l = 1, \ldots, N - 1$.

We conclude that $u_\epsilon = P_{\Omega_{\epsilon,P^\epsilon}} w + \Phi_{\epsilon,P^\epsilon}$ is a solution of (5.1).

Now by the maximum principle it follows easily that $u_\epsilon > 0$ in Ω. Moreover, we have that $\epsilon^N J_\epsilon(u_\epsilon) \to \frac{1}{2} I(w)$ and u_ϵ has only one local maximum point Q^ϵ in $\bar{\Omega}$ such that $Q^\epsilon \in \partial \bar{\Lambda}$. The proof of Theorem 5.1 is complete.

5.3 Notes on the Literature

For the derivation of (5.1) as a limiting problem as well as more details on the Gierer-Meinhardt system and its properties, see the review article [168]. In the pioneering papers [134, 171, 172], Lin, Ni and Takagi established the existence of least-energy solutions and showed that for ϵ sufficiently small the least-energy solution has only one local maximum point P^ϵ, where $P^\epsilon \in \partial \Omega$. Moreover, $H(P^\epsilon) \to$

$\max_{P \in \partial \Omega} H(P)$ as $\epsilon \to 0$, where $H(P)$ is the mean curvature of P at $\partial \Omega$. In [173], Ni and Takagi constructed boundary spike solutions for axially symmetric domains. The second author in [246] studied the general domain case and showed that for single boundary spike solutions, the boundary spike must approach a critical point of the mean curvature; on the other hand, he proved that for any nondegenerate critical point of $H(P)$, one can construct boundary spike solutions whose spike approaches that point. In [78] multiple boundary spike layer solutions at multiple local maximum points of $H(P)$ are constructed while in [258] multiple boundary spike layer solutions at multiple nondegenerate critical points of $H(P)$ are derived. Later these results were improved by Y.Y. Li in [132] using a unified approach. For $p = \frac{N+2}{N-2}$, similar results for the boundary spike layer solutions have been obtained in [2–4, 41, 79, 176, 199–201, 238].

The existence of spikes using Liapunov-Schmidt reduction has been proved in [249, 253, 256–258].

The proof of Theorem 5.1 can be adapted to the case where instead of (5.3) we require

$$\min_{P \in \partial \Lambda} H(P) < \min_{P \in \Lambda} H(P). \tag{5.36}$$

Let $P^0 \in \partial \Omega$ be a nondegenerate critical point of the mean curvature for a smooth domain. Then either (5.3) or (5.36) follow for suitably chosen sets Λ with $P^0 \in \Lambda$.

By adapting the approach in this chapter, multiple boundary spikes can be studied. It is also possible to construct clusters, where multiple spikes approach the same point, see [81]. We recall that one-dimensional clusters for the full Gierer-Meinhardt system (2.2) have been studied in Sect. 2.4.

The stability of boundary spikes is studied in [249]. The degenerate situation when the mean curvature at the boundary is constant at higher order has been considered in [265, 271, 282].

For interior spikes, the interaction is proportional to the distance function to each other and the boundary, respectively [42]. Gui and Wei [80] proved that for any fixed K there exists an $\epsilon_K > 0$ such that for all $0 < \epsilon < \epsilon_K$ problem (5.1) has a steady state with K interior spikes. Lin, Ni and Wei [135] proved that there exists an $\epsilon_0 > 0$ such that for all $0 < \epsilon < \epsilon_0$ and $1 \leq K \leq \frac{c_{\Omega,p,N}}{(\epsilon |\ln \epsilon|)^N}$ problem (5.1) has a steady state with K interior spikes. Recently, Ao, Wei and Zeng [5] proved that there exists an $\epsilon_0 > 0$ such that for all $0 < \epsilon < \epsilon_0$ and $1 \leq K \leq \frac{c_{\Omega,p,M}}{\epsilon^N}$ problem (5.1) has a steady state with K interior spikes, which is optimal. For a survey of further results on the shadow system we refer to [255].

Chapter 6
Existence and Stability of Spikes for the Gierer-Meinhardt System in Two Dimensions

In this chapter, we will give a rigorous proof of the existence and stability of K-solutions ($K = 1, 2, \ldots$) for the Gierer-Meinhardt system in two dimensions. We will consider the weak coupling case: $D(\epsilon) \to \infty$ as $\epsilon \to 0$. In this regime there will be symmetric and asymmetric spikes.

In order to streamline the presentation and improve its transparency, where possible, we will refer to the analysis of the one-dimensional case given in Chaps. 2 and 4.

6.1 Symmetric Multiple Spikes: Existence

We begin by introducing a Green's function G_0 before going on to formulate our main results. Let $G_0(x, \xi)$ be given by

$$
\begin{cases}
\Delta G_0(x, \xi) - \frac{1}{|\Omega|} + \delta_\xi(x) = 0 & \text{in } \Omega, \\
\int_\Omega G_0(x, \xi)dx = 0, \\
\frac{\partial G_0(x, \xi)}{\partial \nu} = 0 & \text{on } \partial\Omega.
\end{cases}
\tag{6.1}
$$

Further, we set

$$
H_0(x, \xi) = \frac{1}{2\pi} \log \frac{1}{|x - \xi|} - G_0(x, \xi)
\tag{6.2}
$$

to be its regular part, where $|\Omega|$ denotes the area of Ω.

For a given number K of spikes, let $\mathbf{P} \in \Omega^K$, where

$$
\mathbf{P} = (P_1, P_2, \ldots, P_K)
$$

with

$$
P_i = (P_{i,1}, P_{i,2}) \quad \text{for } i = 1, \ldots, K.
$$

J. Wei, M. Winter, *Mathematical Aspects of Pattern Formation in Biological Systems*, Applied Mathematical Sciences 189, DOI 10.1007/978-1-4471-5526-3_6, © Springer-Verlag London 2014

For the rest of this chapter we assume that $\mathbf{P} \in \overline{\Lambda_\delta}$, where

$$\Lambda_\delta = \{(P_1, \ldots, P_K) \in \Omega^K : |P_i - P_j| > 4\delta \text{ for } i \neq j$$
$$\text{and } d(P_i, \partial\Omega) > 4\delta \text{ for } i = 1, \ldots, K\} \tag{6.3}$$

and $\delta > 0$ remains fixed and is small enough.

For $\mathbf{P} \in \overline{\Lambda_\delta}$, we set

$$F(\mathbf{P}) = \sum_{k=1}^K H_0(P_k, P_k) - \sum_{i,j=1,\ldots,K, i\neq j} G_0(P_i, P_j) \tag{6.4}$$

and

$$M(\mathbf{P}) = \left(\nabla_{\mathbf{P}}^2 F(\mathbf{P})\right), \tag{6.5}$$

where $M(\mathbf{P})$ is a $(2K) \times (2K)$ matrix with entries $\frac{\partial^2 F(\mathbf{P})}{\partial P_{i,j} \partial P_{k,l}}$, $i, k = 1, \ldots, K$, $j, l = 1, 2$. We remark that $F(\mathbf{P}) \in C^\infty(\overline{\Lambda_\delta})$.

Next, let

$$D = \frac{1}{\beta^2}, \qquad \eta_\epsilon := \frac{\beta^2 |\Omega|}{2\pi} \log \frac{1}{\epsilon}. \tag{6.6}$$

Then $D \to +\infty$ corresponds to $\beta \to 0$.

Steady states of the Gierer-Meinhardt system (1.5) are solutions of the following system of elliptic equations:

$$\begin{cases} \epsilon^2 \Delta A - A + \frac{A^2}{H} = 0, & A > 0 \text{ in } \Omega, \\ \frac{1}{\beta^2} \Delta H - H + A^2 = 0, & H > 0 \text{ in } \Omega, \\ \frac{\partial A}{\partial \nu} = \frac{\partial H}{\partial \nu} = 0 & \text{on } \partial\Omega. \end{cases} \tag{6.7}$$

Now we are ready to state the theorem on the existence of K-spike solutions.

Theorem 6.1 Let $\mathbf{P}^0 = (P_1^0, \ldots, P_K^0) \in \overline{\Lambda_\delta}$ be a nondegenerate critical point of $F(\mathbf{P})$ (defined by (6.4)). Let

$$\lim_{\epsilon \to 0} \eta_\epsilon \neq K \quad \text{if } K > 1, \tag{6.8}$$

where η_ϵ is defined by (6.6).

Then, for ϵ small enough and $D = \frac{1}{\beta^2}$ large enough, problem (6.7) has a solution (A_ϵ, H_ϵ) which satisfies the following properties:

(1) $A_\epsilon(x) = \xi_\epsilon(\sum_{j=1}^K w(\frac{x-P_j^\epsilon}{\epsilon}))(1 + O(k(\epsilon, \beta)))$ uniformly for $x \in \bar{\Omega}$. Here w is the unique solution of the problem

$$\begin{cases} \Delta w - w + w^2 = 0, & w > 0 \text{ in } \mathbb{R}^2, \\ w(0) = \max_{y \in \mathbb{R}^2} w(y), & w(y) \to 0 \text{ as } |y| \to \infty, \end{cases} \tag{6.9}$$

$$\xi_\epsilon = \begin{cases} \dfrac{1}{K}\dfrac{|\Omega|}{\epsilon^2 \int_{\mathbb{R}^2} w^2(y)dy}(1+O(\beta^2 \log \frac{\sqrt{|\Omega|}}{\epsilon})) & \text{if } \eta_\epsilon \to 0, \\[2ex] \dfrac{1}{\eta_\epsilon}\dfrac{|\Omega|}{\epsilon^2 \int_{\mathbb{R}^2} w^2(y)dy}(1+O(\frac{1}{\beta^2 \log(\sqrt{|\Omega|}/\epsilon)})) & \text{if } \eta_\epsilon \to \infty, \\[2ex] \dfrac{1}{K+\eta_0}\dfrac{|\Omega|}{\epsilon^2 \int_{\mathbb{R}^2} w^2(y)dy}(1+O(\beta^2)) & \text{if } \eta_\epsilon \to \eta_0, \end{cases} \qquad (6.10)$$

and

$$k(\epsilon, \beta) := \epsilon^2 \xi_\epsilon \beta^2. \qquad (6.11)$$

(By (6.10), $k(\epsilon, \beta) = O(\min\{\frac{1}{\log(1/\epsilon)}, \beta^2\})$.)

Further, $P_j^\epsilon \to P_j^0$ as $\epsilon \to 0$ for $j = 1, \ldots, K$.

(2) $H_\epsilon(x) = \xi_\epsilon(1 + O(k(\epsilon, \beta)))$ *uniformly for $x \in \bar{\Omega}$.*

Remark 6.2 The technical condition (6.8) in Theorem 6.1 is needed for Liapunov-Schmidt reduction.

Further, we recall that

$$w(y) \sim |y|^{-1/2} e^{-|y|} \quad \text{as } |y| \to \infty. \qquad (6.12)$$

Throughout this chapter, let $C > 0$ be a generic constant which is independent of ϵ and β and may change from line to line. We always assume that $\mathbf{P}, \mathbf{P}^0 \in \overline{\Lambda_\delta}$, where $\overline{\Lambda_\delta}$ was defined in (6.3) and that $|\mathbf{P} - \mathbf{P}^0| < 4\delta$.

6.1.1 The Amplitudes of the Peaks

In this subsection, we formally calculate the amplitudes of the peaks. We will see that in leading order as $\epsilon \to 0$ these amplitudes are independent of their locations but depend on the number of peaks.

For $\beta > 0$, let $G_\beta(x, \xi)$ be the Green's function given by

$$\begin{cases} \Delta G_\beta - \beta^2 G_\beta + \delta_\xi = 0 & \text{in } \Omega, \\ \frac{\partial G_\beta}{\partial \nu} = 0 & \text{on } \partial\Omega \end{cases} \qquad (6.13)$$

and recall that $G_0(x, \xi)$ is given by (6.1). Next we derive a relation between G_β and G_0. Using (6.13), we have

$$\int_\Omega G_\beta(x, \xi)dx = \beta^{-2}.$$

Setting

$$G_\beta(x, \xi) = \frac{\beta^{-2}}{|\Omega|} + \overline{G}_\beta(x, \xi), \qquad (6.14)$$

we get

$$
\begin{cases}
\Delta \bar{G}_\beta - \beta^2 \bar{G}_\beta - \frac{1}{|\Omega|} + \delta_\xi = 0 & \text{in } \Omega, \\
\int_\Omega \bar{G}_\beta(x,\xi) dx = 0, \\
\frac{\partial \bar{G}_\beta}{\partial \nu} = 0 & \text{on } \partial \Omega.
\end{cases}
\tag{6.15}
$$

Then combining (6.1) and (6.15) gives

$$
\bar{G}_\beta(x,\xi) = G_0(x,\xi) + O(\beta^2)
$$

in the operator norm of $L^2(\Omega) \to H^2(\Omega)$. We recall that by Sect. 13.1 the embedding of $H^2(\Omega)$ into $L^\infty(\Omega)$ is compact. Hence we have

$$
G_\beta(x,\xi) = \frac{\beta^{-2}}{|\Omega|} + G_0(x,\xi) + O(\beta^2)
\tag{6.16}
$$

in the operator norm of $L^2(\Omega) \to H^2(\Omega)$.

Next we introduce suitable cut-off functions as follows. Let χ be as defined in (2.51) and set

$$
\chi_{\epsilon,P_j}(x) = \chi\left(\left|\frac{x - P_j}{\delta}\right|\right), \quad x \in \Omega, j = 1, \ldots, K.
\tag{6.17}
$$

We begin the construction of a multiple spike solution (A_ϵ, H_ϵ) of (6.7) with the following ansatz:

$$
\begin{cases}
A_\epsilon(x) \sim \sum_{i=1}^K \xi_{\epsilon,i} w(\frac{x-P_i^\epsilon}{\epsilon}) \chi_{\epsilon,P_i^\epsilon}(x)(1 + O(\beta^2)), \\
H_\epsilon(P_i^\epsilon) = \xi_{\epsilon,i}(1 + O(\beta^2)),
\end{cases}
\tag{6.18}
$$

where w solves (2.3), $\xi_{\epsilon,i}$, $i = 1, \ldots, K$, are the unknown amplitudes of the peaks which will be computed in this section, and $\mathbf{P}^\epsilon = (P_1^\epsilon, \ldots, P_K^\epsilon) \in \overline{\Lambda_\delta}$ are the unknown locations of the peaks which will be determined rigorously in the Liapunov-Schmidt reduction process (see Sect. 6.1.3).

From

$$
\Delta H_\epsilon - \beta^2 H_\epsilon + \beta^2 A_\epsilon^2 = 0
$$

and (6.16) we derive

$$
H_\epsilon(P_i^\epsilon) = \int_\Omega G_\beta(P_i^\epsilon, \xi) \beta^2 A_\epsilon^2(\xi) d\xi
$$

$$
= \int_\Omega \left(\frac{\beta^{-2}}{|\Omega|} + G_0(P_i^\epsilon, \xi) + O(\beta^2)\right) \beta^2
$$

$$
\times \left(\sum_{j=1}^K \xi_{\epsilon,j}^2 w^2\left(\frac{\xi - P_j^\epsilon}{\epsilon}\right)\right) d\xi (1 + O(\beta^2))
$$

$$= \int_{\Omega}\left(\frac{1}{|\Omega|} + \beta^2 G_0(P_i^{\epsilon}, \xi) + O(\beta^4)\right)$$

$$\times \left(\sum_{j=1}^{K}\xi_{\epsilon,j}^2 w^2\left(\frac{\xi - P_j^{\epsilon}}{\epsilon}\right)\right)d\xi\,(1 + O(\beta^2)).$$

Thus we have

$$\xi_{\epsilon,i} = \sum_{j=1}^{K}\xi_{\epsilon,j}^2\frac{\epsilon^2}{|\Omega|}\int_{\mathbb{R}^2}w^2(y)dy$$

$$+\xi_{\epsilon,i}^2\beta^2\int_{\Omega}G_0(P_i^{\epsilon},\xi)w^2\left(\frac{\xi - P_i^{\epsilon}}{\epsilon}\right)d\xi + \sum_{j=1}^{K}\xi_{\epsilon,j}^2 O(\beta^2\epsilon^2). \quad (6.19)$$

Using the expansion for G_0 in (6.19) gives

$$\xi_{\epsilon,i} = \sum_{j=1}^{K}\xi_{\epsilon,j}^2\frac{\epsilon^2}{|\Omega|}\int_{\mathbb{R}^2}w^2(y)dy$$

$$+\xi_{\epsilon,i}^2\beta^2\int_{\Omega}\left(\frac{1}{2\pi}\log\frac{1}{|P_i^{\epsilon}-\xi|} - H_0(P_i^{\epsilon},\xi)\right)w^2\left(\frac{\xi - P_i^{\epsilon}}{\epsilon}\right)d\xi$$

$$+\sum_{j=1}^{K}\xi_{\epsilon,j}^2 O(\beta^2\epsilon^2)$$

$$=\sum_{j=1}^{K}\xi_{\epsilon,j}^2\frac{\epsilon^2}{|\Omega|}\int_{\mathbb{R}^2}w^2(y)dy + \xi_{\epsilon,i}^2\frac{\beta^2}{2\pi}\epsilon^2\log\frac{\sqrt{|\Omega|}}{\epsilon}\int_{\mathbb{R}^2}w^2(y)dy$$

$$+\sum_{j=1}^{K}\xi_{\epsilon,j}^2 O(\beta^2\epsilon^2). \quad (6.20)$$

Note that $H_0 \in C^2(\bar{\Omega} \times \Omega)$.

Defining

$$\hat{\xi}_{\epsilon,i} = \frac{\xi_{\epsilon,i}|\Omega|}{\epsilon^2\int_{\mathbb{R}^2}w^2}, \quad (6.21)$$

we can rewrite (6.20) as

$$\hat{\xi}_{\epsilon,i} = \sum_{j=1}^{K}\hat{\xi}_{\epsilon,j}^2 + \hat{\xi}_{\epsilon,i}^2\eta_{\epsilon} + \sum_{j=1}^{K}\hat{\xi}_{\epsilon,j}^2 O(\beta^2), \quad i = 1,\ldots,K. \quad (6.22)$$

We consider the case that the amplitudes of the spikes are asymptotically equal, i.e. we have

$$\lim_{\epsilon \to 0} \frac{\xi_{\epsilon,i}}{\xi_{\epsilon,j}} = 1, \quad \text{for } i \neq j. \tag{6.23}$$

(In Sect. 6.3, the existence of *asymmetric patterns* will be studied.)
For (6.22) we separately consider three cases.

Case 1: $\eta_\epsilon \to 0$:
Then (6.22) implies

$$\hat{\xi}_{\epsilon,i} = \frac{1}{K} + O(\eta_\epsilon), \quad i = 1, \ldots, K$$

which is equivalent to

$$\xi_{\epsilon,i} = \frac{1}{K} \frac{|\Omega|}{\epsilon^2 \int_{\mathbb{R}^2} w^2(y)dy} (1 + O(\eta_\epsilon)), \quad i = 1, \ldots, K. \tag{6.24}$$

Case 2: $\eta_\epsilon \to \infty$:
Using (6.22), we have

$$\hat{\xi}_{\epsilon,i} = \eta_\epsilon \hat{\xi}_{\epsilon,i}^2 + \sum_{j=1}^{K} \hat{\xi}_{\epsilon,j}^2 O(1)$$

which implies

$$\hat{\xi}_{\epsilon,i} = \frac{1}{\eta_\epsilon}\left(1 + O\left(\frac{1}{\eta_\epsilon}\right)\right), \quad i = 1, \ldots, K$$

and finally

$$\xi_{\epsilon,i} = \frac{|\Omega|}{\eta_\epsilon \epsilon^2 \int_{\mathbb{R}^2} w^2(y)dy}\left(1 + O\left(\frac{1}{\eta_\epsilon}\right)\right), \quad i = 1, \ldots, K. \tag{6.25}$$

Case 3: $\eta_\epsilon \to \eta_0$ $(0 < \eta_0 < \infty)$:
From (6.22) we get

$$\hat{\xi}_{\epsilon,i} = (1 + \eta_0)\hat{\xi}_{\epsilon,i}^2 + \sum_{j \neq i} \hat{\xi}_{\epsilon,j}^2 + \sum_{j=1}^{K} \hat{\xi}_{\epsilon,j}^2 O(\beta^2)$$

which gives

$$\hat{\xi}_{\epsilon,1} = \cdots = \hat{\xi}_{\epsilon,K} = \frac{1}{K + \eta_0}(1 + O(\beta^2)), \quad i = 1, \ldots, K$$

and

$$\xi_{\epsilon,i} = \frac{1}{K + \eta_0} \frac{|\Omega|}{\epsilon^2 \int_{\mathbb{R}^2} w^2} (1 + O(\beta^2)), \quad i = 1, \ldots, K. \tag{6.26}$$

To summarise, in all three cases we have

$$\xi_{\epsilon,i} = \xi_\epsilon (1 + O(h(\epsilon, \beta))), \quad i = 1, \ldots, K,$$

where ξ_ϵ is defined in (6.10) of Theorem 6.1 and

$$h(\epsilon, \beta) = \begin{cases} \eta_\epsilon & \text{if } \eta_\epsilon \to 0, \\ \eta_\epsilon^{-1} & \text{if } \eta_\epsilon \to \infty, \\ \beta^2 & \text{if } \eta_\epsilon \to \eta_0. \end{cases} \tag{6.27}$$

In this subsection we have calculated the amplitudes of the peaks in leading order provided that their shape is given by an ansatz. This will form the basis of a rigorous proof for the existence of equilibrium states in the next two subsections.

6.1.2 Reduction to Finite Dimensions

Let us begin with the proof Theorem 6.1.

Firstly, we will choose a suitable approximation to a steady state. Secondly, using Liapunov-Schmidt reduction, we will reduce the problem to finite dimensions. Thirdly, we will solve the reduced problem. This procedure has been employed to construct steady states for the Gierer-Meinhardt system in one dimension in Chap. 2.

Rescaling

$$x = \epsilon y, \quad x \in \Omega, y \in \Omega_\epsilon = \{y : \epsilon y \in \Omega\},$$

$$\hat{A}(y) = \frac{1}{\xi_\epsilon} A(\epsilon y), \quad y \in \Omega_\epsilon, \tag{6.28}$$

$$\hat{H}(x) = \frac{1}{\xi_\epsilon} H(x), \quad x \in \Omega,$$

where ξ_ϵ is given in (6.10), a steady $(\hat{A}, \hat{H})$ solves the following rescaled Gierer-Meinhardt system:

$$\begin{cases} \Delta_y \hat{A} - \hat{A} + \frac{\hat{A}^2}{\hat{H}} = 0, & y \in \Omega_\epsilon, \\ \Delta_x \hat{H} - \beta^2 \hat{H} + \beta^2 \xi_\epsilon \hat{A}^2 = 0, & x \in \Omega \end{cases} \tag{6.29}$$

together with Neumann boundary conditions.

For given $\hat{A} \in H^2(\Omega_\epsilon)$, let $T[\hat{A}]$ be the unique solution of the problem

$$\Delta T[\hat{A}] - \beta^2 T[\hat{A}] + \beta^2 \xi_\epsilon \hat{A}^2 = 0 \quad \text{in } \Omega, \quad \frac{\partial T[\hat{A}]}{\partial \nu} = 0 \quad \text{on } \partial \Omega \tag{6.30}$$

which can be rewritten as

$$T[\hat{A}](x) = \int_{\Omega} G_{\beta}(x,\xi)\beta^2\xi_{\epsilon}\hat{A}^2\left(\frac{\xi}{\epsilon}\right)d\xi. \tag{6.31}$$

Further, (6.29) is equivalent to

$$S_{\epsilon}(\hat{A}, \hat{H}) = \begin{pmatrix} S_1(\hat{A}, \hat{H}) \\ S_2(\hat{A}, \hat{H}) \end{pmatrix} = 0,$$

$$H_N^2(\Omega_{\epsilon}) \times H_N^2(\Omega) \to L^2(\Omega_{\epsilon}) \times L^2(\Omega), \tag{6.32}$$

where

$$S_1(\hat{A}, \hat{H}) = \Delta_y\hat{A} - \hat{A} + \frac{\hat{A}^2}{\hat{H}} : H_N^2(\Omega_{\epsilon}) \times H_N^2(\Omega) \to L^2(\Omega_{\epsilon}),$$

$$S_2(\hat{A}, \hat{H}) = \Delta_x\hat{H} - \beta^2\hat{H} + \beta^2\xi_{\epsilon}\hat{A}^2 : H_N^2(\Omega_{\epsilon}) \times H_N^2(\Omega) \to L^2(\Omega).$$

For $\mathbf{P} = (P_1, \ldots, P_K) \in \overline{\Lambda_{\delta}}$ and

$$w_{\epsilon,j}(y) := w\left(y - \frac{P_j}{\epsilon}\right)\chi_{\epsilon,P_j}(\epsilon y), \quad y \in \Omega_{\epsilon}, j = 1, \ldots, K, \tag{6.33}$$

where w is the unique solution of (6.9) and χ_{ϵ,P_j} has been defined in (6.17), we choose the approximation $(\hat{A}, \hat{H})$ to a steady state as

$$A_{\epsilon,\mathbf{P}}(y) := \sum_{j=1}^{K} w_{\epsilon,j}(y), \qquad H_{\epsilon,\mathbf{P}}(x) := T[A_{\epsilon,\mathbf{P}}](x), \quad x = \epsilon y \in \Omega. \tag{6.34}$$

Hence

$$H_{\epsilon,\mathbf{P}}(P_j) = \beta^2\xi_{\epsilon}\int_{\Omega} G_{\beta}(x,\xi)\sum_{j=1}^{K} w_{\epsilon,j}^2\left(\frac{\xi}{\epsilon}\right)d\xi + \text{e.s.t.}$$

Similarly to Sect. 6.1.1, we compute

$$H_{\epsilon,\mathbf{P}}(P_j) = 1 + O\big(h(\epsilon, \beta)\big), \quad j = 1, \ldots, K.$$

Inserting the ansatz (6.34) into (6.32), we have

$$S_1(A_{\epsilon,\mathbf{P}}, H_{\epsilon,\mathbf{P}}) = \Delta_y A_{\epsilon,\mathbf{P}} - A_{\epsilon,\mathbf{P}} + \frac{A_{\epsilon,\mathbf{P}}^2}{H_{\epsilon,\mathbf{P}}}$$

$$= \sum_{j=1}^{K}\left[\Delta_y w\left(y - \frac{P_j}{\epsilon}\right) - w\left(y - \frac{P_j}{\epsilon}\right)\right]$$

$$+ \sum_{j=1}^{K} w^2 \left(y - \frac{P_j}{\epsilon} \right) H_{\epsilon,\mathbf{P}}^{-1} + \text{e.s.t.}$$

$$= \sum_{j=1}^{K} w^2 \left(y - \frac{P_j}{\epsilon} \right) \left(H_{\epsilon,\mathbf{P}}^{-1} - 1 \right) + \text{e.s.t.}$$

$$= \sum_{j=1}^{K} w^2 \left(y - \frac{P_j}{\epsilon} \right) \left(H_{\epsilon,\mathbf{P}}^{-1}(P_j) - 1 \right)$$

$$+ \sum_{j=1}^{K} w^2 \left(y - \frac{P_j}{\epsilon} \right) \left(H_{\epsilon,\mathbf{P}}^{-1}(x) - H_{\epsilon,\mathbf{P}}^{-1}(P_j) \right) + \text{e.s.t.} \quad (6.35)$$

$$S_2(A_{\epsilon,\mathbf{P}}, H_{\epsilon,\mathbf{P}}) = 0. \quad (6.36)$$

Next, we compute for $j = 1, \dots, K$ and $x = P_j + \epsilon z$, $|\epsilon z| < \delta$:

$$H_{\epsilon,\mathbf{P}}(P_j + \epsilon z) - H_{\epsilon,\mathbf{P}}(P_j)$$

$$= \beta^2 \int_{\Omega} [G_\beta(P_j + \epsilon z, \xi) - G_\beta(P_j, \xi)]\xi_\epsilon A_{\epsilon,\mathbf{P}}^2 d\xi$$

$$= \beta^2 \xi_\epsilon \int_{\Omega} [G_\beta(P_j + \epsilon z, \xi) - G_\beta(P_j, \xi)] w_{\epsilon,j}^2 d\xi$$

$$+ \beta^2 \xi_\epsilon \int_{\Omega} [G_\beta(P_j + \epsilon z, \xi) - G_\beta(P_j, \xi)] \sum_{l \neq j} w_{\epsilon,l}^2 d\xi + \text{e.s.t.}$$

$$= k(\epsilon, \beta) \int_{\mathbb{R}^2} \frac{1}{2\pi} \log \frac{|\zeta|}{|z - \zeta|} w^2(\zeta) d\zeta$$

$$- k(\epsilon, \beta) \left(\sum_{k=1}^{2} \frac{\partial F(\mathbf{P})}{\partial P_{j,k}} \epsilon z_k \int_{\mathbb{R}^2} w^2 \right) + O\left(\epsilon \beta^2 k(\epsilon, \beta)|z|\right), \quad (6.37)$$

where $k(\epsilon, \beta)$ is given by (6.11), and $F(\mathbf{P})$ is defined at (6.4).

We substitute (6.37) into (6.35) and derive the following key result:

Lemma 6.3 *For $x = P_j + \epsilon z$, $|\epsilon z| < \delta$ we can decompose*

$$S_1(A_{\epsilon,\mathbf{P}}, H_{\epsilon,\mathbf{P}}) = S_{1,1} + S_{1,2}, \quad (6.38)$$

where

$$S_{1,1}(z) = k(\epsilon, \beta) \left(H_{\epsilon,P_j}(P_j) \right)^{-2} \left(\int_{\mathbb{R}^2} w^2 \right) w^2(z) \left(\epsilon \nabla_{\mathbf{P}_j} F(\mathbf{P}) \cdot z + O\left(\epsilon \beta^2 |z|\right) \right)$$

and

$$S_{1,2}(z) = k(\epsilon, \beta) w^2(z) R(|z|) + O\left(\epsilon \beta^2 k(\epsilon, \beta)|z|\right).$$

Further, $R(|z|)$ is a radially symmetric function which satisfies the estimate $R(|z|) = O(\log(1 + |z|))$. Finally, we have $S_1(A_{\epsilon,\mathbf{P}}, H_{\epsilon,\mathbf{P}}) = e.s.t.$ for $|x - P_j| \geq \delta$, $j = 1, \ldots, K$.

Next we study a linearised operator which is given by

$$\tilde{L}_{\epsilon,\mathbf{P}} : H_N^2(\Omega_\epsilon) \times H_N^2(\Omega) \to L^2(\Omega_\epsilon) \times L^2(\Omega), \quad \tilde{L}_{\epsilon,\mathbf{P}} := S_\epsilon' \begin{pmatrix} A_{\epsilon,\mathbf{P}} \\ H_{\epsilon,\mathbf{P}} \end{pmatrix},$$

where $\mathbf{P} \in \overline{\Lambda_\delta}$ and $\epsilon > 0$ is small enough.

Setting

$$K_{\epsilon,\mathbf{P}} := \operatorname{span}\left\{ \frac{\partial A_{\epsilon,\mathbf{P}}}{\partial P_{j,l}} : j = 1, \ldots, K, l = 1, 2 \right\} \subset H_N^2(\Omega_\epsilon)$$

and

$$C_{\epsilon,\mathbf{P}} := \operatorname{span}\left\{ \frac{\partial A_{\epsilon,\mathbf{P}}}{\partial P_{j,l}} : j = 1, \ldots, K, l = 1, 2 \right\} \subset L^2(\Omega_\epsilon),$$

we note that $\tilde{L}_{\epsilon,\mathbf{P}}$ is not uniformly invertible in ϵ and β since it possesses the approximate kernel

$$\mathcal{K}_{\epsilon,\mathbf{P}} := K_{\epsilon,\mathbf{P}} \oplus \{0\} \subset H_N^2(\Omega_\epsilon) \times H_N^2(\Omega). \tag{6.39}$$

With the approximate cokernel

$$\mathcal{C}_{\epsilon,\mathbf{P}} := C_{\epsilon,\mathbf{P}} \oplus \{0\} \subset L^2(\Omega_\epsilon) \times L^2(\Omega) \tag{6.40}$$

we set

$$\mathcal{K}_{\epsilon,\mathbf{P}}^{\perp} := K_{\epsilon,\mathbf{P}}^{\perp} \oplus H_N^2(\Omega) \subset H_N^2(\Omega_\epsilon) \times H_N^2(\Omega), \tag{6.41}$$

$$\mathcal{C}_{\epsilon,\mathbf{P}}^{\perp} := C_{\epsilon,\mathbf{P}}^{\perp} \oplus L^2(\Omega) \subset L^2(\Omega_\epsilon) \times L^2(\Omega). \tag{6.42}$$

Here $C_{\epsilon,\mathbf{P}}^{\perp}$ and $K_{\epsilon,\mathbf{P}}^{\perp}$ are orthogonal complements with the $L^2(\Omega_\epsilon)$ scalar product taken within the spaces $H_N^2(\Omega_\epsilon)$ and $L^2(\Omega_\epsilon)$, respectively.

Using the projection $\pi_{\epsilon,\mathbf{P}}$ in $L^2(\Omega_\epsilon) \times L^2(\Omega)$ onto $\mathcal{C}_{\epsilon,\mathbf{P}}^{\perp}$, where the second component of the projection is the identity map, we will prove that

$$\pi_{\epsilon,\mathbf{P}} \circ S_\epsilon \begin{pmatrix} A_{\epsilon,\mathbf{P}} + \Phi_{\epsilon,\mathbf{P}} \\ H_{\epsilon,\mathbf{P}} + \Psi_{\epsilon,\mathbf{P}} \end{pmatrix} = 0 \tag{6.43}$$

has the unique solution $\Sigma_{\epsilon,\mathbf{P}} = \begin{pmatrix} \Phi_{\epsilon,\mathbf{P}}(y) \\ \Psi_{\epsilon,\mathbf{P}}(x) \end{pmatrix} \in \mathcal{K}_{\epsilon,\mathbf{P}}^{\perp}$ provided that ϵ and β are chosen small enough.

Setting

$$\mathcal{L}_{\epsilon,\mathbf{P}} = \pi_{\epsilon,\mathbf{P}} \circ \tilde{L}_{\epsilon,\mathbf{P}} : \mathcal{K}_{\epsilon,\mathbf{P}}^{\perp} \to \mathcal{C}_{\epsilon,\mathbf{P}}^{\perp}, \tag{6.44}$$

we have the following two propositions on the invertibility of the linearised operator $\mathcal{L}_{\epsilon,\mathbf{P}}$.

Proposition 6.4 *Assume that (6.8) holds and let $\mathcal{L}_{\epsilon,\mathbf{P}}$ be defined in (6.44). Then there are positive constants $\overline{\epsilon}, \overline{\beta}, C$ such that for all $\epsilon \in (0, \overline{\epsilon}), \beta \in (0, \overline{\beta})$ we have*

$$\|\mathcal{L}_{\epsilon,\mathbf{P}}\Sigma\|_{L^2(\Omega_\epsilon) \times L^2(\Omega)} \geq C \|\Sigma\|_{H^2(\Omega_\epsilon) \times H^2(\Omega)} \tag{6.45}$$

for arbitrary $\mathbf{P} \in \overline{\Lambda_\delta}$ and $\Sigma \in \mathcal{K}^{\perp}_{\epsilon,\mathbf{P}}$.

Proposition 6.5 *Assume that (6.8) holds. Then there exist positive constants $\overline{\overline{\epsilon}}$ and $\overline{\overline{\beta}}$ such that for all $\epsilon \in (0, \overline{\overline{\epsilon}})$ and $\beta \in (0, \overline{\overline{\beta}})$ the map $\mathcal{L}_{\epsilon,\mathbf{P}}$ is surjective for arbitrary $\mathbf{P} \in \overline{\Lambda_\delta}$.*

The proofs of Propositions 6.4 and 6.5 are similar to those in Sect. 2.3.3. Technical details can be found there and are omitted here. The main new feature stems from the limiting problem of the linearised operator as we take $\epsilon \to 0$, where

$$\phi_i \in \left\{ \phi \in H^2(\mathbb{R}^2) : \int_{\mathbb{R}^2} \phi \frac{\partial w}{\partial y_j} dy = 0, i = 1, \ldots, Kj = 1, 2 \right\} = K_0^{\perp}$$

and $\phi_i, i = 1, \ldots, K$ has to satisfy the following nonlocal linear problem:

Case 1: $\eta_\epsilon \to 0$

$$\Delta \phi_i - \phi_i + 2w\phi_i - \frac{2 \sum_{j=1}^{K} \int_{\mathbb{R}^2} w\phi_j}{K \int_{\mathbb{R}^2} w^2} w^2 \in C_0^{\perp}. \tag{6.46}$$

Case 2: $\eta_\epsilon \to \infty$

$$\Delta \phi_i - \phi_i + 2w\phi_i - \frac{2 \int_{\mathbb{R}^2} w\phi_i}{\int_{\mathbb{R}^2} w^2} w^2 \in C_0^{\perp}. \tag{6.47}$$

Case 3: $\eta_\epsilon \to \eta_0$

$$\Delta \phi_i - \phi_i + 2w\phi_i - \frac{2[(1 + \eta_0) \int_{\mathbb{R}^2} w\phi_i + \sum_{j \neq i} \int_{\mathbb{R}^2} w\phi_j]}{(K + \eta_0) \int_{\mathbb{R}^2} w^2} w^2 \in C_0^{\perp}, \tag{6.48}$$

where

$$C_0 := \text{span} \left\{ \frac{\partial w}{\partial y_j}, j = 1, 2 \right\}$$

and $K_0^{\perp}, C_0^{\perp}$ is the orthogonal complement with the $L^2(\mathbb{R}^2)$ scalar product taken within the spaces $H^2(\mathbb{R}^2)$ and $L^2(\mathbb{R}^2)$, respectively.

Diagonalisation gives for the transformed functions $(\phi_1, \ldots, \phi_K)$ the following decoupled equations:

$$\Delta \phi_i - \phi_i + 2w\phi_i - 2\rho_i \frac{\int_{\mathbb{R}^2} w\phi_i}{\int_{\mathbb{R}^2} w^2} w \in C_0^{\perp}, \qquad (6.49)$$

where

$$\rho_i = \begin{cases} 0, \ldots, 0, K & \text{in Case 1,} \\ 1, \ldots, 1 & \text{in Case 2,} \\ \frac{\eta_0}{K+\eta_0}, \ldots, \frac{\eta_0}{K+\eta_0}, 1 & \text{in Case 3.} \end{cases}$$

Since $L_0 w = w^2$, (6.49) can be written as

$$(\Delta_y - 1 + 2w)\left(\phi_i - 2\rho_i \frac{\int_{\mathbb{R}^2} w\phi_i}{\int_{\mathbb{R}^2} w^2} w\right) \in C_0^{\perp}.$$

By Lemma 13.4 the operator

$$L_0 = \Delta_y - 1 + 2w : K_0^{\perp} \to C_0^{\perp}$$

is a one-to-one and invertible mapping and we get

$$\phi_i - 2\rho_i \frac{\int_{\mathbb{R}^2} w\phi_i}{\int_{\mathbb{R}^2} w} = 0. \qquad (6.50)$$

Multiplying by w and integrating, we get

$$(1 - 2\rho_i) \int_{\mathbb{R}^2} w\phi_i = 0. \qquad (6.51)$$

If $\rho_i \neq \frac{1}{2}$, then by (6.51) we have

$$\int_{\mathbb{R}^2} w\phi_i = 0.$$

Thus we get

$$L_0 \phi_i = 0, \quad i = 1, \ldots, K$$

and by Lemma 13.4 we conclude

$$\phi_i \in K_0, \quad i = 1, \ldots, K.$$

Therefore, by (6.50), we get

$$\phi_i = 0, \quad i = 1, \ldots, K.$$

Let us explain why Remark 6.2 plays a crucial role: We have $\rho_i = \frac{1}{2}$ for some i if and only $K > 1$ and $\eta_0 = K$. Then the dimension of the kernel in the limiting problem changes and Liapunov-Schmidt reduction is not easily applicable.

We consider the limit $\epsilon \to 0$ and $\psi_i \to 0$ in $H^2(\Omega)$, where

$$\sum_{i=1}^{K}\left(\|\phi_i\|^2_{H^2(\mathbb{R}^2)} + \|\psi_i\|^2_{H^2(\Omega)}\right) = 1.$$

The last statement contradicts the assumption $\phi_i = \psi_i = 0$ and Proposition 5.11 follows.

Next we consider the nonlinear problem

$$\pi_{\epsilon,\mathbf{P}} \circ S_\epsilon \begin{pmatrix} A_{\epsilon,\mathbf{P}} + \phi \\ H_{\epsilon,\mathbf{P}} + \psi \end{pmatrix} = 0. \tag{6.52}$$

Using the fact that $\mathcal{L}_{\epsilon,\mathbf{P}}|_{\mathcal{K}^{\perp}_{\epsilon,\mathbf{P}}}$ is invertible and denoting the inverse by $\mathcal{L}^{-1}_{\epsilon,\mathbf{P}}$, we can write (6.52) as follows:

$$\Sigma = -\left(\mathcal{L}^{-1}_{\epsilon,\mathbf{P}} \circ \pi_{\epsilon,\mathbf{P}}\right)\left(S_\epsilon\begin{pmatrix} A_{\epsilon,\mathbf{P}} \\ H_{\epsilon,\mathbf{P}} \end{pmatrix}\right) - \left(\mathcal{L}^{-1}_{\epsilon,\mathbf{P}} \circ \pi_{\epsilon,\mathbf{P}}\right)\left(N_{\epsilon,\mathbf{P}}(\Sigma)\right) =: M_{\epsilon,\mathbf{P}}(\Sigma), \tag{6.53}$$

where

$$N_{\epsilon,\mathbf{P}}(\Sigma) = S_\epsilon\begin{pmatrix} A_{\epsilon,\mathbf{P}} + \phi \\ H_{\epsilon,\mathbf{P}} + \psi \end{pmatrix} - S_\epsilon\begin{pmatrix} A_{\epsilon,\mathbf{P}} \\ H_{\epsilon,\mathbf{P}} \end{pmatrix} - S'_\epsilon\begin{pmatrix} A_{\epsilon,\mathbf{P}} \\ H_{\epsilon,\mathbf{P}} \end{pmatrix}\begin{bmatrix} \phi \\ \psi \end{bmatrix}.$$

Note that the operator $M_{\epsilon,\mathbf{P}}$ in (6.53) is defined for all $\Sigma = (\phi, \psi)^T \in H^2_N(\Omega_\epsilon) \times H^2_N(\Omega)$. In the next step we prove that the operator $M_{\epsilon,\mathbf{P}}$ is a contraction mapping defined on the set

$$B_{\epsilon,\delta} \equiv \left\{\Sigma \in H^2(\Omega_\epsilon) \times H^2(\Omega) : \|\Sigma\|_{H^2(\Omega_\epsilon) \times H^2(\Omega)} < \delta\right\} \tag{6.54}$$

provided that the constant δ is chosen suitably.

By Lemma 6.3, we get

$$\left\|S_1(A_{\epsilon,\mathbf{P}}, H_{\epsilon,\mathbf{P}})\right\|_{H^2(\Omega_\epsilon)} \le C_0 k(\epsilon, \beta). \tag{6.55}$$

Using (6.55) as well as Propositions 5.10 and 5.11, we derive

$$\left\|M_{\epsilon,\mathbf{P}}(\Sigma)\right\|_{H^2(\Omega_\epsilon) \times H^2(\Omega)} \le \lambda^{-1}\left(\left\|\pi_{\epsilon,\mathbf{P}} \circ N_{\epsilon,\mathbf{P}}(\Sigma)\right\|_{L^2(\Omega_\epsilon) \times L^2(\Omega)}\right.$$

$$\left. + \left\|\pi_{\epsilon,\mathbf{P}} \circ S_\epsilon\begin{pmatrix} A_{\epsilon,\mathbf{P}} \\ H_{\epsilon,\mathbf{P}} \end{pmatrix}\right\|_{L^2(\Omega_\epsilon) \times L^2(\Omega)}\right)$$

$$\le \lambda^{-1} C_0\left(c(\delta)\delta + k(\epsilon, \beta)\right),$$

where $\lambda > 0$ is independent of $\delta > 0$ and $c(\delta) \to 0$ as $\delta \to 0$. Similarly, we prove

$$\left\| M_{\epsilon,\mathbf{P}}(\Sigma) - M_{\epsilon,\mathbf{P}}(\Sigma') \right\|_{H^2(\Omega_\epsilon) \times H^2(\Omega)} \leq \lambda^{-1} c(\delta) \left\| \Sigma - \Sigma' \right\|_{H^2(\Omega_\epsilon) \times H^2(\Omega)},$$

where $c(\delta) \to 0$ as $\delta \to 0$. Taking

$$\delta = C_1 k(\epsilon, \beta) \quad \text{for } \lambda^{-1} C_0 < C_1 \text{ and } \epsilon \text{ small enough,} \tag{6.56}$$

it follows that the operator $M_{\epsilon,\mathbf{P}}$ is a contraction mapping of $B_{\epsilon,\delta}$ into itself. Employing the Contraction Mapping Principle, there exists a unique fixed point $\Sigma_{\epsilon,\mathbf{P}} \in B_{\epsilon,\delta}$. Then $\Sigma_{\epsilon,\mathbf{P}}$ solves (6.53).

We have proved the following result:

Lemma 6.6 *There are $\bar{\epsilon} > 0$, $\bar{\beta} > 0$ such that for every triple $(\epsilon, \beta, \mathbf{P})$ with $0 < \epsilon < \bar{\epsilon}$, $0 < \beta < \bar{\beta}$, $\mathbf{P} \in \overline{\Lambda_\delta}$ there exists a unique $(\Phi_{\epsilon,\mathbf{P}}, \Psi_{\epsilon,\mathbf{P}}) \in K_{\epsilon,\mathbf{P}}^\perp$ which solves $S_\epsilon \left(\begin{smallmatrix} A_{\epsilon,\mathbf{P}} + \Phi_{\epsilon,\mathbf{P}} \\ H_{\epsilon,\mathbf{P}} + \Psi_{\epsilon,\mathbf{P}} \end{smallmatrix} \right) \in C_{\epsilon,\mathbf{P}}$ and*

$$\left\| (\Phi_{\epsilon,\mathbf{P}}, \Psi_{\epsilon,\mathbf{P}}) \right\|_{H^2(\Omega_\epsilon) \times H^2(\Omega)} \leq C k(\epsilon, \beta). \tag{6.57}$$

Next we derive more precise estimates for $\Phi_{\epsilon,\mathbf{P}}$. Lemma 6.3 implies that $S_1 = S_{1,1} + S_{1,2}$, where $S_{1,1}$ in leading order is an odd function and $S_{1,2}$ in leading order is a radially symmetric function. Similarly, we can decompose $\Phi_{\epsilon,\mathbf{P}}$:

Lemma 6.7 *For $\Phi_{\epsilon,\mathbf{P}}$ given in Lemma 6.6 and for $x = P_i + \epsilon z$, $|\epsilon z| < \delta$, we derive*

$$\Phi_{\epsilon,\mathbf{P}} = \Phi_{\epsilon,\mathbf{P},1} + \Phi_{\epsilon,\mathbf{P},2}, \tag{6.58}$$

where

$$\Phi_{\epsilon,\mathbf{P},1} = O\big(\epsilon k(\epsilon, \beta)\big) \quad \text{in } H_N^2(\Omega_\epsilon) \tag{6.59}$$

and $\Phi_{\epsilon,\mathbf{P},2}$ is a radially symmetric function in z which solves

$$\Phi_{\epsilon,\mathbf{P},2} = O\big(k(\epsilon, \beta)\big) \quad \text{in } H_N^2(\Omega_\epsilon). \tag{6.60}$$

Proof Let $S[v] := S_1(v, T[v])$. We consider the solutions of the problems

$$S[A_{\epsilon,\mathbf{P}} + \Phi_{\epsilon,\mathbf{P},2}] - S[A_{\epsilon,\mathbf{P}}] + \sum_{j=1}^{K} S_{1,2}\left(y - \frac{P_j}{\epsilon}\right) \in C_{\epsilon,\mathbf{P}} \tag{6.61}$$

for $\Phi_{\epsilon,\mathbf{P},2} \in K_{\epsilon,\mathbf{P}}^\perp$ and

$$S[A_{\epsilon,\mathbf{P}} + \Phi_{\epsilon,\mathbf{P},2} + \Phi_{\epsilon,\mathbf{P},1}] - S[A_{\epsilon,\mathbf{P}} + \Phi_{\epsilon,\mathbf{P},2}] + \sum_{j=1}^{K} S_{1,1}\left(y - \frac{P_j}{\epsilon}\right) \in C_{\epsilon,\mathbf{P}} \tag{6.62}$$

for $\Phi_{\epsilon,\mathbf{P},1} \in K_{\epsilon,\mathbf{P}}^{\perp}$. By the argument in the proof of Lemma 6.6, (6.61) and (6.62) each have a unique solution provided that ϵ and β are small enough. Thus we get $\Phi_{\epsilon,\mathbf{P}} = \Phi_{\epsilon,\mathbf{P},1} + \Phi_{\epsilon,\mathbf{P},2}$. Since $S_{1,1} = S_{1,1}^0 + S_{1,1}^{\perp}$, where $\|S_{1,1}^0\|_{H^2(\Omega_\epsilon)} = O(\epsilon k(\epsilon, \beta))$ and $S_{1,1}^{\perp} \in C_{\epsilon,\mathbf{P}}^{\perp}$, it follows that $\Phi_{\epsilon,\mathbf{P},1}$ and $\Phi_{\epsilon,\mathbf{P},2}$ possess the properties stated in the lemma. $\qquad\square$

6.1.3 The Reduced Problem

In this subsection, we solve the reduced problem to complete the proof of Theorem 6.1. Let $\mathbf{P}^0$ be a nondegenerate critical point of the function $F(\mathbf{P})$. By Lemma 6.6, for each $\mathbf{P} \in B_\delta(\mathbf{P}^0)$ and δ small enough, there is a unique solution $(\Phi_{\epsilon,\mathbf{P}}, \psi_{\epsilon,\mathbf{P}}) \in \mathcal{K}_{\epsilon,\mathbf{P}}^{\perp}$ of the problem

$$S_\epsilon \begin{pmatrix} A_{\epsilon,\mathbf{P}} + \Phi_{\epsilon,\mathbf{P}} \\ H_{\epsilon,\mathbf{P}} + \Psi_{\epsilon,\mathbf{P}} \end{pmatrix} = \begin{pmatrix} v_{\epsilon,\mathbf{P}} \\ 0 \end{pmatrix} \in C_{\epsilon,\mathbf{P}}.$$

Next we determine $\mathbf{P} = \mathbf{P}^\epsilon \in B_\delta(\mathbf{P}^0)$ such that

$$S_\epsilon \begin{pmatrix} A_{\epsilon,\mathbf{P}} + \Phi_{\epsilon,\mathbf{P}} \\ H_{\epsilon,\mathbf{P}} + \Psi_{\epsilon,\mathbf{P}} \end{pmatrix} \perp C_{\epsilon,\mathbf{P}}. \tag{6.63}$$

For $\mathbf{P} \in \overline{\Lambda_\delta}$, let

$$W_{\epsilon,j,i}(\mathbf{P}) := \frac{1}{k(\epsilon, \beta)} \int_{\Omega_\epsilon} \left(S_1(A_{\epsilon,\mathbf{P}} + \Phi_{\epsilon,\mathbf{P}}, H_{\epsilon,\mathbf{P}} + \Psi_{\epsilon,\mathbf{P}}) \frac{\partial A_{\epsilon,\mathbf{P}}}{\partial P_{j,i}} \right),$$

$$j = 1, \ldots, K, i = 1, 2, \tag{6.64}$$

$$W_\epsilon(\mathbf{P}) := \left(W_{\epsilon,1,1}(\mathbf{P}), \ldots, W_{\epsilon,K,2}(\mathbf{P}) \right), \tag{6.65}$$

where $k(\epsilon, \beta) = \epsilon^2 \beta \xi_\epsilon$.

Note that the mapping $W_\epsilon(\mathbf{P})$ is continuous in $\mathbf{P}$ and all that remains is to determine a root of the vector field $W_\epsilon(\mathbf{P})$.

Let

$$\Omega_{\epsilon,P_j} = \{y : \epsilon y + P_j \in \Omega\}.$$

The asymptotic expansion of $W_{\epsilon,j,i}(\mathbf{P})$ is computed as

$$\frac{1}{k(\epsilon, \beta)} \int_{\Omega_\epsilon} S_1(A_{\epsilon,\mathbf{P}} + \Phi_{\epsilon,\mathbf{P}}, H_{\epsilon,\mathbf{P}} + \Psi_{\epsilon,\mathbf{P}}) \frac{\partial A_{\epsilon,\mathbf{P}}}{\partial P_{j,i}}$$

$$= \frac{1}{k(\epsilon, \beta)} \int_{\Omega_\epsilon} \left[\Delta(A_{\epsilon,\mathbf{P}} + \Phi_{\epsilon,\mathbf{P}}) - (A_{\epsilon,\mathbf{P}} + \Phi_{\epsilon,\mathbf{P}}) + \frac{(A_{\epsilon,\mathbf{P}} + \Phi_{\epsilon,\mathbf{P}})^2}{H_{\epsilon,\mathbf{P}} + \Psi_{\epsilon,\mathbf{P}}} \right] \frac{\partial A_{\epsilon,\mathbf{P}}}{\partial P_{j,i}}$$

$$= \frac{1}{k(\epsilon, \beta)} \int_{\Omega_\epsilon} \left[\Delta(A_{\epsilon,\mathbf{P}} + \Phi_{\epsilon,\mathbf{P}}) - (A_{\epsilon,\mathbf{P}} + \Phi_{\epsilon,\mathbf{P}}) + \frac{(A_{\epsilon,\mathbf{P}} + \Phi_{\epsilon,\mathbf{P}})^2}{H_{\epsilon,\mathbf{P}}} \right] \frac{\partial A_{\epsilon,\mathbf{P}}}{\partial P_{j,i}}$$

$$+ \frac{1}{k(\epsilon, \beta)} \int_{\Omega_\epsilon} \left[\frac{(A_{\epsilon, \mathbf{P}} + \Phi_{\epsilon, \mathbf{P}})^2}{H_{\epsilon, \mathbf{P}} + \Psi_{\epsilon, \mathbf{P}}} - \frac{(A_{\epsilon, \mathbf{P}} + \Phi_{\epsilon, \mathbf{P}})^2}{H_{\epsilon, \mathbf{P}}} \right] \frac{\partial A_{\epsilon, \mathbf{P}}}{\partial P_{j,i}}$$

$$=: I_1 + I_2.$$

By Lemma 6.7 we get

$$I_1 = \frac{1}{k(\epsilon, \beta)} \left(\int_{\Omega_\epsilon} \left[\Delta(A_{\epsilon, \mathbf{P}} + \Phi_{\epsilon, \mathbf{P}}) - (A_{\epsilon, \mathbf{P}} + \Phi_{\epsilon, \mathbf{P}}) + \frac{(A_{\epsilon, \mathbf{P}} + \Phi_{\epsilon, \mathbf{P}})^2}{H_{\epsilon, \mathbf{P}}(P_j)} \right] \frac{\partial A_{\epsilon, \mathbf{P}}}{\partial P_{j,i}} \right.$$

$$\left. - \int_{\Omega_\epsilon} \frac{(A_{\epsilon, \mathbf{P}} + \Phi_{\epsilon, \mathbf{P}})^2}{H^2_{\epsilon, \mathbf{P}}(P_j)} \left(H_{\epsilon, \mathbf{P}} - H_{\epsilon, \mathbf{P}}(P_j) \right) \frac{\partial A_{\epsilon, \mathbf{P}}}{\partial P_{j,i}} \right) + o(1)$$

$$= \frac{1}{\epsilon k(\epsilon, \beta)} \left(- \int_{\Omega_{\epsilon, P_j}} \left[\Delta(w_{\epsilon, j} + \Phi_{\epsilon, \mathbf{P}}) - (w_{\epsilon, j} + \Phi_{\epsilon, \mathbf{P}}) + (w_{\epsilon, j} + \Phi_{\epsilon, \mathbf{P}})^2 \right] \frac{\partial w_{\epsilon, j}}{\partial y_i} \right.$$

$$\left. + \int_{\Omega_{\epsilon, P_j}} \frac{(w_{\epsilon, j} + \Phi_{\epsilon, \mathbf{P}, 2})^2(y)}{(H_{\epsilon, \mathbf{P}}(P_j))^2} \left(H_{\epsilon, \mathbf{P}}(P_j + \epsilon y) - H_{\epsilon, \mathbf{P}}(P_j) \right) \frac{\partial w_{\epsilon, j}(y)}{\partial y_i} dy \right)$$

$$+ o(1).$$

Lemma 6.7 implies

$$\int_{\Omega_{\epsilon, P_j}} [\Delta \Phi_{\epsilon, \mathbf{P}} - \Phi_{\epsilon, \mathbf{P}} + 2 w_{\epsilon, j} \Phi_{\epsilon, \mathbf{P}}] \frac{\partial w_{\epsilon, j}}{\partial y_i}$$

$$= \int_{\Omega_{\epsilon, P_j}} \Phi_{\epsilon, \mathbf{P}, 1} \frac{\partial}{\partial y_i} [\Delta w - w + w^2] + o(\epsilon k(\epsilon, \beta)) = o(\epsilon k(\epsilon, \beta)), \quad (6.66)$$

$$\int_{\Omega_{\epsilon, P_j}} (\Phi_{\epsilon, \mathbf{P}})^2 \frac{\partial w_{\epsilon, j}}{\partial y_i}$$

$$= \int_{\Omega_{\epsilon, P_j}} (\Phi_{\epsilon, \mathbf{P}, 1})^2 \frac{\partial w_{\epsilon, j}}{\partial y_i} + o(\epsilon k(\epsilon, \beta)) = o(\epsilon k(\epsilon, \beta)). \quad (6.67)$$

Next from (6.37), (6.66) and (6.67) we get

$$I_1 = o(1) - \frac{1}{\epsilon k(\epsilon, \beta)} \int_{\Omega_{\epsilon, P_j}} w^2_{\epsilon, j}(y) \left(H_{\epsilon, \mathbf{P}}(P_j + \epsilon y) - H_{\epsilon, \mathbf{P}}(P_j) \right) \frac{\partial w_{\epsilon, j}(y)}{\partial y_i} dy$$

$$= o(1) + \sum_{k=1}^{2} \frac{\partial F(\mathbf{P})}{\partial P_{j,k}} \int_{\mathbb{R}^2} w^2 y_k \frac{\partial w}{\partial y_i} \int_{\mathbb{R}^2} w^2$$

$$= o(1) + \frac{\partial F(\mathbf{P})}{\partial P_{j,i}} \int_{\mathbb{R}^2} w^2 y_i \frac{\partial w}{\partial y_i} \int_{\mathbb{R}^2} w^2$$

$$= o(1) - \frac{1}{3} \int_{\mathbb{R}^2} w^3 \int_{\mathbb{R}^2} w^2 \frac{\partial F(\mathbf{P})}{\partial P_{j,i}}. \quad (6.68)$$

Next we compute

$$
I_2 = \frac{1}{k(\epsilon, \beta)} \int_{\Omega_\epsilon} \left[\frac{(A_{\epsilon,\mathbf{P}} + \Phi_{\epsilon,\mathbf{P}})^2}{H_{\epsilon,\mathbf{P}} + \Psi_{\epsilon,\mathbf{P}}} - \frac{(A_{\epsilon,\mathbf{P}} + \Phi_{\epsilon,\mathbf{P}})^2}{H_{\epsilon,\mathbf{P}}} \right] \frac{\partial A_{\epsilon,\mathbf{P}}}{\partial P_{j,i}}
$$

$$
= -\frac{1}{k(\epsilon, \beta)} \int_{\Omega_\epsilon} \frac{(A_{\epsilon,\mathbf{P}} + \Phi_{\epsilon,\mathbf{P}})^2}{H_{\epsilon,\mathbf{P}}^2} \Psi_{\epsilon,\mathbf{P}} \frac{\partial A_{\epsilon,\mathbf{P}}}{\partial P_{j,i}} + o(1)
$$

$$
= -\frac{1}{\epsilon k(\epsilon, \beta)} \int_{\Omega_{\epsilon,P_j}} \frac{1}{3} \frac{\partial w_{\epsilon,j}^3}{\partial y_i} \left(\Psi_{\epsilon,\mathbf{P}} - \Psi_{\epsilon,\mathbf{P}}(P_j) \right) + o(1). \tag{6.69}
$$

Recalling that $\Psi_{\epsilon,\mathbf{P}}$ satisfies

$$
\Delta \Psi_{\epsilon,\mathbf{P}} - \beta^2 \Psi_{\epsilon,\mathbf{P}} + 2\beta^2 \xi_\epsilon A_{\epsilon,\mathbf{P}} \Phi_{\epsilon,\mathbf{P}} + \beta^2 \xi_\epsilon \Phi_{\epsilon,\mathbf{P}}^2 = 0,
$$

computations such as those leading to (6.37) give

$$
\Psi_{\epsilon,\mathbf{P}}(P_j + \epsilon y) - \Psi_{\epsilon,\mathbf{P}}(P_j)
$$

$$
= \int_\Omega \left(G_\beta(P_j + \epsilon y, \xi) - G_\beta(P_j, \xi) \right) \beta^2
$$

$$
\times \xi_\epsilon \left(2 A_{\epsilon,\mathbf{P}} \left(\frac{\xi}{\epsilon} \right) \Phi_{\epsilon,\mathbf{P}} \left(\frac{\xi}{\epsilon} \right) + \Phi_{\epsilon,\mathbf{P}}^2 \left(\frac{\xi}{\epsilon} \right) \right) d\xi
$$

$$
= o\left(\epsilon k(\epsilon, \beta) \left| \nabla_{P_j} F(\mathbf{P}) \right| |y| \right) + k(\epsilon, \beta) R_1(|y|), \tag{6.70}
$$

where $R_1(|y|)$ is radially symmetric.

We substitute (6.70) into (6.69) and estimate

$$
I_2 = o(1). \tag{6.71}
$$

A combination of the estimates for I_1 and I_2 gives

$$
W_\epsilon(\mathbf{P}) = c_0 \nabla_{\mathbf{P}} F(\mathbf{P}) + E_\epsilon(\mathbf{P}), \tag{6.72}
$$

where $c_0 = -\frac{1}{3} \int_{\mathbb{R}^2} w^3 \int_{\mathbb{R}^2} w^2 < 0$ and the function $E_\epsilon(\mathbf{P})$ is continuous and tends to 0 as $\epsilon \to 0$ uniformly for $\mathbf{P} \in \overline{\Lambda_\delta}$.

At $\mathbf{P}^0$, we have $\nabla_{\mathbf{P}}|_{\mathbf{P}=\mathbf{P}^0} F(\mathbf{P}^0) = 0$, $\det(\nabla_{\mathbf{P}}^2|_{\mathbf{P}=\mathbf{P}^0}(F(\mathbf{P}^0))) \neq 0$.

We assume that (i) $\delta = \delta(\epsilon) \to 0$ as $\epsilon \to 0$ and (ii) W_ϵ has exactly one zero in $B_\delta(\mathbf{P}^0)$ for ϵ small enough (which is possible by (6.72)) and compute the mapping degree of $W_\epsilon(\mathbf{P})$ for the set B_δ and the value 0 as follows:

$$
\deg(W_\epsilon, 0, B_\delta) = \operatorname{sign} \det\left(-\nabla_{\mathbf{P}}^2|_{\mathbf{P}=\mathbf{P}^0}(F(\mathbf{P})) \right)
$$

$$
= \operatorname{sign} \det\left(-M(\mathbf{P}^0) \right) \neq 0
$$

since by assumption $\mathbf{P}^0$ is a nondegenerate critical point of $F(\mathbf{P})$. By standard degree theory (see Sect. 13.1), for ϵ small enough there is a point $\mathbf{P}^\epsilon \in B_\delta$ such that $W_\epsilon(\mathbf{P}^\epsilon) = 0$. Further, by (6.72) we have $\mathbf{P}^\epsilon \to \mathbf{P}^0$.

We have proved the following result:

Proposition 6.8 *For ϵ small enough there are points $\mathbf{P}^\epsilon$ such that $\mathbf{P}^\epsilon \to \mathbf{P}^0$ and $W_\epsilon(\mathbf{P}^\epsilon) = 0$.*

Now we complete the proof of Theorem 6.1.

Proof of Theorem 6.1 By Proposition 6.8, there are points such that $\mathbf{P}^\epsilon \to \mathbf{P}^0$ and $W_\epsilon(\mathbf{P}^\epsilon) = 0$. Equivalently, we have $S_1(A_{\epsilon,\mathbf{P}^\epsilon} + \Phi_{\epsilon,\mathbf{P}^\epsilon}, H_{\epsilon,\mathbf{P}^\epsilon} + \Psi_{\epsilon,\mathbf{P}^\epsilon}) = 0$. Letting $A_\epsilon = \xi_\epsilon(A_{\epsilon,\mathbf{P}^\epsilon} + \Phi_{\epsilon,\mathbf{P}^\epsilon})$, $H_\epsilon = \xi_\epsilon(H_{\epsilon,\mathbf{P}^\epsilon} + \Psi_{\epsilon,\mathbf{P}^\epsilon})$, it follows that $H_\epsilon = \xi_\epsilon T[A_{\epsilon,\mathbf{P}^\epsilon} + \Phi_{\epsilon,\mathbf{P}^\epsilon}] > 0$. This implies $A_\epsilon \geq 0$. By the Maximum Principle, we get $A_\epsilon > 0$. Thus (A_ϵ, H_ϵ) satisfies all the properties stated in Theorem 6.1. $\qquad\square$

6.2 Symmetric Multiple Spikes: Stability

For the stability we study separately the eigenvalues of order $O(1)$, called large eigenvalues, and the eigenvalues of order $o(1)$, called small eigenvalues. We will see that the small eigenvalues are related to the derivatives of the Green's function G_0 at the spike locations.

Studying the (linearised) stability of the K-spike solutions constructed in Theorem 6.1 amounts to considering the eigenvalue problem

$$\mathcal{L}_\epsilon \begin{pmatrix} \phi_\epsilon \\ \psi_\epsilon \end{pmatrix} = \begin{pmatrix} \epsilon^2 \Delta \phi_\epsilon - \phi_\epsilon + 2\frac{A_\epsilon}{H_\epsilon}\phi_\epsilon - \frac{A_\epsilon^2}{H_\epsilon^2}\psi_\epsilon \\ \frac{1}{\tau}(\frac{1}{\beta^2}\Delta\psi_\epsilon - \psi_\epsilon + 2A_\epsilon\phi_\epsilon) \end{pmatrix} = \lambda_\epsilon \begin{pmatrix} \phi_\epsilon \\ \psi_\epsilon \end{pmatrix}, \tag{6.73}$$

where the steady state (A_ϵ, H_ϵ) is given in Theorem 6.1 and the eigenvalues satisfies $\lambda_\epsilon \in \mathbb{C}$.

Next we state the main result on stability:

Theorem 6.9 *Assume that ϵ is sufficiently small and $D = \frac{1}{\beta^2}$ sufficiently large. Further, assume that (6.8) holds and*

$$\mathbf{P}^0 \in \overline{\Lambda_\delta} \text{ is a nondegenerate local maximum point of } F(\mathbf{P}). \tag{$*$}$$

Let (A_ϵ, H_ϵ) be the K-spike solution given in Theorem 6.1 with spike locations converging to $\mathbf{P}^0$. Then we have

Case 1. $\eta_\epsilon \to 0$ (i.e., $\frac{2\pi D}{|\Omega|} \gg \log\frac{\sqrt{|\Omega|}}{\epsilon}$).
If $K = 1$, there is a unique $\tau_1 > 0$ such that for $\tau < \tau_1$ the solution (A_ϵ, H_ϵ) is linearly stable and for $\tau > \tau_1$ it is linearly unstable.
If $K > 1$, then (A_ϵ, H_ϵ) is linearly unstable for any $\tau \geq 0$.

Case 2. $\eta_\epsilon \to +\infty$ (i.e., $\frac{2\pi D}{|\Omega|} \ll \log\frac{\sqrt{|\Omega|}}{\epsilon}$).
(A_ϵ, H_ϵ) is linearly stable for any $\tau \geq 0$.

Case 3. $\eta_\epsilon \to \eta_0 \in (0, +\infty)$ (i.e., $\frac{2\pi D}{|\Omega|} \sim \frac{1}{\eta_0} \log \frac{\sqrt{|\Omega|}}{\epsilon}$).

If $K > 1$ and $\eta_0 < K$, then the solution (A_ϵ, H_ϵ) is linearly unstable for any $\tau \geq 0$.

If $\eta_0 > K$, then there are $0 < \tau_2 \leq \tau_3$ such that (A_ϵ, H_ϵ) is linearly stable for $\tau < \tau_2$ and $\tau > \tau_3$.

If $K = 1$ and $\eta_0 < 1$, then there are $0 < \tau_4 \leq \tau_5$ such that (A_ϵ, H_ϵ) is linearly stable for $\tau < \tau_4$ and linearly unstable for $\tau > \tau_5$.

Let us explain the statements of Theorem 6.9 by the following remarks.

Remark 6.10 We assume that the condition $(*)$ holds and ϵ is small enough. Then the stability properties of (A_ϵ, H_ϵ) are as follows:

	Case 1	Case 2	Case 3 ($\eta_0 < K$)	Case 3 ($\eta_0 > K$)
$K = 1$, τ small	stable	stable	stable	stable
$K = 1$, τ finite	?	stable	?	?
$K = 1$, τ large	unstable	stable	unstable	stable
$K > 1$, τ small	unstable	stable	unstable	stable
$K > 1$, τ finite	unstable	stable	unstable	?
$K > 1$, τ large	unstable	stable	unstable	stable

Remark 6.11 If condition $(*)$ holds, ϵ is small enough and τ is small enough or large enough, then for $K \geq 2$ there are stability thresholds

$$D_2(\epsilon) > D_3(\epsilon) > \cdots > D_K(\epsilon) > \cdots$$

which have the following properties:

$$\text{If } \lim_{\epsilon \to 0} \frac{D_K(\epsilon)}{D} > 1, \text{ then the } K\text{-spike solution is stable.}$$

$$\text{If } \lim_{\epsilon \to 0} \frac{D_K(\epsilon)}{D} < 1, \text{ then the } K\text{-spike solution is unstable.}$$

In leading order, the stability thresholds are characterised by

$$D_K(\epsilon) = \frac{|\Omega|}{2\pi K} \log \frac{\sqrt{|\Omega|}}{\epsilon} \quad \text{as } \epsilon \to 0.$$

In particular, if

$$\lim_{\epsilon \to 0} \frac{D}{\log(\sqrt{|\Omega|}/\epsilon)} = 0 \quad \text{as } \epsilon \to 0$$

then for $\epsilon = \epsilon(K)$ small enough the K-spike solution is stable ($K = 1, 2, \ldots$).

Remark 6.12 Cases 1 and 3 with $\eta_0 < K$ behave like the shadow system $D = \infty$, whereas Cases 2 and 3 with $\eta_0 > K$ resemble the strong coupling case $D = O(1)$ as $\epsilon \to 0$.

Remark 6.13 Condition (∗) is generic with respect to the positions $\mathbf{P}^0 \in \overline{\Lambda_\delta}$ which can be seen as follows: The functional $F(\mathbf{P})$ has a global maximum at some $\mathbf{P}^0 \in \overline{\Lambda_\delta}$ in any bounded smooth domain Ω since then $F(\mathbf{P})$ goes to $-\infty$ as $|P_i - P_j|$ or $d(P_i, \partial\Omega)$ tends to 0. The global maximum point $\mathbf{P}^0$ is a critical point of $F(\mathbf{P})$ and supposing in addition that $\mathbf{P}^0$ is a nondegenerate critical point of $F(\mathbf{P})$, which is a generic property, the matrix $M(\mathbf{P}^0)$ has only negative eigenvalues. Numerical computations for the critical points of $F(\mathbf{P})$ can be found in [111].

Let us compare the two-dimensional results with the one-dimensional case (see Chap. 4): Then in leading order the critical thresholds $D_K(\epsilon) = D_K$ ($K = 2, 3, \dots$) are independent of ϵ. Moreover, they are derived from the small eigenvalues and are coupled to the peak locations. In contrast, in two dimensions we have that $D_K(\epsilon) \to +\infty$ as $\epsilon \to 0$ and the thresholds are obtained from the large eigenvalues. Further, in two dimensions these thresholds are independent of the peak locations.

6.2.1 Large Eigenvalues

We consider the (linearised) stability of the solution (A_ϵ, H_ϵ) given in Theorem 6.1.

Linearising the system (1.5) around the equilibrium states (A_ϵ, H_ϵ) we get the following eigenvalue problem for $(\phi_\epsilon, \psi_\epsilon) \in H^2_N(\Omega_\epsilon) \times H^2_N(\Omega)$:

$$\begin{cases} \Delta_y \phi_\epsilon - \phi_\epsilon + 2\frac{A_\epsilon}{H_\epsilon}\phi_\epsilon - \frac{A_\epsilon^2}{H_\epsilon^2}\psi_\epsilon = \lambda_\epsilon \phi_\epsilon, \\ \frac{1}{\beta^2}\Delta\psi_\epsilon - \psi_\epsilon + 2A_\epsilon\phi_\epsilon = \tau\lambda_\epsilon\psi_\epsilon, \end{cases} \tag{6.74}$$

where $D = \frac{1}{\beta^2}$ and $\lambda_\epsilon \in \mathbb{C}$.

Setting

$$\hat{A}_\epsilon = \xi_\epsilon^{-1} A_\epsilon = A_{\epsilon, \mathbf{P}^\epsilon} + \Phi_{\epsilon, \mathbf{P}^\epsilon}, \qquad \hat{H}_\epsilon = \xi_\epsilon^{-1} H_\epsilon = H_{\epsilon, \mathbf{P}^\epsilon} + \Psi_{\epsilon, \mathbf{P}^\epsilon}, \tag{6.75}$$

(6.74) can be written as

$$\begin{cases} \Delta_y \phi_\epsilon - \phi_\epsilon + 2\frac{A_{\epsilon, \mathbf{P}^\epsilon}}{H_{\epsilon, \mathbf{P}^\epsilon}}\phi_\epsilon - \frac{A_{\epsilon, \mathbf{P}^\epsilon}^2}{H_{\epsilon, \mathbf{P}^\epsilon}^2}\psi_\epsilon = \lambda_\epsilon \phi_\epsilon, \\ \frac{1}{\beta^2}\Delta\psi_\epsilon - \psi_\epsilon + 2\xi_\epsilon A_{\epsilon, \mathbf{P}^\epsilon}\phi_\epsilon = \tau\lambda_\epsilon\psi_\epsilon. \end{cases} \tag{6.76}$$

In this subsection, we study the large eigenvalues, i.e., we assume that $|\lambda_\epsilon| \geq c > 0$ for ϵ small. Choosing c small enough, we may assume without loss of generality that $(1 + \tau)c < \frac{1}{2}$. If $\text{Re}(\lambda_\epsilon) \leq -c$, we are done, and so we assume without loss of generality that $\text{Re}(\lambda_\epsilon) \geq -c$. Choosing a suitable subsequence $\epsilon \to 0$, we have $\lambda_\epsilon \to \lambda_0 \neq 0$. The limiting eigenvalue problem consists of a system of NLEPs.

The second equation of (6.76) can be written as

$$\Delta\psi_\epsilon - \beta^2(1 + \tau\lambda_\epsilon)\psi_\epsilon + 2\beta^2\xi_\epsilon\hat{A}_\epsilon\phi_\epsilon = 0. \tag{6.77}$$

Next we define

$$\beta_{\lambda_\epsilon} = \beta\sqrt{1+\tau\lambda_\epsilon},$$

taking the principal part of the square root. This implies that the real part of $\sqrt{1+\tau\lambda_\epsilon}$ is positive, which is possible since $\mathrm{Re}(1+\tau\lambda_\epsilon) \geq \frac{1}{2}$.

By scaling the eigenfunction we have

$$\|\phi_\epsilon\|_{H^2(\Omega_\epsilon)} = 1. \tag{6.78}$$

Using cut-off functions, we define

$$\phi_{\epsilon,j}(y) = \phi_\epsilon(y)\chi_{\epsilon,P_j^\epsilon}(\epsilon y), \tag{6.79}$$

where $\chi_{\epsilon,P_j^\epsilon}(x)$ is given in (6.17).

From (6.76) and Lemma 6.6, using the facts that $\mathrm{Re}(\lambda_\epsilon) \geq -c$ and w decays exponentially (see (6.12)), we derive

$$\phi_\epsilon = \sum_{j=1}^K \phi_{\epsilon,j} + \text{e.s.t.} \quad \text{in } H^2(\Omega_\epsilon). \tag{6.80}$$

Using a standard procedure (see for example Sect. 7.12 in [74]), we extend $\phi_{\epsilon,j}$ to a function defined on $\mathbb{R}^2$ such that

$$\|\phi_{\epsilon,j}\|_{H^2(\mathbb{R}^2)} \leq C\|\phi_{\epsilon,j}\|_{H^2(\Omega_\epsilon)}, \quad j = 1, \ldots, K.$$

Since $\|\phi_\epsilon\|_{H^2(\Omega_\epsilon)} = 1$, $\|\phi_{\epsilon,j}\|_{H^2(\Omega_\epsilon)} \leq C$, for a subsequence of ϵ we have $\phi_{\epsilon,j} \to \phi_j$ as $\epsilon \to 0$ in $H^2(E)$ for $j = 1, \ldots, K$ uniformly for any compact subset E of $\mathbb{R}^2$. This implies

$$w\phi_{\epsilon,j} \to w\phi_j \text{ strongly in } L^\infty(\mathbb{R}^2) \quad \text{as } \epsilon \to 0.$$

We have by (6.77)

$$\psi_\epsilon(x) = 2\beta^2 \xi_\epsilon \int_\Omega G_{\beta_{\lambda_\epsilon}}(x,\xi)\hat{A}_\epsilon\left(\frac{\xi}{\epsilon}\right)\phi_\epsilon\left(\frac{\xi}{\epsilon}\right)d\xi. \tag{6.81}$$

For $x = P_i^\epsilon$ ($i = 1, \ldots, K$), we get

$$\psi_\epsilon(P_j^\epsilon) = 2\beta^2 \int_\Omega G_{\beta_{\lambda_\epsilon}}(P_i^\epsilon,\xi)\sum_{j=1}^K \xi_\epsilon w\left(\frac{\xi-P_j^\epsilon}{\epsilon}\right)\phi_{\epsilon,j}\left(\frac{\xi}{\epsilon}\right)d\xi + o(1)$$

$$= 2\beta^2 \int_\Omega\left(\frac{(\beta_{\lambda_\epsilon})^{-2}}{|\Omega|} + G_0(P_i^\epsilon,\xi) + O(|\beta_{\lambda_\epsilon}|^2)\right)$$

$$\times \sum_{j=1}^K \xi_\epsilon w\left(\frac{\xi-P_j^\epsilon}{\epsilon}\right)\phi_{\epsilon,j}\left(\frac{\xi}{\epsilon}\right)d\xi + o(1)$$

$$= 2 \int_{\Omega} \left(\frac{1}{|\Omega|(1+\tau\lambda_\epsilon)} + \beta^2 G_0(P_i^\epsilon, \xi) + O\left(|\beta_{\lambda_\epsilon}|^4\right) \right)$$

$$\times \xi_\epsilon w\left(\frac{x - P_i^\epsilon}{\epsilon} \right) \phi_{\epsilon,i}\left(\frac{\xi}{\epsilon} \right) d\xi + o(1)$$

$$+ 2 \sum_{j \neq i} \int_{\Omega} \left(\frac{1}{|\Omega|(1+\tau\lambda_\epsilon)} + \beta^2 G_0(P_i^\epsilon, P_j^\epsilon) + O\left(|\beta_{\lambda_\epsilon}|^4\right) \right)$$

$$\times \xi_\epsilon w\left(\frac{\xi - P_j^\epsilon}{\epsilon} \right) \phi_{\epsilon,j}\left(\frac{\xi}{\epsilon} \right) d\xi$$

$$= 2 \sum_{j=1}^{K} \frac{1}{|\Omega|(1+\tau\lambda_\epsilon)} \xi_\epsilon \epsilon^2 \int_{\mathbb{R}^2} w(y) \phi_{\epsilon,j}(y) dy$$

$$+ 2\xi_\epsilon \frac{\beta^2}{2\pi} \epsilon^2 \log \frac{\sqrt{|\Omega|}}{\epsilon} \int_{\mathbb{R}^2} w(y) \phi_{\epsilon,i}(y) dy + o(1). \tag{6.82}$$

We consider three cases (compare Sect. 6.1):

Case 1: $\eta_\epsilon \to 0$

Using (6.82), we have

$$\psi_\epsilon\left(P_i^\epsilon\right) = 2 \sum_{j=1}^{K} \frac{1}{|\Omega|(1+\tau\lambda_\epsilon)} \xi_\epsilon \epsilon^2 \int_{\mathbb{R}^2} w\phi_{\epsilon,j}(1+o(1)). \tag{6.83}$$

We put (6.83) into (6.76). We employ (6.24). Then for $\epsilon \to 0$ we get the nonlocal eigenvalue problem (NLEP)

$$\Delta\phi_i - \phi_i + 2w\phi_i - \frac{2 \sum_{j=1}^{K} \int_{\mathbb{R}^2} w\phi_j}{K(1+\tau\lambda_0) \int_{\mathbb{R}^2} w^2} w^2 = \lambda_0\phi_i, \quad i = 1, \ldots, K. \tag{6.84}$$

If $K = 1$ and $\tau < \tau_1$, Theorem 3.6 implies that problem (6.84) is stable and so the large eigenvalues of (6.76) have negative real part.

If $\tau > \tau_1$, Theorem 3.6 implies that problem (6.84) possesses an eigenvalue λ_0 with $\text{Re}(\lambda_0) \geq a_0 > 0$. We claim that problem (6.76) has an eigenvalue λ_ϵ with $\lambda_\epsilon = \lambda_0 + o(1)$, which implies that problem (6.76) is unstable. To this end, we follow the argument given in Sect. 2 of [37] similarly to Chap. 4.

Then case $K = 1$ follows immediately.

If $K > 1$, we see that problem (6.84) has a positive eigenvalue $\lambda_0 = \mu_1$. We can construct an eigenfunction as follows:

$$\phi_1 = -\phi_2 = \Phi_0, \qquad \phi_3 = \cdots = \phi_K = 0,$$

where Φ_0 is the principal eigenfunction of L_0 given in Lemma 13.5.

We conclude that there is an eigenvalue of (6.76) with positive real part. The behaviour in this case is the same as for the shadow system: All multi-spike solutions are unstable.

Case 2: $\eta_\epsilon \to \infty$

From (6.82), we get

$$\psi_\epsilon\left(P_i^\epsilon\right) = 2\xi_\epsilon \frac{\eta_\epsilon}{|\Omega|}\epsilon^2 \int_{\mathbb{R}^2} w\phi_{\epsilon,i}\left(1 + o(1)\right). \tag{6.85}$$

Taking the limit $\epsilon \to 0$ (for any $\tau \geq 0$ fixed), we derive the NLEP:

$$\Delta\phi_i - \phi_i + 2w\phi_i - \frac{2\int_{\mathbb{R}^2} w\phi_i}{\int_{\mathbb{R}^2} w^2}w^2 = \lambda_0\phi_i, \quad i = 1, \ldots, K. \tag{6.86}$$

By Theorem 3.6, all eigenvalues of (6.86) are stable.

To summarise, assuming that $\eta_\epsilon \to \infty$, then for a K-spike solution all large eigenvalues have negative real part.

Case 3: $\eta_\epsilon \to \eta_0$

From (6.82), we have

$$\psi_\epsilon\left(P_i^\epsilon\right) = \left(2\sum_{j=1}^K \frac{1}{|\Omega|(1 + \tau\lambda_0)}\xi_\epsilon\epsilon^2 \int_{\mathbb{R}^2} w\phi_{\epsilon,j} + 2\xi_\epsilon\frac{\eta_0}{|\Omega|}\epsilon^2 \int_{\mathbb{R}^2} w\phi_{\epsilon,i}\right)$$
$$\times \left(1 + o(1)\right). \tag{6.87}$$

Taking the limit $\epsilon \to 0$ we obtain the NLEP

$$\Delta\phi_i - \phi_i + 2w\phi_i - \frac{2[(1 + \eta_0(1 + \tau\lambda_0))\int_{\mathbb{R}^2} w\phi_i + \sum_{j\neq i}\int_{\mathbb{R}^2} w\phi_j]}{(K + \eta_0)(1 + \tau\lambda_0)\int_{\mathbb{R}^2} w^2}w^2$$
$$= \lambda_0\phi_i, \quad i = 1, \ldots, K. \tag{6.88}$$

We introduce the matrix

$$\mathcal{G} = \begin{pmatrix} 1 + \eta_0(1 + \tau\lambda_0) & 1 & \cdots & 1 \\ 1 & 1 + \eta_0(1 + \tau\lambda_0) & \cdots & 1 \\ \vdots & & & \vdots \\ 1 & \cdots & \cdots & 1 + \eta_0(1 + \tau\lambda_0) \end{pmatrix}$$

and note that $\mathcal{G}$ is symmetric and the eigenvalues of $\mathcal{G}$ are given by

$$\rho_1 = \cdots = \rho_{K-1} = \eta_0(1 + \tau\lambda_0), \qquad \rho_K = K + \eta_0(1 + \tau\lambda_0).$$

Let P be an orthogonal matrix which satisfies

$$PGP^{-1} = \begin{pmatrix} \eta_0(1+\tau\lambda_0) & 0 & \cdots & 0 \\ 0 & \eta_0(1+\tau\lambda_0) & \cdots & 0 \\ 0 & \cdots & \eta_0(1+\tau\lambda_0) & 0 \\ 0 & \cdots & 0 & K+\eta_0(1+\tau\lambda_0) \end{pmatrix}.$$

We introduce the notation

$$\Phi = \begin{pmatrix} \phi_1 \\ \vdots \\ \phi_K \end{pmatrix}.$$

Then (6.88) implies

$$\Delta\Phi - \Phi + 2w\Phi - \frac{2\mathcal{G}\int_{\mathbb{R}^2}\Phi w}{(K+\eta_0)(1+\tau\lambda_0)\int_{\mathbb{R}^2}w^2}w^2 = \lambda_0\Phi.$$

Letting $P\Phi = \bar{\Phi}$, we have

$$\Delta\bar{\Phi} - \bar{\Phi} + 2w\bar{\Phi} - \frac{2}{(K+\eta_0)(1+\tau\lambda_0)\int_{\mathbb{R}^2}w^2}$$

$$\times \begin{pmatrix} \eta_0(1+\tau\lambda_0) & 0 & \cdots & 0 \\ 0 & \eta_0(1+\tau\lambda_0) & \cdots & 0 \\ 0 & \cdots & \eta_0(1+\tau\lambda_0) & 0 \\ 0 & \cdots & 0 & K+\eta_0(1+\tau\lambda_0) \end{pmatrix}$$

$$\times \left(\int_{\mathbb{R}^2}w\bar{\Phi}\right)w^2 = \lambda_0\bar{\Phi}.$$

In scalar notation we get

$$\Delta\bar{\Phi}_i - \bar{\Phi}_i + 2w\bar{\Phi}_i - \frac{2\rho_i}{(K+\eta_0)(1+\tau\lambda_0)\int_{\mathbb{R}^2}w^2}\left(\int_{\mathbb{R}^2}w(y)\bar{\Phi}_i(y)dy\right)w^2$$

$$= \lambda_0\bar{\Phi}_i, \quad i=1,\dots,K. \tag{6.89}$$

In case $i=1,\dots,K-1$ we have

$$\Delta\bar{\Phi}_i - \bar{\Phi}_i + 2w\bar{\Phi}_i - \frac{2\eta_0}{(K+\eta_0)\int_{\mathbb{R}^2}w^2}\left(\int_{\mathbb{R}^2}w(y)\bar{\Phi}_i(y)dy\right)w^2$$

$$= \lambda_0\bar{\Phi}_i, \quad i=1,\dots,K-1. \tag{6.90}$$

In case $i=K$ we get

$$\Delta\bar{\Phi}_K - \bar{\Phi}_K + 2w\bar{\Phi}_K - \frac{2(K+\eta_0(1+\tau\lambda_0))}{(K+\eta_0)(1+\tau\lambda_0)\int_{\mathbb{R}^2}w^2}\left(\int_{\mathbb{R}^2}w(y)\bar{\Phi}_K(y)dy\right)w^2$$

$$= \lambda_0\bar{\Phi}_K. \tag{6.91}$$

If $K > 1$ and $\frac{2\eta_0}{K+\eta_0} < 1$ (i.e. $\eta_0 < K$), then by Theorem 3.6, problem (6.90) is unstable for all $\tau \geq 0$. Thus problem (6.76) is (linearly) unstable for all $\tau \geq 0$.

If $K \geq 1$ and $\frac{2\eta_0}{K+\eta_0} > 1$, or, equivalently, $\eta_0 > K$, Theorem 3.6 implies that problem (6.90) is stable. If $0 \leq \tau < \tau_2$ or $\tau > \tau_3$ for suitable $\tau_2 \leq \tau_3$, it is a consequence of Theorem 3.9 that (6.91) is stable.

If $K = 1$ and $\eta_0 < 1$, the only problem to consider is (6.91). Theorem 3.9 implies that there are positive constants $\tau_4 \leq \tau_5$ such that problem (6.91) is stable for $0 \leq \tau < \tau_4$ and unstable for $\tau > \tau_5$,

Thus we have proved the part of Theorem 6.1 concerning large eigenvalues.

6.2.2 Small Eigenvalues

In this subsection, we study (6.76) in the case of small eigenvalues, i.e. we assume that they are of order $o(1)$ as $\epsilon \to 0$. We will show that the small eigenvalues are related to the matrix $M(\mathbf{P}^0)$ introduced in (6.5).

Let us assume that condition $(*)$ is true, i.e. the matrix $M(\mathbf{P}^0)$ has only negative eigenvalues. We will show that if $\lambda_\epsilon \to 0$ then

$$\lambda_\epsilon \sim \epsilon^2 k(\epsilon, \beta)\sigma_0, \tag{6.92}$$

where σ_0 is an eigenvalue of $M(\mathbf{P}^0)$. Then (6.92) implies that all small eigenvalues of $\mathcal{L}_\epsilon$ have negative real part.

Let (A_ϵ, H_ϵ) be the steady state of (6.7) given in Theorem 6.1. We also consider the rescaled steady state $(\hat{A}_\epsilon, \hat{H}_\epsilon)$ stated in (6.75). Using cut-off functions, we introduce

$$\hat{A}_{\epsilon,j}(y) = \chi_{\epsilon,P_j^\epsilon}(\epsilon y)\hat{A}_\epsilon(y), \quad j = 1, \ldots, K, \tag{6.93}$$

where $\chi_{\epsilon,P_j^\epsilon}$ is given in (6.17). Then we have

$$\hat{A}_\epsilon(y) = \sum_{j=1}^{K} \hat{A}_{\epsilon,j}(y) + \text{e.s.t.} \quad \text{in } H_N^2(\Omega_\epsilon). \tag{6.94}$$

Motivated by the decomposition in Liapunov-Schmidt reduction, we compute

$$\phi_\epsilon = \sum_{j=1}^{K}\sum_{k=1}^{2} a_{j,k}^\epsilon \frac{\partial \hat{A}_{\epsilon,j}}{\partial y_k} + \phi_\epsilon^\perp, \tag{6.95}$$

where $a_{j,k}^\epsilon$ are complex numbers and

$$\phi_\epsilon^\perp \perp \tilde{\mathcal{K}}_\epsilon := \text{span}\left\{\frac{\partial \hat{A}_{\epsilon,j}}{\partial y_k} : j = 1, \ldots, K, k = 1, 2\right\} \subset H_N^2(\Omega_\epsilon). \tag{6.96}$$

Similarly,

$$\psi_\epsilon(x) = \sum_{j=1}^{K}\sum_{k=1}^{2} a_{j,k}^\epsilon \psi_{\epsilon,j,k} + \psi_\epsilon^\perp, \tag{6.97}$$

where $\psi_{\epsilon,j,k}$ is the unique solution of the problem

$$\begin{cases} \frac{1}{\beta^2}\Delta_x \psi_{\epsilon,j,k} - (1+\tau\lambda_\epsilon)\psi_{\epsilon,j,k} + 2\xi_\epsilon \hat{A}_\epsilon, \quad \frac{\partial \hat{A}_{\epsilon,j}}{\partial y_k} = 0 \text{ in } \Omega, \\ \frac{\partial \psi_{\epsilon,j,k}}{\partial \nu} = 0 \quad \text{on } \partial\Omega, \end{cases} \tag{6.98}$$

and $\psi_\epsilon^\perp$ is given by

$$\begin{cases} \frac{1}{\beta^2}\Delta_x \psi_\epsilon^\perp - (1+\tau\lambda_\epsilon)\psi_\epsilon^\perp + 2\xi_\epsilon \hat{A}_\epsilon \phi_\epsilon^\perp = 0 \quad \text{in } \Omega, \\ \frac{\partial \psi_\epsilon^\perp}{\partial \nu} = 0 \quad \text{on } \partial\Omega. \end{cases} \tag{6.99}$$

Suppose that $\|\phi_\epsilon\|_{H^2(\Omega_\epsilon)} = 1$. Then $|a_{j,k}^\epsilon| \leq C$ since

$$a_{j,k}^\epsilon = \frac{\int_{\Omega_\epsilon} \phi_\epsilon(\partial \hat{A}_{\epsilon,j}/\partial y_k)}{\int_{\mathbb{R}^2}(\partial w/\partial y_1)^2} + o(1).$$

We put the decompositions of ϕ_ϵ and ψ_ϵ into (6.76) and get

$$\epsilon \sum_{j=1}^{K}\sum_{k=1}^{2} a_{j,k}^\epsilon \frac{(\hat{A}_{\epsilon,j})^2}{(\hat{H}_\epsilon)^2}\left[-\frac{1}{\epsilon}\psi_{\epsilon,j,k} + \frac{\partial \hat{H}_\epsilon}{\partial x_k}\right] + \text{e.s.t.}$$

$$+ \Delta_y \phi_\epsilon^\perp - \phi_\epsilon^\perp + 2\frac{\hat{A}_\epsilon}{\hat{H}_\epsilon}\phi_\epsilon^\perp - \frac{(\hat{A}_\epsilon)^2}{(\hat{H}_\epsilon)^2}\psi_\epsilon^\perp - \lambda_\epsilon \phi_\epsilon^\perp$$

$$= \lambda_\epsilon \sum_{j=1}^{K}\sum_{k=1}^{2} a_{j,k}^\epsilon \frac{\partial \hat{A}_{\epsilon,j}}{\partial y_k} \quad \text{in } \Omega_\epsilon. \tag{6.100}$$

We define

$$I_3 := \epsilon \sum_{j=1}^{K}\sum_{k=1}^{2} a_{j,k}^\epsilon \frac{(\hat{A}_{\epsilon,j})^2}{(\hat{H}_\epsilon)^2}\left[-\frac{1}{\epsilon}\psi_{\epsilon,j,k} + \frac{\partial \hat{H}_\epsilon}{\partial x_k}\right] \tag{6.101}$$

and

$$I_4 := \Delta_y \phi_\epsilon^\perp - \phi_\epsilon^\perp + 2\frac{\hat{A}_\epsilon}{\hat{H}_\epsilon}\phi_\epsilon^\perp - \frac{(\hat{A}_\epsilon)^2}{(\hat{H}_\epsilon)^2}\psi_\epsilon^\perp - \lambda_\epsilon \phi_\epsilon^\perp. \tag{6.102}$$

Now the proof will proceed in two steps. Firstly, we will derive an error estimate for $\phi_\epsilon^\perp$. Secondly, we will obtain algebraic equations for $a_{j,k}^\epsilon$ using the matrix $M(\mathbf{P}^0)$.

Step 1: Estimates for $\phi_\epsilon^\perp$.

We will derive an error bound for $\phi_\epsilon^\perp$.

Using equation (6.100) and the fact that $\phi_\epsilon^\perp \perp \tilde{\mathcal{K}}_\epsilon$, then similarly to the proof of Proposition 2.7 we get

$$\|\phi_\epsilon^\perp\|_{H^2(\Omega_\epsilon)} \leq C \|I_3\|_{L^2(\Omega_\epsilon)}. \tag{6.103}$$

Next we compute I_3. Recalling that ξ_ϵ and $k(\epsilon, \beta)$ have been defined in Theorem 6.1, then for $x \in B_\delta(P_l^\epsilon)$ we get

$$\frac{\partial \hat{H}_\epsilon}{\partial x_k}(x) = \xi_\epsilon \beta^2 \int_\Omega \frac{\partial}{\partial x_k} G_\beta(x, \xi) \left(\hat{A}_\epsilon\left(\frac{\xi}{\epsilon}\right)\right)^2 d\xi$$

$$= \xi_\epsilon \beta^2 \left(\int_\Omega \frac{\partial}{\partial x_k} (K_0(|x - \xi|) - H_0(x, \xi)) \left(\hat{A}_{\epsilon,l}\left(\frac{\xi}{\epsilon}\right)\right)^2 d\xi\right.$$

$$\left. + \int_\Omega \sum_{s \neq l} \frac{\partial}{\partial x_k} G_0(x, \xi) \left(\hat{A}_{\epsilon,s}\left(\frac{\xi}{\epsilon}\right)\right)^2 d\xi + O(\beta^4 \epsilon^2)\right).$$

Using (6.16), we have

$$\psi_{\epsilon,l,k}(x) = 2\beta^2 \xi_\epsilon \int_\Omega G_{\beta_{\lambda_\epsilon}}(x, z) \hat{A}_{\epsilon,l} \frac{\partial \hat{A}_{\epsilon,l}}{\partial y_k} dz$$

$$= \epsilon \xi_\epsilon \beta^2 \int_\Omega (K_0(|x - \xi|) - H_0(x, \xi) + O(\beta^2)) \frac{\partial}{\partial \xi_k} (\hat{A}_{\epsilon,l})^2 d\xi.$$

Hence, for $x \in B_\delta(P_l^\epsilon)$, we get

$$\frac{\partial \hat{H}_\epsilon}{\partial x_k}(x) - \frac{1}{\epsilon} \psi_{\epsilon,l,k}(x)$$

$$= \xi_\epsilon \beta^2 \left[\left(\int_\Omega \left[\frac{\partial}{\partial x_k} K_0(|x - \xi|) \left(\hat{A}_{\epsilon,l}\left(\frac{\xi}{\epsilon}\right)\right)^2\right.\right.\right.$$

$$\left.\left. - K_0(|x - \xi|) \frac{\partial}{\partial \xi_k} \left(\hat{A}_{\epsilon,l}\left(\frac{\xi}{\epsilon}\right)\right)^2\right] d\xi\right)$$

$$- \int_\Omega \left[\frac{\partial}{\partial x_k} H_0(x, \xi) \left(\hat{A}_{\epsilon,l}\left(\frac{\xi}{\epsilon}\right)\right)^2 - H_0(x, \xi) \frac{\partial}{\partial \xi_k} \left(\hat{A}_{\epsilon,l}\left(\frac{\xi}{\epsilon}\right)\right)^2\right] d\xi$$

$$\left. + \int_\Omega \sum_{s \neq l} \frac{\partial}{\partial x_k} G_0(x, \xi) \left(\hat{A}_{\epsilon,s}\left(\frac{\xi}{\epsilon}\right)\right)^2 d\xi + O(\epsilon^2 \beta^4)\right].$$

By the radial symmetry of $K(|x|)$, we have

$$\frac{\partial \hat{H}_\epsilon}{\partial x_k}(x) - \frac{1}{\epsilon} \psi_{\epsilon,l,k}(x) = k(\epsilon, \beta) \int_{\mathbb{R}^2} w^2 \left(-\frac{\partial}{\partial x_k} F_l(x) + o(\epsilon)\right), \tag{6.104}$$

where

$$F_l(x) = H_0(x, P_l^\epsilon) - \sum_{j \neq l} G_0(x, P_j^\epsilon).\tag{6.105}$$

Since $\mathbf{P}^\epsilon \to \mathbf{P}^0$ and $\mathbf{P}^0$ is a critical point of $F(\mathbf{P})$, we have

$$\frac{\partial}{\partial x_m} F_l(x)\bigg|_{x=P_l^\epsilon} = o(1).$$

Thus we get

$$\|I_3\|_{L^2(\Omega_\epsilon)} = o\left(\epsilon k(\epsilon, \beta) \sum_{j=1}^K \sum_{k=1}^2 |a_{j,k}^\epsilon|\right)\tag{6.106}$$

and

$$\|\phi_\epsilon^\perp\|_{H^2(\Omega_\epsilon)} \leq C \|I_3\|_{L^2(\Omega_\epsilon)} = o\left(\epsilon k(\epsilon, \beta) \sum_{j=1}^K \sum_{k=1}^2 |a_{j,k}^\epsilon|\right).\tag{6.107}$$

By the equation for $\psi_\epsilon^\perp$ and relation (6.107), we have

$$\psi_\epsilon^\perp(x) = o\left(\epsilon k(\epsilon, \beta) \sum_{j=1}^K \sum_{k=1}^2 |a_{j,k}^\epsilon|\right).\tag{6.108}$$

We compute

$$\int_{\Omega_\epsilon} I_4 \frac{\partial \hat{A}_{\epsilon,l}}{\partial y_m} d\xi = \int_{\Omega_\epsilon} \frac{\hat{A}_{\epsilon,l}^2}{H_\epsilon^2}\left(\epsilon \frac{\partial \hat{H}_\epsilon}{\partial x_m} \phi_\epsilon^\perp - \frac{\partial \hat{A}_{\epsilon,l}}{\partial y_m} \psi_\epsilon^\perp\right) d\xi - \lambda_\epsilon \int_{\Omega_\epsilon} \phi_\epsilon^\perp \frac{\partial \hat{A}_{\epsilon,l}}{\partial y_m} d\xi$$

$$= \int_{\Omega_{\epsilon,P_l^\epsilon}} \frac{\hat{A}_{\epsilon,l}^2}{\hat{H}_\epsilon^2}\left(\epsilon \frac{\partial \hat{H}_\epsilon}{\partial x_m}(P_l^\epsilon + \epsilon y) - \epsilon \frac{\partial \hat{H}_\epsilon}{\partial x_m}(P_l^\epsilon)\right)\phi_\epsilon^\perp$$

$$+ \int_{\Omega_{\epsilon,P_l^\epsilon}} \frac{\hat{A}_{\epsilon,l}^2}{\hat{H}_\epsilon^2}\left(\epsilon \frac{\partial \hat{H}_\epsilon}{\partial x_m}(P_l^\epsilon)\right)\phi_\epsilon^\perp$$

$$- \int_{\Omega_{\epsilon,P_l^\epsilon}} \frac{\hat{A}_{\epsilon,l}^2}{\hat{H}_\epsilon^2} \frac{\partial \hat{A}_{\epsilon,l}}{\partial y_m}\left(\psi_\epsilon^\perp(P_l^\epsilon + \epsilon y) - \psi_\epsilon^\perp(P_l^\epsilon)\right)$$

$$- \lambda_\epsilon \int_{\Omega_{\epsilon,P_l^\epsilon}} \phi_\epsilon^\perp \frac{\partial \hat{A}_{\epsilon,l}}{\partial y_m}$$

$$= o\left(\epsilon^2 k(\epsilon, \beta) \sum_{j=1}^K \sum_{k=1}^2 |a_{j,k}^\epsilon|\right).\tag{6.109}$$

Here we have utilised (6.99) and

$$\frac{\partial \hat{H}_\epsilon}{\partial x_m} = O\big(k(\epsilon, \beta)\big) \quad \text{in } \Omega.$$

Step 2: Algebraic equations for $a^\epsilon_{j,k}$.

In this step we derive algebraic equations for $a^\epsilon_{j,k}$.

We multiply both sides of (6.100) by $\frac{\partial \hat{A}_{\epsilon,l}}{\partial y_m}$ and integrate over Ω_ϵ. This gives

$$\text{r.h.s.} = \lambda_\epsilon \sum_{j=1}^{K} \sum_{k=1}^{2} a^\epsilon_{j,k} \int_{\Omega_{\epsilon,P^\epsilon_l}} \frac{\partial \hat{A}_{\epsilon,j}}{\partial y_k} \frac{\partial \hat{A}_{\epsilon,l}}{\partial y_m}$$

$$= \lambda_\epsilon \sum_{j=1}^{K} \sum_{k=1}^{2} a^\epsilon_{j,k} \delta_{jl} \delta_{km} \int_{\mathbb{R}^2} \left(\frac{\partial w}{\partial y_1} \right)^2 dy (1 + o(1))$$

$$= \lambda_\epsilon a^\epsilon_{l,m} \int_{\mathbb{R}^2} \left(\frac{\partial w}{\partial y_1} \right)^2 dy (1 + o(1)).$$

From (6.104) and (6.109) we have

$$\text{l.h.s.} = \epsilon \sum_{j=1}^{K} \sum_{k=1}^{2} a^\epsilon_{j,k} \int_{\Omega_{\epsilon,P^\epsilon_l}} \frac{(\hat{A}_{\epsilon,j})^2}{(\hat{H}_\epsilon)^2} \left[-\frac{1}{\epsilon} \psi_{\epsilon,j,k} + \frac{\partial \hat{H}_\epsilon}{\partial x_k} \right] \frac{\partial \hat{A}_{\epsilon,l}}{\partial y_m} + \int_{\Omega_{\epsilon,P^\epsilon_l}} \left(I_4 \frac{\partial \hat{A}_{\epsilon,l}}{\partial y_m} \right)$$

$$= \epsilon \sum_{j=1}^{K} \sum_{k=1}^{2} a^\epsilon_{j,k} \int_{\Omega_{\epsilon,P^\epsilon_l}} \frac{(\hat{A}_{\epsilon,j})^2}{(\hat{H}_\epsilon)^2} \left[-\frac{1}{\epsilon} \psi_{\epsilon,j,k} + \frac{\partial \hat{H}_\epsilon}{\partial x_k} \right] \frac{\partial \hat{A}_{\epsilon,l}}{\partial y_m}$$

$$+ o\left(\epsilon^2 k(\epsilon, \beta) \sum_{j=1}^{K} \sum_{k=1}^{2} |a^\epsilon_{j,k}| \right). \tag{6.110}$$

By (6.104), we get

$$\text{l.h.s.} = \epsilon k(\epsilon, \beta) \sum_{j=1}^{K} \sum_{k=1}^{2} a^\epsilon_{j,k} \int_{\Omega_{\epsilon,P^\epsilon_l}} \frac{(\hat{A}_{\epsilon,j})^2}{(\hat{H}_\epsilon)^2} \left(-\frac{\partial}{\partial x_k} F_j(x) \right) \frac{\partial \hat{A}_{\epsilon,l}}{\partial y_m} \int w^2$$

$$+ o\left(\epsilon^2 k(\epsilon, \beta) \sum_{j=1}^{K} \sum_{k=1}^{2} |a^\epsilon_{j,k}| \right)$$

$$= \epsilon^2 k(\epsilon, \beta) \int_{\mathbb{R}^2} w^2 \frac{\partial w}{\partial y_m} y_m \sum_{j=1}^{K} \sum_{k=1}^{2} a^\epsilon_{j,k} \left(-\frac{\partial}{\partial P^\epsilon_{l,m}} \frac{\partial}{\partial P^\epsilon_{j,k}} F(\mathbf{P}^\epsilon) \right)$$

$$+ o\left(\epsilon^2 k(\epsilon, \beta) \sum_{j=1}^{K} \sum_{k=1}^{2} |a^\epsilon_{j,k}| \right). \tag{6.111}$$

Using the identity

$$\int_{\mathbb{R}^2} w^2 \frac{\partial w}{\partial y_m} y_m = -\frac{1}{3} \int_{\mathbb{R}^2} w^3,$$

we compute

$$\text{l.h.s.} = \frac{\epsilon^2 k(\epsilon, \beta)}{3} \left(\int_{\mathbb{R}^2} w^3 \right) \sum_{j=1}^{K} \sum_{k=1}^{2} a_{j,k}^{\epsilon} \left(\frac{\partial}{\partial P_{l,m}^{\epsilon}} \frac{\partial}{\partial P_{j,k}^{\epsilon}} F(\mathbf{P}^{\epsilon}) \right)$$

$$+ o\left(\epsilon^2 k(\epsilon, \beta) \sum_{j=1}^{K} \sum_{k=1}^{2} |a_{j,k}^{\epsilon}| \right). \tag{6.112}$$

Combining l.h.s. and r.h.s., we get

$$\frac{\epsilon^2 k(\epsilon, \beta)}{3} \left(\int_{\mathbb{R}^2} w^3 \right) \sum_{j=1}^{K} \sum_{k=1}^{2} a_{j,k}^{\epsilon} \left(\frac{\partial}{\partial P_{l,m}^{\epsilon}} \frac{\partial}{\partial P_{j,k}^{\epsilon}} F(\mathbf{P}^{\epsilon}) \right)$$

$$+ o\left(\epsilon^2 k(\epsilon, \beta) \sum_{j=1}^{K} \sum_{k=1}^{2} |a_{j,k}^{\epsilon}| \right)$$

$$= \lambda_{\epsilon} a_{l,m}^{\epsilon} \int_{\mathbb{R}^2} \left(\frac{\partial w}{\partial y_1} \right)^2 dy (1 + o(1)). \tag{6.113}$$

Finally (6.113) implies that the small eigenvalues with $\lambda_{\epsilon} \to 0$ satisfy $|\lambda_{\epsilon}| \sim \epsilon^2 k(\epsilon, \beta)$ and we have

$$\frac{\lambda_{\epsilon}}{\epsilon^2 k(\epsilon, \beta)} \to \frac{\int_{\mathbb{R}^2} w^3}{3 \int_{\mathbb{R}^2} (\partial w / \partial y_1)^2 dy} \sigma_0$$

as $\epsilon \to 0$, where σ_0 is an eigenvalue of the matrix $M(\mathbf{P}^0) = \nabla_{\mathbf{P}}^2|_{\mathbf{P}=\mathbf{P}^0}(F(\mathbf{P}))$, and $\mathbf{P}^{\epsilon} \to \mathbf{P}^0$ as $\epsilon \to 0$. Note that the vector $a^{\epsilon} = (a_{1,1}^{\epsilon}, a_{1,2}^{\epsilon}, \ldots, a_{K,2}^{\epsilon})^T$ approaches an eigenvector of $M(\mathbf{P}^0)$ corresponding to σ_0. From condition $(*)$ it follows that the matrix $M(\mathbf{P}^0)$ is negative definite and $\text{Re}(\lambda_{\epsilon}) \leq -c\epsilon^2 k(\epsilon, \beta)$ for some $c > 0$. Thus all the small eigenvalues λ_{ϵ} for (6.76) have negative real part for ϵ small enough.

This concludes the proof of Theorem 6.1.

6.3 Asymmetric Multiple Spikes: Existence

We consider the existence of asymmetric multiple spikes. Recall that

$$\beta^2 = \frac{1}{D}, \quad \eta_{\epsilon} = \frac{\beta^2 |\Omega|}{2\pi} \log \frac{\sqrt{|\Omega|}}{\epsilon},$$

where $|\Omega|$ denotes the area of Ω. For asymmetric multiple spikes, we assume that

$$\lim_{\epsilon \to 0} \eta_\epsilon = \eta_0 \in (0, +\infty). \tag{6.114}$$

It will be shown that this is a necessary condition for asymmetric multiple spikes for ϵ small enough. Note that (6.114) implies that

$$D \to \infty \quad \text{and} \quad \beta \to 0 \quad \text{as } \epsilon \to 0.$$

Explicitly, we assume

$$D \sim \frac{|\Omega| \log(\sqrt{|\Omega|}/\epsilon)}{2\pi \eta_0} \tag{6.115}$$

and

$$\beta^2 \sim \frac{2\pi \eta_0}{|\Omega| \log(\sqrt{|\Omega|}/\epsilon)}. \tag{6.116}$$

We will show that the amplitudes $(\xi_{\epsilon,1}, \ldots, \xi_{\epsilon,K})$ satisfy a nonlinear algebraic system which can be solved explicitly. As a consequence, we will see that the asymmetric patterns are generated by multiple spikes which have exactly two different amplitudes and that more than two different amplitudes are not possible. Using the method of Liapunov-Schmidt reduction we give a rigorous proof of the existence of asymmetric K-spike stationary states. We remark that this will enable us to reduce the infinite-dimensional problem of finding a steady state to (1.5) to the finite-dimensional problem of locating the K points at which the spikes concentrate. We will state a condition for these K points in terms of a Green's function and its derivatives.

For an integer $K \geq 2$, let $k_1, k_2 \geq 1$ be two integers such that

$$k_1 + k_2 = K. \tag{6.117}$$

Further, we assume that η_0 given in (6.114) satisfies

$$\eta_0 > 2\sqrt{k_1 k_2}. \tag{6.118}$$

Setting

$$\rho_+ = \frac{2k_2 + \eta_0 + \sqrt{\eta_0^2 - 4k_1 k_2}}{2\eta_0(\eta_0 + K)}, \qquad \rho_- = \frac{2k_2 + \eta_0 - \sqrt{\eta_0^2 - 4k_1 k_2}}{2\eta_0(\eta_0 + K)}, \tag{6.119}$$

$$\eta_+ = \frac{2k_1 + \eta_0 - \sqrt{\eta_0^2 - 4k_1 k_2}}{2\eta_0(\eta_0 + K)}, \qquad \eta_- = \frac{2k_1 + \eta_0 + \sqrt{\eta_0^2 - 4k_1 k_2}}{2\eta_0(\eta_0 + K)}, \tag{6.120}$$

we have

$$\rho_+ + \eta_+ = \frac{1}{\eta_0}, \qquad \rho_- + \eta_- = \frac{1}{\eta_0}. \tag{6.121}$$

We select $(\rho, \eta) = (\rho_+, \eta_+)$ or $(\rho, \eta) = (\rho_-, \eta_-)$ and drop the subscript "$\pm$" if this can be done without causing confusion.

Next we set $(\hat{\xi}_1, \ldots, \hat{\xi}_K) \in R_+^K$, where

$$\hat{\xi}_j \in \{\rho, \eta\}, \text{ and } \rho \text{ appears in } (\hat{\xi}_1, \ldots, \hat{\xi}_K) \ k_1 \text{ times.} \tag{6.122}$$

Note that η appears in $(\hat{\xi}_1, \ldots, \hat{\xi}_K)$ k_2 times.

For $\mathbf{P} \in \overline{\Lambda_\delta}$ let

$$F(\mathbf{P}) = \sum_{k=1}^{K} H_0(P_k, P_k)\hat{\xi}_k^4 - \sum_{i,j=1,\ldots,K, i\neq j} G_0(P_i, P_j)\hat{\xi}_i^2\hat{\xi}_j^2 \tag{6.123}$$

and

$$M(\mathbf{P}) = \nabla_{\mathbf{P}}^2 F(\mathbf{P}). \tag{6.124}$$

We remark that $F(\mathbf{P}) \in C^\infty(\overline{\Lambda_\delta})$.

The result on the existence of asymmetric K-spike solutions can be stated as follows:

Theorem 6.14 *For an integer $K \geq 2$, let $k_1, k_2 \geq 1$ be two integers such that $k_1 + k_2 = K$. Let*

$$\beta^2 = \frac{1}{D}, \quad \eta_\epsilon = \frac{\beta^2 |\Omega|}{2\pi} \log \frac{\sqrt{|\Omega|}}{\epsilon}.$$

Assume that (6.114) and (6.118) hold. Further, assume that

$$\eta_0 \neq K \tag{6.125}$$

and let

$$\mathbf{P}^0 = \left(P_1^0, \ldots, P_K^0 \right) \in \overline{\Lambda_\delta} \text{ be a nondegenerate critical point of } F(\mathbf{P}), \tag{6.126}$$

where $F(\mathbf{P})$ is given in (6.123).

Then, for ϵ small enough, problem (2.2) has a steady state (A_ϵ, H_ϵ) with the following properties:

(1) $A_\epsilon(x) = \sum_{j=1}^{K} \xi_{\epsilon,j} w(\frac{x-P_j^\epsilon}{\epsilon})(1 + O(\frac{1}{D}))$ *uniformly for $x \in \bar{\Omega}$, where w is the unique solution of (6.9) and*

$$\xi_{\epsilon,j} = \xi_\epsilon \hat{\xi}_{\epsilon,j}, \quad \xi_\epsilon = \frac{|\Omega|}{\epsilon^2 \int_{\mathbb{R}^2} w^2}. \tag{6.127}$$

Further, $(\hat{\xi}_{\epsilon,1}, \ldots, \hat{\xi}_{\epsilon,K}) \to (\hat{\xi}_1, \ldots, \hat{\xi}_K)$ which has been introduced in (6.122).

(2) $H_\epsilon(P_j^\epsilon) = \xi_{\epsilon,j}(1 + \frac{1}{D})$ *for $j = 1, \ldots, K$.*

(3) $P_j^\epsilon \to P_j^0$ *as $\epsilon \to 0$ for $j = 1, \ldots, K$.*

Remark 6.15 The technical condition (6.125) is needed for Liapunov-Schmidt reduction to control the kernels of the linearised operator. The technical condition (6.126) will be essential in the Liapunov-Schmidt reduction to guarantee the solvability of the reduced problem.

6.3.1 Analysing the Algebraic System for the Amplitudes

Taking the limit $\epsilon \to 0$ in (6.22), we have

$$\hat{\xi}_i = \sum_{j=1}^{K} \hat{\xi}_j^2 + \hat{\xi}_i^2 \eta_0, \quad i = 1, \ldots, K. \tag{6.128}$$

In this subsection, we compute solutions of $\hat{\xi}_i$, $i = 1, \ldots, K$, of (6.128). Using the notation

$$\rho(t) = t - \eta_0 t^2, \tag{6.129}$$

(6.128) can be rewritten as

$$\rho(\hat{\xi}_i) = \sum_{j=1}^{K} \hat{\xi}_j^2, \quad i = 1, \ldots, K \tag{6.130}$$

which implies

$$\rho(\hat{\xi}_i) = \rho(\hat{\xi}_j) \quad \text{for } i \neq j. \tag{6.131}$$

Further, we get

$$(\hat{\xi}_i - \hat{\xi}_j)(1 - \eta_0(\hat{\xi}_i + \hat{\xi}_j)) = 0. \tag{6.132}$$

Hence for $i \neq j$ we have

$$\hat{\xi}_i - \hat{\xi}_j = 0 \quad \text{or} \quad \hat{\xi}_i + \hat{\xi}_j = \frac{1}{\eta_0}. \tag{6.133}$$

Now we assume that the solutions are asymmetric, i.e., there is an $i \in \{2, \ldots, N\}$ such that $\hat{\xi}_i \neq \hat{\xi}_1$. Without loss of generality, we may choose

$$\hat{\xi}_2 \neq \hat{\xi}_1$$

and so

$$\hat{\xi}_1 + \hat{\xi}_2 = \frac{1}{\eta_0}. \tag{6.134}$$

We calculate $\hat{\xi}_j$, $j = 3, \ldots, K$. If $\hat{\xi}_j \neq \hat{\xi}_1$, then by (6.133), $\hat{\xi}_j + \hat{\xi}_1 = \frac{1}{\eta_0}$, which implies that $\hat{\xi}_j = \hat{\xi}_2$.

Thus for $j \geq 3$, we have either $\hat{\hat{\xi}}_j = \hat{\hat{\xi}}_1$ or $\hat{\hat{\xi}}_j = \hat{\hat{\xi}}_2$.

Let $\hat{\hat{\xi}}_1$ and $\hat{\hat{\xi}}_2$ appear in $\{\hat{\hat{\xi}}_1, \ldots, \hat{\hat{\xi}}_K\}$ k_1 and k_2 times, respectively, where $k_1 \geq 1$, $k_2 \geq 1$, $k_1 + k_2 = K$.

This implies

$$\hat{\hat{\xi}}_1 - \eta_0 \hat{\hat{\xi}}_1^2 = \sum_{j=1}^{K} \hat{\hat{\xi}}_j^2 = k_1 \hat{\hat{\xi}}_1^2 + k_2 \hat{\hat{\xi}}_2^2, \tag{6.135}$$

$$\hat{\hat{\xi}}_2 = \frac{1}{\eta_0} - \hat{\hat{\xi}}_1. \tag{6.136}$$

We put (6.136) into (6.135) and get

$$\hat{\hat{\xi}}_1 - \eta_0 \hat{\hat{\xi}}_1^2 = k_1 \hat{\hat{\xi}}_1^2 + k_2 \left(\frac{1}{\eta_0} - \hat{\hat{\xi}}_1 \right)^2.$$

Therefore we have

$$(k_1 + k_2 + \eta_0) \hat{\hat{\xi}}_1^2 - \frac{2k_2 + \eta_0}{\eta_0} \hat{\hat{\xi}}_1 + \frac{k_2}{\eta_0^2} = 0. \tag{6.137}$$

Equation (6.137) has a solution if and only if

$$(2k_2 + \eta_0)^2 \geq 4k_2(k_1 + k_2 + \eta_0). \tag{6.138}$$

The inequality (6.138) in the strict sense is the same as (6.118).

If (6.118) holds, then (6.137) has two different solutions which are given by $(\rho_\pm, \eta_\pm)$.

Finally, we have the following conclusion:

Lemma 6.16 *Let $\eta_0 > 2\sqrt{k_1 k_2}$. Then the solutions of (6.128) are given by $(\hat{\hat{\xi}}_1, \ldots, \hat{\hat{\xi}}_N) \in (\{\rho_\pm, \eta_\pm\})^K$ where the $\rho_\pm$ appears k_1 times and $\eta_\pm$ appears k_2 times.*

If $\eta_0 > 2\sqrt{k_1 k_2}$, then (6.128) has two solutions $(\rho_\pm, \eta_\pm)$.
If $\eta_0 = 2\sqrt{k_1 k_2}$, then (6.128) has one solution $(\rho_\pm, \rho_\pm)$.
If $\eta_0 < 2\sqrt{k_1 k_2}$, then (6.128) has no solutions.

From now on, we assume that (6.118) is true. As in Lemma 6.16 we choose the amplitudes $(\hat{\hat{\xi}}_1, \ldots, \hat{\hat{\xi}}_K)$.

6.3.2 The Reduced Problem

The existence proof uses Liapunov-Schmidt reduction in the same way as for symmetric spikes. We only have to be careful to extend the computations to the case where the variable amplitudes $\xi_{\epsilon,i}$ vary with i. Finally, we will get another reduced problem.

Combining the estimates for I_1 and I_2, we get

$$W_\epsilon(\mathbf{P}) = -\frac{\pi \eta_0}{6} \mathcal{D}^{-1} \nabla_\mathbf{P} F(\mathbf{P}) + E_\epsilon(\mathbf{P}). \tag{6.139}$$

Here the matrix $\mathcal{D}$ is defined by

$$(\mathcal{D})_{ij} = \hat{\xi}_{\epsilon,j} \delta_{ij}, \tag{6.140}$$

where δ_{ij} denotes the Kronecker symbol. Finally, the function $E_\epsilon(\mathbf{P}) = o(1)$ depends on $\mathbf{P}$ continuously and tends to 0 as $\epsilon \to 0$ uniformly for $\mathbf{P} \in \overline{\Lambda_\delta}$.

Computing at $\mathbf{P}^0$, we get $\nabla_\mathbf{P}|_{\mathbf{P}=\mathbf{P}^0} F(\mathbf{P}^0) = 0$, $\det(\nabla^2_\mathbf{P}|_{\mathbf{P}=\mathbf{P}^0}(F(\mathbf{P}))) \neq 0$. Thus if (i) $\delta = \delta(\epsilon) \to 0$ as $\epsilon \to 0$ and (ii) W_ϵ has exactly one zero in $B_\delta(\mathbf{P}^0)$ (which is possible by (6.139) for ϵ small enough), we compute the mapping degree of $W_\epsilon(\mathbf{P})$ for the set B_δ and the value 0 as follows:

$$\deg(W_\epsilon, 0, B_\delta) = \operatorname{sign} \det\left(-\mathcal{D}^{-1} \nabla^2_\mathbf{P}|_{\mathbf{P}=\mathbf{P}^0}(F(\mathbf{P}))\right)$$
$$= \operatorname{sign} \det\left(-\mathcal{D}^{-1} M(\mathbf{P}^0)\right) \neq 0$$

by condition (6.126) in Theorem 6.14. Thus, using standard degree theory, for ϵ small enough, there is a $\mathbf{P}^\epsilon \in B_\delta$ such that $W_\epsilon(\mathbf{P}^\epsilon) = 0$. Further, from (6.139) we get $P^\epsilon \to \mathbf{P}^0$.

We summarise this result as follows:

Proposition 6.17 *For ϵ small enough, there are points $\mathbf{P}^\epsilon$ such that $\mathbf{P}^\epsilon \to \mathbf{P}^0$ and $W_\epsilon(\mathbf{P}^\epsilon) = 0$.*

Finally, we finish the proof of Theorem 6.14 in the same way as for symmetric spikes.

6.4 Asymmetric Multiple Spikes: Stability

In this section, we study the stability properties of the asymmetric K-spike given in Theorem 6.14.

We will study the large eigenvalues of order $O(1)$ and the small eigenvalues of order $o(1)$ separately. We will link the small eigenvalues to a Green's function and its derivatives. Under the assumption that the small eigenvalues all have negative real part, we will show that stable asymmetric K-spike solutions exist only for

$$\frac{1}{2\pi K} \log \frac{\sqrt{|\Omega|}}{\epsilon} < \frac{D}{|\Omega|} < \frac{1}{4\pi \sqrt{k_1 k_2}} \log \frac{\sqrt{|\Omega|}}{\epsilon}. \tag{6.141}$$

Here ϵ must be small enough and k_1, k_2 are two integers satisfying $k_1 + k_2 = K$, $k_1 \geq 1, k_2 \geq 1$.

Theorem 6.18 *Let ϵ be small enough. Assume that (6.114) and (6.118) hold. Further, assume that (6.125) and (6.126) of Theorem 6.14 are true. We consider the K-spike solutions (A_ϵ, H_ϵ) given in Theorem 6.14 whose peaks converge to $\mathbf{P}^0 \in \overline{\Lambda_\delta}$ as $\epsilon \to 0$.*

Assume that

$$\mathbf{P}^0 \text{ is a nondegenerate local maximum point of } F(\mathbf{P}). \qquad (**)$$

Then we have:

(a) (*Stability*)

   Assume that

$$2\sqrt{k_1 k_2} < \eta_0 < K \qquad (6.142)$$

   and

$$k_1 > k_2 = 1, \qquad (\rho, \eta) = (\rho_+, \eta_+).$$

   Then, for τ small enough, (A_ϵ, H_ϵ) is linearly stable.

(b) (*Instability*)

   Assume that either

$$\eta_0 > K$$

   or

$$\tau \text{ is large enough}$$

   or

$$k_1 \le k_2, \qquad (\rho, \eta) = (\rho_+, \eta_+)$$

   or

$$k_2 > 1.$$

Then (A_ϵ, H_ϵ) is linearly unstable.

Remark 6.19 (a) The theorem implies that for asymmetric multiple spikes to be stable it is necessary that there is only one small spike and there are more than one large spikes.

6.4.1 Large Eigenvalues

We study the stability of the asymmetric solution (A_ϵ, H_ϵ) given in Theorem 6.14. We begin by considering large eigenvalues of order $O(1)$.

Similarly to the symmetric case, we derive

$$\psi_\epsilon\left(P_i^\epsilon\right) = \left(2\sum_{j=1}^{K}\frac{1}{|\Omega|(1+\tau\lambda_0)}\xi_\epsilon\hat{\xi}_{\epsilon,j}\epsilon^2\int_{\mathbb{R}^2}w\phi_{\epsilon,j} + 2\xi_\epsilon\hat{\xi}_{\epsilon,i}\frac{\eta_0}{|\Omega|}\epsilon^2\int_{\mathbb{R}^2}w\phi_{\epsilon,i}\right)$$

$$\times \left(1+o(1)\right). \tag{6.143}$$

Putting (6.143) into (6.74), we have for $\epsilon \to 0$

$$\Delta\phi_i - \phi_i + 2w\phi_i - \frac{2}{1+\tau\lambda_0}\sum_{j=1}^{K}\hat{\xi}_j\frac{\int w\phi_j}{\int w^2} - 2\eta_0\hat{\xi}_i\frac{\int w\phi_i}{\int w^2}w^2$$

$$= \lambda_0\phi_i, \quad i = 1,\dots,K. \tag{6.144}$$

Using the notation

$$\Phi = \begin{pmatrix}\phi_1\\ \vdots \\ \phi_K\end{pmatrix},$$

we can rewrite (6.144) as

$$\Delta\Phi - \Phi + 2w\Phi - \frac{2\int_{\mathbb{R}^2}w\mathcal{B}\Phi}{\int_{\mathbb{R}^2}w^2}w^2 = \lambda_0\Phi, \tag{6.145}$$

where

$$\mathcal{B} = \begin{pmatrix}\eta_0\hat{\xi}_1 & & \\ & \ddots & \\ & & \eta_0\hat{\xi}_K\end{pmatrix} + \frac{1}{1+\tau\lambda_0}\begin{pmatrix}\hat{\xi}_1 & \cdots & \hat{\xi}_K \\ \vdots & & \vdots \\ \hat{\xi}_1 & \cdots & \hat{\xi}_K\end{pmatrix} \tag{6.146}$$

Note that in general $\mathcal{B}$ is not self-adjoint since $\lambda_0 \in \mathbb{C}$.

Let us now compute the eigenvalues of $\mathcal{B}$ in two special cases. We claim that

Lemma 6.20 *Let $(\hat{\xi}_1,\dots,\hat{\xi}_K)$ be given by Lemma 6.16. Then the eigenvalues of $\mathcal{B}$ solve*

$$\frac{k_1\rho}{\eta_0\rho - \lambda} + \frac{k_2\eta}{\eta_0\eta - \lambda} + 1 + \tau\lambda_0 = 0, \tag{6.147}$$

or, if $k_1 > 1$, they satisfy

$$\lambda = \eta_0\rho, \tag{6.148}$$

or, if $k_2 > 1$, they satisfy

$$\lambda = \eta_0\eta, \tag{6.149}$$

where ρ and η are given by (6.122).

If $\tau = 0$, the eigenvalues of $\mathcal{B}$ are given by

$$\lambda_1 = 1, \qquad \lambda_2 = k_1 \rho + k_2 \eta,$$
$$\lambda_3 = \eta_0 \rho \quad (\text{if } k_1 > 1), \qquad \lambda_4 = \eta_0 \eta \quad (\text{if } k_2 > 1). \tag{6.150}$$

If $\tau = +\infty$, then the eigenvalues of $\mathcal{B}$ are given by

$$\lambda_1 = \eta_0 \rho, \qquad \lambda_2 = \eta_0 \eta. \tag{6.151}$$

Proof Let $\mathbf{q} = (q_1, \ldots, q_K)^T$ be an eigenvector of $\mathcal{B}$ with corresponding eigenvalue λ. Then we get

$$\mathcal{B}\mathbf{q} = \lambda \mathbf{q}. \tag{6.152}$$

Writing (6.152) in components, we have

$$\eta_0 \hat{\xi}_i q_i + \frac{1}{1 + \tau \lambda_0} \sum_{j=1}^{N} q_j \hat{\xi}_j = \lambda q_i, \quad i = 1, \ldots, K.$$

Hence, we get

$$(\eta_0 \hat{\xi}_i - \lambda) q_i = -\frac{1}{1 + \tau \lambda_0} \sum_{j=1}^{N} q_j \hat{\xi}_j = c, \tag{6.153}$$

where

$$q_i = \frac{c}{\eta_0 \hat{\xi}_i - \lambda}, \tag{6.154}$$

or

$$\lambda = \eta_0 \rho \quad (\text{if } k_1 > 1)$$

or

$$\lambda = \eta_0 \eta \quad (\text{if } k_2 > 1).$$

Substituting (6.154) into (6.153), we obtain that

$$\sum_{j=1}^{K} \frac{\hat{\xi}_j}{\eta_0 \hat{\xi}_j - \lambda} + 1 + \tau \lambda_0 = 0. \tag{6.155}$$

Using (6.122), we get

$$\frac{k_1 \rho}{\eta_0 \rho - \lambda} + \frac{k_2 \eta}{\eta_0 \eta - \lambda} + 1 + \tau \lambda_0 = 0,$$

which is precisely (6.147).

For $\tau = 0$ we use $\rho + \eta = \frac{1}{\eta_0}$ and have

$$\lambda^2 - \lambda(k_1\rho + k_2\eta + 1) + \eta_0(K + \eta_0)\rho\eta = 0. \tag{6.156}$$

Now (6.156) has two solutions which are given in (6.150).

Finally, we consider the case $\tau = +\infty$. Then the matrix $\mathcal{B}$ is diagonal and the result is trivial. □

Selecting a basis for $\mathbb{R}^K$ such that $\mathcal{B}$ becomes a diagonal matrix, the eigenvalue problem (6.145) is reduced to the following two NLEPs:

$$\Delta \Phi_i - \Phi_i + 2w\Phi_i - \frac{2\lambda_i \int_{\mathbb{R}^2} w\Phi_i}{\int_{\mathbb{R}^2} w^2}w^2$$
$$= \lambda_0\Phi_i, \quad i = 1, 2, \Phi_i \in H^2(\mathbb{R}^2), \tag{6.157}$$

where λ_i are the two eigenvalues of $\mathcal{B}$ satisfying (6.147). Using results of Chap. 3 we study the stability properties of these two eigenvalue problems.

For $\tau = 0$, we have $\lambda_1 = 1$, $\lambda_2 = k_1\rho + k_2\eta$. By Theorem 3.1, the first eigenvalue does not result in an instability of (6.157). For the second eigenvalue, we compute in case $(\rho, \eta) = (\rho_\pm, \eta_\pm)$ that

$$2\lambda_2 - 1 = \frac{4k_1k_2 - \eta_0^2 \pm (k_1 - k_2)\sqrt{\eta_0^2 - 4k_1k_2}}{\eta_0(\eta_0 + K)}. \tag{6.158}$$

If $\eta_0 > K$, we have

$$\eta_0^2 > (k_1 + k_2)^2$$

and so

$$\eta_0^2 - 4k_1k_2 > (k_1 - k_2)^2.$$

This implies

$$\lambda_2 < \frac{1}{2} \quad \text{if } \eta_0 > K.$$

Theorem 3.6 implies that for all $\tau > 0$ there is a positive eigenvalue λ_0 of (6.157). Together with a perturbation argument based on the compactness of operators [37] it follows that the instability of (6.74) with respect to the $O(1)$ eigenvalues carries over from the limiting eigenvalue problem ($\epsilon = 0$) to the original eigenvalue problem ($\epsilon > 0$ small enough).

However, for the case $2\sqrt{k_1k_2} < \eta_0 < K$, we necessarily have

$$0 < \eta_0^2 - 4k_1k_2 < (k_1 - k_2)^2$$

which implies that $k_1 \neq k_2$. Choosing $k_1 > k_2$, $(\rho, \eta) = (\rho_+, \eta_+)$, we conclude that $\lambda_2 > 1/2$. Further, if $k_2 > 1$, then by (6.149), we have $\lambda_4 = \eta_0\eta$. Then, since $\rho + \eta =$

$\frac{1}{\eta_0}$ and using $\eta < \rho$, we derive $\lambda_4 < \frac{1}{2}$ (and similarly, if $k_1 > 1$ by (6.148), we get $\lambda_3 > \frac{1}{2}$, which does not cause an instability). Therefore, for stability, we need to assume that $k_2 = 1$. By Theorem 3.1, we have stability of (6.74) with respect to large eigenvalues, for τ small enough.

Finally, when $\tau = +\infty$, we have $\lambda_1 = \eta_0 \rho$, $\lambda_2 = \eta_0 \eta$ which again implies that $\lambda_2 < \frac{1}{2}$ and the solution is unstable.

Thus by Theorems 3.6, 3.9, if τ is large enough, we have instability of (6.157) and also of (6.74) with respect to $O(1)$ eigenvalues.

Thus we have proved the part of Theorem 6.9 concerning large eigenvalues.

In the next subsection we will study the small eigenvalues λ_ϵ which tend to zero as $\epsilon \to 0$.

6.4.2 Small Eigenvalues

In this subsection we study (6.74) in the case of small eigenvalues, i.e. we assume that $\lambda_\epsilon \to 0$ as $\epsilon \to 0$. The analysis is similar to the case of symmetric spikes. We will show that the small eigenvalues are related to the matrix $M(\mathbf{P}^0)$ introduced in (6.124).

We assume that condition $(**)$ of Theorem 6.18 holds. We will show that if $\lambda_\epsilon \to 0$ then

$$\lambda_\epsilon \sim \frac{\epsilon^2}{\log(\sqrt{|\Omega|}/\epsilon)} 2\pi \eta_0 \frac{1}{\int_{\mathbb{R}^2} w^2} \sigma_0, \tag{6.159}$$

where σ_0 is an eigenvalue of $\mathcal{D}^{-1} M(\mathbf{P}^0) \mathcal{D}^{-2}$ and $\mathcal{D}$ is given in (6.140). Then (6.159) implies that all the small eigenvalues of $\mathcal{L}_\epsilon$ have negative real part.

The computations of the small eigenvalues change due to inclusion of varying amplitudes and the special choice of the asymptotic behaviour for $D(\epsilon)$ as $\epsilon \to 0$. Therefore we include a sketch of the computations.

We state the algebraic equations for $a^\epsilon_{j,k}$, whose derivation follows the same steps as for symmetric spikes:

$$\frac{\epsilon^4 \xi_\epsilon \beta^2}{6} \frac{\hat{\xi}_{\epsilon,l}}{(\hat{\xi}_{\epsilon,j})^2} \int_{\mathbb{R}^2} w^3 dy \int_{\mathbb{R}^2} w^2 dy \sum_{j=1}^{K} \sum_{k=1}^{2} a^\epsilon_{j,k} \left(\frac{\partial}{\partial P^\epsilon_{l,m}} \frac{\partial}{\partial P^\epsilon_{j,k}} F(\mathbf{P}^\epsilon) \right)$$

$$+ o\left(\frac{\epsilon^2}{\log(\sqrt{|\Omega|}/\epsilon)} \sum_{j=1}^{K} \sum_{k=1}^{2} |a^\epsilon_{j,k}| \right)$$

$$= \lambda_\epsilon a^\epsilon_{l,m} \hat{\xi}^2_{\epsilon,l} \int_{\mathbb{R}^2} \left(\frac{\partial w}{\partial y_1} \right)^2 dy (1 + o(1)). \tag{6.160}$$

Letting $\epsilon \to 0$ in (6.160), for the small eigenvalues we have

$$\frac{\lambda_\epsilon}{\epsilon^4 \xi_\epsilon \beta^2} \to \frac{\int_{\mathbb{R}^2} w^3 \int_{\mathbb{R}^2} w^2}{6 \int_{\mathbb{R}^2} (\partial w / \partial y_1)^2 dy} \sigma_0 \quad \text{as } \epsilon \to 0,$$

where σ_0 is an eigenvalue of the matrix $\mathcal{D}^{-1} M(\mathbf{P}^0) \mathcal{D}^{-2}$, $\mathcal{D}$ is given by (6.140), $M(\mathbf{P}^0) = \nabla^2_{\mathbf{P}} |_{\mathbf{P}=\mathbf{P}^0} (F(\mathbf{P}))$ and $\mathbf{P}^\epsilon \to \mathbf{P}^0$ as $\epsilon \to 0$. Using (**) of Theorem 6.18, it follows that the matrix $M(\mathbf{P}^0)$ is negative definite. Thus we have $\mathrm{Re}(\sigma_0) < 0$ and, in particular, $\mathrm{Re}(\lambda_\epsilon) \leq -c\epsilon^4 \xi_\epsilon \beta^2$ for some $c > 0$ if ϵ is small enough. We conclude that all small eigenvalues λ_ϵ for (6.74) have negative real part if ϵ is small enough.

Theorem 6.18 now follows by combining the results for large and small eigenvalues.

6.5 Notes on the Literature

The existence of spikes for the two-dimensional Gierer-Meinhardt system has been proved in [259, 262] for the strong coupling case

$$D = O(1) \quad \text{as } \epsilon \to 0.$$

The weak coupling case

$$D \to \infty \quad \text{as } \epsilon \to 0$$

has been considered in [261, 269]. In [261] multiple symmetric spikes are studied and in [269] multiple asymmetric spikes are investigated.

More information on deriving the position of the spikes from the Green's function has been obtained in [110, 111]. Roughly speaking, assuming that condition (*) holds and that τ is small, then for $\epsilon \ll 1$, $D_K(\epsilon) = \frac{|\Omega|}{2\pi K} \log \frac{1}{\epsilon}$ is the critical threshold for the asymptotic behaviour of the inhibitor diffusivity which determines the stability of K-spike solutions. This result is similar to the one-dimensional case, [99, 273]. In higher dimensions the analysis is very different since it has to reflect the geometry of the domain.

In the strong coupling case an arbitrary number of spikes is stable if ϵ is small enough. The results and the approach are similar, but a different Green's function is required. In [261] single interior spike solutions have been constructed.

The stability in the strong coupling case has been investigated in [262]. After constructing interior K-spike solutions the stability for $\tau = 0$ is studied provided that the limiting peaks $\mathbf{P}^0 = (P_1^0, \ldots, P_K^0)$ form a nondegenerate local maximum point of the functional

$$F_1(\mathbf{P}) = \sum_{k=1}^{K} H_1(P_k, P_k) - \sum_{i,j=1,\ldots,K, i \neq j} G_1(P_i, P_j), \qquad (6.161)$$

where the Green's function $G_1(P, x)$ satisfies

$$\begin{cases} -\Delta G_1 + G_1 = \delta_\xi & \text{in } \Omega, \\ \frac{\partial G_1}{\partial v} = 0 & \text{on } \partial\Omega \end{cases}$$

and

$$H_1(\xi, x) = \frac{1}{2\pi} \log \frac{1}{|x - \xi|} - G_1(\xi, x).$$

Hence, for any finite $D = O(1)$, the stability of K-spike solutions is independent of D and depends on the peak locations only.

For the uniqueness and stability of spikes for the Gierer-Meinhardt system in the limit $D \to \infty$ (shadow system) see [249].

The stability of stripes for the Gierer-Meinhardt system has been studied in [96] using a SLEP approach and in [45] with NLEP methods. These studies have been extended to the Gierer-Meinhardt system with saturation in [114] based on the latter approach. Ermentrout [64] considered the problem of stripes versus spots in general reaction-diffusion systems using amplitude equations and identified stripes with one-dimensional profiles, and spots with two-dimensional profiles, and showed that quadratic nonlinearities lead to spots and cubic nonlinearities to stripes. A system preferring stripes has been considered in [136].

More generally speaking, there are solutions to the Gierer-Meinhardt system which do not concentrate near points (zero-dimensional sets), but near manifolds of arbitrary dimension. Particular cases include solutions concentrating near a sphere ($N - 1$ dimensional subset of N-dimensional space) [175], ring (one-dimensional set in two-dimensional space) or smoke ring (one-dimensional set in space of dimension three or higher) [109, 116]. However, we expect most of these patterns to be unstable.

A priori estimates have been derived in [39, 70, 103, 104, 170, 210]. Breathing and wiggling motions have been studied in [224].

Chapter 7
The Gierer-Meinhardt System with Inhomogeneous Coefficients

Turing systems are usually studied for coefficients which are constant in space. In the previous chapters we have always made this assumption. However, in real-life biological systems many effects influencing their behaviour are spatially dependent. One particularly important example is precursor gradients in reaction-diffusion models which represent pre-existing spatial structures within a living organism.

We consider two cases of inhomogeneities. The first concerns a spatially variable coefficient for the decay term of the inhibitor which we study by rescaling the spike. In the analysis we have to consider extra spatially dependent terms which interact with the Green's function. The second concerns a discontinuous diffusion coefficient of the inhibitor which can be handled by studying a new Green's function and an inner-outer expansion.

7.1 Precursors

Holloway [92] introduced precursor gradients into (2.1) by considering spatially varying coefficients. This results in a Turing bifurcation of the second kind: There are no longer any homogeneous uniform steady states and so it is not possible to use standard Turing instability analysis to understand the formation of patterns.

In [92] the author uses numerical methods to investigate the following Gierer-Meinhardt system with precursor gradient (inhomogeneity) $\mu(x)$ for the morphogen A:

$$\begin{cases} A_t = \epsilon^2 A'' - \mu(x)A + \frac{A^2}{H} & \text{in } (-1, 1), \\ \tau H_t = DH'' - H + A^2 & \text{in } (-1, 1), \\ A'(-1) = A'(1) = H'(-1) = H'(1) = 0. \end{cases} \quad (7.1)$$

Two cases of precursors have been studied. The first is linear precursors which are given by $\mu(x) = Ax + B$, where A and B are some real constants. The second concerns exponential precursors for which we have $\mu(x) = \sum_{i=1}^{m} A_i e^{-|x-x_i|}$ for given constants $A_i > 0$ and points $x_i \in (-1, 1)$. Precursor gradients play an important role

J. Wei, M. Winter, *Mathematical Aspects of Pattern Formation in Biological Systems*, Applied Mathematical Sciences 189, DOI 10.1007/978-1-4471-5526-3_7, © Springer-Verlag London 2014

in the existence and stability of spiky patterns. They change their properties, such as their amplitude, position and stability, to name a few. In particular, precursor gradients can result in stable asymmetric multi-spike patterns for which the distances and amplitudes are not uniform. This behaviour is often observed in real-world biological systems, e.g. in seashell patterns and on fish skins. In the next subsection we will begin a rigorous analysis of multi-spike patterns with precursor gradients.

7.1.1 Results on Existence and Stability

The main goal of this section is to rigorously state and prove results for the Gierer-Meinhardt system in case of a spatially varying decay rate $\mu(x)$ of the activator concerning the existence and stability of multi-spike steady states. In [244], Ward et al. considered pinning phenomena for the problem

$$
\begin{cases}
A_t = \epsilon^2 \Delta A - \mu_1(x)A + \frac{A^2}{H} & \text{in } \Omega, \\
\tau H_t = D\Delta H - \mu_2(x)H + A^2 & \text{in } \Omega, \\
\frac{\partial A}{\partial \nu} = \frac{\partial H}{\partial \nu} = 0 & \text{on } \partial \Omega
\end{cases}
\tag{7.2}
$$

where Ω is a bounded smooth domain in $\mathbb{R}^1$ or $\mathbb{R}^2$. They restricted their attention to the consideration of one-spike solutions. We shall consider multi-spike steady states of (7.1) in the one-dimensional case.

The stationary solutions to (7.1) solve

$$
\begin{cases}
\epsilon^2 A'' - \mu(x)A + \frac{A^2}{H} = 0 & \text{in } (-1, 1), \\
DH'' - H + A^2 = 0 & \text{in } (-1, 1), \\
A'(-1) = A'(1) = H'(-1) = H'(1) = 0.
\end{cases}
\tag{7.3}
$$

The proof of the existence of solutions is more difficult than for the constant coefficient case. It is not possible to construct one spike on a smaller interval and then extend it to multiple spikes on the whole interval by periodic continuation since in general the precursor $\mu(x)$ is not periodic. Instead, we construct a multiple-spike solution directly. We extend Liapunov-Schmidt reduction to this case. Now we introduce the technical framework required for this approach.

Let $\mu(x)$ satisfy

$$
\mu(x) \in C^3(\Omega), \quad \mu(x) > 0 \text{ in } \Omega.
\tag{7.4}
$$

Let w be the ground state given by (2.3). Let $G_D(x, z)$ be the Green's function given in (2.13). Let $-1 < t_1^0 < \cdots < t_j^0 < \cdots < t_N^0 < 1$ be N points in $(-1, 1)$ and set $\mu_i^0 = \mu(t_i^0)$, $i = 1, \ldots, N$. Let $D > 0$ be a constant which does not depend on ϵ.

Using rescaling, the function

$$
w_a(y) = aw(a^{1/2}y),
\tag{7.5}
$$

solves

$$\begin{cases} w_a'' - aw_a + w_a^2 = 0 & \text{in } \mathbb{R}, \\ w_a > 0, \quad w_a(0) = \max_{y \in \mathbb{R}} w_a(y), \quad w_a(y) \to 0 \text{ as } |y| \to \infty, \end{cases} \tag{7.6}$$

where w is given by (2.3). Then we have

$$\int_{\mathbb{R}} w_a^2(y) dy = a^{3/2} \int_{\mathbb{R}} w^2(z) dz, \qquad \int_{\mathbb{R}} w_a^3(y) dy = a^{5/2} \int_{\mathbb{R}} w^3(z) dz,$$
$$\int_{\mathbb{R}} (w_a')^2(y) dy = a^{5/2} \int_{\mathbb{R}} (w')^2(z) dz. \tag{7.7}$$

Recall from (2.18) that

$$\xi_\epsilon := \left(\epsilon \int_{\mathbb{R}} w^2(z) dz \right)^{-1}.$$

The following three assumptions are obvious modifications of those in the constant coefficient case:

(H1b) There exists a solution $(\hat{\xi}_1^0, \ldots, \hat{\xi}_N^0)$ of the equation

$$\sum_{j=1}^{N} G_D(t_i^0, t_j^0)(\hat{\xi}_j^0)^2 (\mu_j^0)^{3/2} = \hat{\xi}_i^0, \quad i = 1, \ldots, N. \tag{7.8}$$

(H2b) We have

$$\frac{1}{2} \notin \sigma(\mathcal{B}), \tag{7.9}$$

where $\sigma(\mathcal{B})$ is the spectrum of the matrix $\mathcal{B}$ which is given by

$$\mathcal{B} = (b_{ij}), \quad b_{ij} = G_D(t_i^0, t_j^0) \hat{\xi}_j^0 (\mu_j^0)^{3/2}.$$

We introduce

$$\mathcal{H}(\mathbf{t}) = (\hat{\xi}_i(\mathbf{t}) \delta_{ij}), \quad \mu(\mathbf{t}) = (\mu(t_i) \delta_{ij}), \quad \mu'(\mathbf{t}) = (\mu'(t_i) \delta_{ij}) \tag{7.10}$$

and define

$$F(\mathbf{t}) = (F_1(\mathbf{t}), \ldots, F_N(\mathbf{t}))$$

with

$$F_i(\mathbf{t}) = \frac{5}{4} \hat{\xi}_i \frac{\mu'(t_i)}{\mu(t_i)} + \sum_{l=1}^{N} \nabla_{t_i} G_D(t_i, t_l) \hat{\xi}_l^2 \mu_l^{3/2}, \quad i = 1, \ldots, N. \tag{7.11}$$

Then we let

$$\mathcal{M}(\mathbf{t}) = \left(\frac{\partial F_i(\mathbf{t})}{\partial t_j} \right). \tag{7.12}$$

Finally, we state our third condition.

(H3b) At the point $\mathbf{t}^0 = (t_1^0, \ldots, t_N^0)$ we have

$$F(\mathbf{t}^0) = 0, \qquad \det(\mathcal{M}(\mathbf{t}^0)) \neq 0. \tag{7.13}$$

Now we present the main existence theorem.

Theorem 7.1 *Suppose that* (H1b), (H2b) *and* (H3b) *are valid. Then for $\epsilon \ll 1$, problem* (7.1) *has a steady state with multiple spikes located at the points $t_1^\epsilon, \ldots, t_N^\epsilon$ which approach limiting positions. Further, we have for $\epsilon \ll 1$ that*

$$A_\epsilon(x) = \sum_{i=1}^{N} \xi_\epsilon \hat{\xi}_i^0 w_{a_i}\left(\frac{x - t_i^\epsilon}{\epsilon}\right)(1 + o(1)), \tag{7.14}$$

where w_{a_i} is given by (7.6) *for $a_i = \mu(t_i^0)$, ξ_ϵ has been defined in* (2.18), $\hat{\xi}_i^0$ *has been introduced in* (H1b),

$$H_\epsilon(t_i^\epsilon) = \xi_\epsilon \hat{\xi}_i^0 (1 + o(1)), \quad i = 1, \ldots, N, \tag{7.15}$$

$$t_i^\epsilon \to t_i^0, \quad i = 1, \ldots, N. \tag{7.16}$$

Next we state the main result on stability.

Theorem 7.2 *Let (A_ϵ, H_ϵ) be the multi-peak steady states given in Theorem 7.1. Then for $\epsilon \ll 1$ we have the following:*

(1) (*Stability*) *Suppose that*

$$\min_{\sigma \in \sigma(\mathcal{B})} \sigma > \frac{1}{2} \tag{7.17}$$

and

$$\sigma(\mathcal{M}(\mathbf{t}^0)) \subseteq \{\sigma \mid \mathrm{Re}(\sigma) \geq c\} \quad \text{for some } c > 0. \tag{7.18}$$

Then there exists a $\tau_0 > 0$ such that (A_ϵ, H_ϵ) is linearly stable for $0 \leq \tau < \tau_0$.

(2) (*Instability*) *Suppose that*

$$\min_{\sigma \in \sigma(\mathcal{B})} \sigma < \frac{1}{2}. \tag{7.19}$$

Then (A_ϵ, H_ϵ) is linearly unstable for all $\tau \geq 0$.

(3) (*Instability*) *Suppose that there exists*

$$\sigma \in \sigma(\mathcal{M}(\mathbf{t}^0)), \quad \mathrm{Re}(\sigma) < 0. \tag{7.20}$$

Then (A_ϵ, H_ϵ) is linearly unstable for all $\tau \geq 0$.

Let us now calculate $\mathcal{M}(\mathbf{t}^0)$. This computation includes all the contributions from the constant coefficient case. Additionally we get extra terms which stem from the spatially varying coefficient $\mu(t)$.

We begin by computing the derivatives of $\hat{\xi}(\mathbf{t})$. By the Implicit Function Theorem, the function $\hat{\xi}(\mathbf{t})$ is C^1 in $\mathbf{t}$, and by (7.9) we get

$$
\nabla_{t_j}\hat{\xi}_i = 2\sum_{l=1}^{N} G_D(t_i, t_l)\hat{\xi}_l\mu_l^{3/2}\nabla_{t_j}\hat{\xi}_l + \frac{\partial}{\partial t_j}\left(G_D(t_i, t_j)\right)\hat{\xi}_j^2\mu_j^{3/2}
$$

$$
+ \frac{3}{2}G_D(t_i, t_j)\hat{\xi}_j^2\mu_j^{1/2}\mu_j' \quad \text{for } i \neq j,
$$

$$
\nabla_{t_i}\hat{\xi}_i = 2\sum_{l=1}^{N} G_D(t_i, t_l)\hat{\xi}_l\mu_l^{3/2}\nabla_{t_i}\hat{\xi}_l + \sum_{l=1}^{N}\frac{\partial}{\partial t_i}\left(G_D(t_i, t_l)\right)\hat{\xi}_l^2\mu_l^{3/2}
$$

$$
+ \frac{3}{2}G_D(t_i, t_i)\hat{\xi}_i^2\mu_i^{1/2}\mu_i'
$$

$$
= 2\sum_{l=1}^{N} G_D(t_i, t_l)\hat{\xi}_l\mu_l^{3/2}\nabla_{t_i}\hat{\xi}_l + \nabla_{t_i}G_D(t_i, t_i)\hat{\xi}_i^2\mu_i^{3/2} - \frac{5}{4}\hat{\xi}_i\frac{\mu_i'}{\mu_i}
$$

$$
+ \frac{3}{2}G_D(t_i, t_i)\hat{\xi}_i^2\mu_i^{1/2}\mu_i' + O\left(\sum_{j=1}^{N}|F_j(\mathbf{t})|\right) \quad \text{for } i = j,
$$

where we have used (7.11).

Note that

$$
\left(\nabla_{t_j}G_D(t_i, t_j)\right) = (\nabla\mathcal{G}_D)^T.
$$

Therefore, introducing matrix notation

$$
\nabla\xi = (\nabla_{t_j}\hat{\xi}_i), \qquad \mathcal{P} = \left(I - 2\mathcal{G}_D\mathcal{H}\mu^{3/2}\right)^{-1}, \tag{7.21}
$$

we have

$$
\nabla\xi(\mathbf{t}) = \mathcal{P}\left[(\nabla\mathcal{G}_D)^T\mathcal{H}^2\mu^{3/2} - \frac{5}{4}\mathcal{H}\mu^{-1}\mu' + \frac{3}{2}\mathcal{G}_D\mathcal{H}^2\mu^{1/2}\mu'\right]
$$

$$
+ O\left(\sum_{j=1}^{N}|F_j(\mathbf{t})|\right). \tag{7.22}
$$

Let

$$
\mathcal{Q} = (q_{ij}) = \left(\left(\frac{1}{D}\hat{\xi}_i^{-1}\mu_i^{-3/2} - \frac{1}{2D^{3/2}}\right)\delta_{ij}\right). \tag{7.23}
$$

We compute $\mathcal{M}(\mathbf{t}^0)$ using (7.22). This gives for $i \neq j$

$$\sum_{l=1}^{N} (\nabla_{t_j} \nabla_{t_i} G_D(t_i, t_l)) \hat{\xi}_l^2 \mu_l^{3/2} = (\nabla_{t_j} \nabla_{t_i} G_D(t_i, t_j)) \hat{\xi}_j^2 \mu_j^{3/2}. \qquad (7.24)$$

In case $i = j$ we have

$$\sum_{l=1}^{N} (\nabla_{t_i} \nabla_{t_i} G_D(t_i, t_l)) \hat{\xi}_l^2 \mu_l^{3/2}$$

$$= \sum_{l=1,\dots,N, l \neq i} \nabla_{t_i} \nabla_{t_i} G_D(t_i, t_l) \hat{\xi}_l^2 \mu_l^{3/2}$$

$$- \left(\frac{\partial^2}{\partial x^2} \Big|_{x=t_i} H_D(x, t_i) \right) \hat{\xi}_i^2 \mu_i^{3/2} - \left(\frac{\partial^2}{\partial x \partial y} \Big|_{x=t_i, y=t_i} H_D(x, y) \right) \hat{\xi}_i^2 \mu_i^{3/2}$$

$$= \frac{1}{D} \sum_{l=1,\dots,N, l \neq i} G_D(t_i, t_l) \hat{\xi}_l^2 \mu_l^{3/2} - \frac{1}{D} H_D(t_i, t_i) \hat{\xi}_i^2 \mu_i^{3/2}$$

$$- \left(\frac{\partial^2}{\partial x \partial y} \Big|_{x=t_i, y=t_i} H_D(x, t_i) \right) \hat{\xi}_i^2 \mu_i^{3/2}$$

$$= \frac{1}{D} \sum_{l=1}^{N} G_D(t_i, t_l) \hat{\xi}_l^2 \mu_l^{3/2} - \frac{1}{D} K_D(0) \hat{\xi}_i^2 \mu_i^{3/2} + (\nabla_{t_i} \nabla_{t_i} G_D(t_i, t_i)) \hat{\xi}_i^2 \mu_i^{3/2}$$

$$= \frac{1}{D} \hat{\xi}_i - \frac{1}{D} K_D(0) \hat{\xi}_i^2 \mu_i^{3/2} + (\nabla_{t_i} \nabla_{t_i} G_D(t_i, t_i)) \hat{\xi}_i^2 \mu_i^{3/2} \qquad (7.25)$$

by (7.8), and hence

$$\mathcal{M}(\mathbf{t}^0) = (\nabla^2 \mathcal{G}_D + \mathcal{Q}) \mathcal{H}^2 \mu^{3/2} + 2 \nabla \mathcal{G}_D \mathcal{H} \nabla \boldsymbol{\xi} \mu^{3/2}$$

$$+ \frac{5}{4} [\nabla \hat{\boldsymbol{\xi}} \mu^{-1} \mu' - \mathcal{H} \mu^{-2} (\mu')^2 + \mathcal{H} \mu^{-1} \mu'']$$

$$+ \frac{3}{2} \nabla \mathcal{G}_D \mathcal{H}^2 \mu^{1/2} \mu'. \qquad (7.26)$$

Using

$$\nabla \boldsymbol{\xi}(\mathbf{t}^0) = \mathcal{P} \left[(\nabla \mathcal{G}_D)^T \mathcal{H}^2 \mu^{3/2} - \frac{5}{4} \mathcal{H} \mu^{-1} \mu' + \frac{3}{2} \mathcal{G}_D \mathcal{H}^2 \mu^{1/2} \mu' \right], \qquad (7.27)$$

which follows from (H3b) and (7.22), we obtain

$$\mathcal{M}(\mathbf{t}^0) = (\nabla^2 \mathcal{G}_D + \mathcal{Q}) \mathcal{H}^2 \mu^3 + 2 \nabla \mathcal{G}_D \mathcal{H} \mathcal{P} (\nabla \mathcal{G}_D)^T \mathcal{H}^2 \mu^3$$

$$+ \frac{5}{4} \mathcal{P} \left[(\nabla \mathcal{G}_D)^T \mathcal{H}^2 \mu^{1/2} \mu' - \frac{5}{4} \mathcal{H} \mu^{-2} (\mu')^2 + \frac{3}{2} \mathcal{G}_D \mathcal{H}^2 \mu^{-1/2} (\mu')^2 \right]$$

$$+ \frac{5}{4}\mathcal{H}[\mu^{-1}\mu'' - \mu^{-2}(\mu')^2] - \frac{5}{2}\nabla\mathcal{G}_D\mathcal{H}\mathcal{P}\mathcal{H}\mu^{1/2}\mu'$$
$$+ 3\nabla\mathcal{G}_D\mathcal{H}\mathcal{P}\mathcal{G}_D\mathcal{H}^2\mu^2\mu'.$$

Remark 7.3 Let us consider symmetric N-spikes for

$$\mu(x) = A + \sum_{j=1}^{N} B_j(x - t_j^0)^2, \quad t_j^0 = -1 + \frac{(2j-1)}{N}, j = 1, \ldots, N$$

with $\xi_1^0 = \xi_2^0 = \cdots = \xi_N^0$. For this choice of t_j^0 and ξ_j^0 the assumptions (H1b) and (H2b) are valid. Further, we get $\mu_i^0 = A$ and the matrix $\mathcal{M}$ can be computed as

$$\mathcal{M} = \left(m_{ij}^1 + m_{ij}^2\right) = \mathcal{M}^1 + \mathcal{M}^2,$$

where

$$\mathcal{M}^1 = \frac{5}{4}\mathcal{H}\mu^{-1}\mu'',$$
$$\mathcal{M}^2 = (\nabla^2\mathcal{G}_D + \mathcal{Q})\mathcal{H}^2\mu^{3/2} + 2\nabla\mathcal{G}_D\mathcal{H}\mathcal{P}(\nabla\mathcal{G}_D)^T\mathcal{H}^2\mu^3.$$

Note that $\mathcal{H} = \xi_0 I$, $\mu = AI$. So

$$m_{ij}^1 = c_0 B_i \delta_{ij}.$$

Furthermore, $\mathcal{M}^2 = (m_{ij}^2)$ is independent of B_i. Note that the corresponding eigenvalues have been computed in Sect. 4.1.3. These results imply the following: Suppose that we fix A and set $B_i = -B < 0$ with varying constant B. Then, if we choose B large enough, the eigenvalue problem has unstable eigenvalues. To summarise, we have shown that precursors *can destabilise* multi-spike patterns.

To give a rigorous proof of Theorem 7.1 we follow the unified approach introduced in Sect. 2.3. To fill in the details, the conditions and computations presented in this subsection are needed. Then the rest is very similar as before. Here we only remark that for the reduced problem we get

$$W_{\epsilon,i}(\mathbf{t}) = 2.4\mu_i^{5/2}F_i(\mathbf{t}) + O(\epsilon), \quad i = 1, \ldots, N,$$

where the functions F_i have been introduced in (7.11).

The proof of Theorem 7.2 goes along the lines of Chap. 4. Here we only state that in the limit $\epsilon \to 0$ the large eigenvalues solve

$$L\phi_i = (\phi_i)'' - \phi_i + 2w\phi_i - 2\sigma_i\left(\int_{\mathbb{R}} w\phi_i dy\right)\left(\int_{\mathbb{R}} w^2 dy\right)^{-1} w^2$$
$$= \lambda_0\phi_i, \quad i = 1, \ldots, N, \tag{7.28}$$

Fig. 7.1 Two Spikes for (7.1) with $\epsilon^2 = 0.001$, $D = 0.1$, $\mu \equiv 1$ (i.e. no precursor). We get symmetric spikes with the same amplitudes and regular spacing

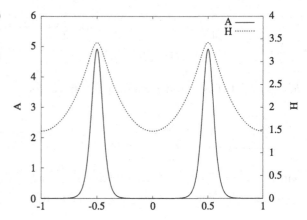

where

$$\phi_i \in H^2(\mathbb{R}), \quad i = 1, \ldots, N,$$

σ_i, $i = 1, \ldots, N$ are the eigenvalues of the matrix $\mathcal{B}$. Further, the small eigenvalues λ_ϵ of (4.20) are given by

$$\lambda_\epsilon = -2\epsilon^2 \sigma \left(\mathcal{H}^{-1} \mathcal{M}(\mathbf{t}^0) \right) \left(1 + o(1) \right),$$

where $\mathcal{H}$ has been defined in (7.10). We omit the details.

7.1.2 Numerical Computations

We present some numerical computations to illustrate the influence of precursors on multi-spike patterns for system (7.1). We set $\Omega = (-1, 1)$, $\tau = 0.1$ and choose varying diffusivities ($\epsilon^2 = 0.001$, $D = 0.1$ and $\epsilon^2 = 0.0001$, $D = 0.01$).

We first show A, then H.

We begin by considering the system with constant coefficients: $\mu(x) \equiv 1$, $\epsilon^2 = 0.001$, $D = 0.1$ (Fig. 7.1).

Next we consider a linearly increasing precursor gradient. We observe that now the spikes are shifted in the direction of smaller values of the precursor (Fig. 7.2).

Finally, we compute multi-spike steady states for a precursor with oscillating profile which is given by a trigonometric function. Again the spikes are shifted towards smaller values of μ (Fig. 7.3).

These simulations conclude our study of nontrivial precursor gradients. We now turn to another type of spatially varying coefficient.

Fig. 7.2 Multiple spikes for (7.1) with $\epsilon^2 = 0.001$, $D = 0.1$ and $\mu = 1 + 0.1x$

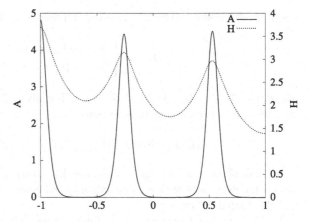

Fig. 7.3 Asymmetric two-spike steady states spikes for (7.1) with $\mu(x) = 1 + 0.1\cos(4\pi x)$

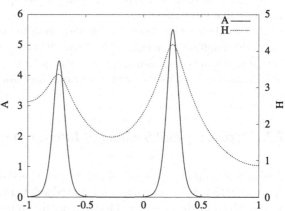

7.2 Discontinuous Diffusivities

We consider the Gierer-Meinhardt system in case the inhibitor diffusivity is piecewise constant but has a jump discontinuity. Further, the activator diffusivity is constant. We will explore the effect of this jump discontinuity on spiky patterns.

Spatially inhomogeneous diffusivities arising in real-world biological systems are often induced by a control biochemical regulating the diffusion processes of morphogens in reaction-diffusion systems. A class of very simple, yet conceptually very intriguing, inhomogeneities of diffusion coefficients are jump discontinuities. Frequently this modelling assumption approximates real-life situations very well, for example in the case where two different types of tissues such as groups of differentiated cells make contact at a joint border. It can also be used to approach more general inhomogeneities.

The Gierer-Meinhardt system adapted to this setting is given by

$$\begin{cases} a_t = \epsilon^2 a_{xx} - a + \frac{a^2}{h}, \\ \tau h_t = (D(x)h_x)_x - h + a^2. \end{cases} \tag{7.29}$$

Let $0 < \epsilon \ll 1$ and $\tau \geq 0$ both be constant. Further, assume that

$$D(x) = \begin{cases} D_1, & -1 < x < x_b, \\ D_2, & x_b < x < 1, \end{cases} \tag{7.30}$$

where $0 < D_1$, $0 < D_2$ and $D_1 \neq D_2$. We will study (7.29) in the interval $(-1, 1)$. We choose Neumann boundary conditions

$$a_x(-1) = a_x(1) = 0, \qquad h_x(-1) = h_x(1) = 0. \tag{7.31}$$

We will show that there are different spiky steady states: (i) an interior spike located far away from the jump discontinuity and boundary; and (ii) a spike located near the jump discontinuity.

We shall characterise the position of the interior spike by some Green's function and show that it is stable.

Next we derive a condition for the position of the spike near the jump discontinuity. This implies that in general there are either two possible locations or there are none. Further, this condition is most easily satisfied if the discontinuity is located near the centre of the interval and if the two diffusivities differ by a large amount.

7.2.1 Existence and Stability of Interior Spike

Due to the jump discontinuity of the inhibitor diffusivity the smoothness of the inhibitor is lost. More precisely, for classical solutions $D(x)h_x(x)$ is continuous at $x = x_b$ which implies that $h_x(x)$ has a jump discontinuity at $x = x_b$. To handle the resulting discontinuity of h, the function spaces for the solutions have to be chosen accordingly.

Thus we set

$$(a, h) \in H_N^2(-1, 1) \times H_N^{2,*}(-1, 1),$$

where

$$H_N^2(-1, 1) := \left\{ a \in H^2(-1, 1) : a_x(-1) = a_x(1) = 0 \right\},$$

$$H^{2,*}(-1, 1) := \left\{ h \in H^1(-1, 1) : \left(D(x)h_x \right)_x \in L^2(-1, 1) \right\},$$

$$H_N^{2,*}(-1, 1) := \left\{ h \in H^{2,*}(-1, 1) : h_x(-1) = h_x(1) = 0 \right\}$$

and

$$\left\| (a, h) \right\|_{2,*}^2 := \| a \|_{H^2(-1,1)}^2 + \| h \|_{2,*}^2,$$

where

$$\| h \|_{2,*}^2 := \| h \|_{H^1(-1,1)}^2 + \left\| \left(D(x)h_x \right)_x \right\|_{L^2(-1,1)}.$$

Let the function w be given by (2.3). We set

$$\rho(y) := \int_0^y w^2(z)dz. \tag{7.32}$$

Elementary calculations give

$$\alpha := \int_0^\infty w^2(y)dy = \int_0^\infty w(y)dy = 3,$$

$$\int_0^\infty w^3(y)dy = 3.6,$$

$$\rho(y) = \frac{9}{2}\tanh\frac{y}{2} - \frac{3}{2}\tanh^3\frac{y}{2}, \tag{7.33}$$

$$\int_0^\infty w^3(y)\rho(y)dy = \frac{297}{64} = 4.640625,$$

$$\int_0^\infty (w')^2 dy = \int_0^\infty w^3 dy - \int_0^\infty w^2 dy = 0.6.$$

Recall the notation $\Omega_\epsilon = (-\frac{1}{\epsilon}, \frac{1}{\epsilon})$ and define $u(x) \in H_\epsilon^2(\Omega)$ if and only if $u(\frac{x}{\epsilon}) \in H^2(\Omega_\epsilon)$, where the norm of the former space is defined by the norm of the latter, i.e.

$$\|u\|_{H_\epsilon^2(\Omega)} := \|u(\cdot/\epsilon)\|_{H^2(\Omega_\epsilon)}.$$

First we have an existence result.

Theorem 7.4 (Existence of an interior-spike solution) *Assume that*

$$\frac{1}{\theta_1}\tanh\theta_1(1 + x_b) > \frac{1}{\theta_2}\tanh\theta_2(1 - x_b), \tag{7.34}$$

where $\theta_i = D_i^{-1/2}$. Then there is a steady state of (7.29)–(7.31) which has an interior spike profile in the subinterval $(-1, x_b)$. Further,

$$a_\epsilon(x) = \xi_0 w\left(\frac{x - t^\epsilon}{\epsilon}\right) + o(1) \quad in\ H_\epsilon^2(\Omega), \tag{7.35}$$

where $t^\epsilon \to t_0 \in (-1, x_b)$ and $\xi_0/h(t^\epsilon) \to 1$ as $\epsilon \to 0$,

$$\frac{1}{\theta_1}\tanh\big(\theta_1(2t_0 + 1 - x_b)\big) = \frac{1}{\theta_2}\tanh\big(\theta_2(1 - x_b)\big). \tag{7.36}$$

Under condition (7.34) there are no steady states (7.29)–(7.31) which have an interior spike profile in the subinterval $(x_b, 1)$.

Remark 7.5

(i) In the special case $x_b = 0$ there exists a steady state with a spike in the subinterval with the larger diffusivity D_1 but there is none with a spike in the subinterval with smaller diffusivity. This follows from Condition (7.34) using the strict monotonicity of the function $\tanh \alpha / \alpha$ with respect to $\alpha > 0$.

(ii) The results for (7.34) with reverse sign follow by a symmetry argument: Reflection of the solution about the centre $x = 0$ of the interval results in exchange of θ_1 and θ_2 and change of sign for x_b.

The proof of Theorem 7.4 uses the method of Liapunov-Schmidt reduction as in Chap. 2. We begin with the following: Let $G(x, t_0)$ be the Green's function given by

$$
\begin{cases}
(D(x)G(x, t_0)_x)_x - G(x, t_0) + \delta_{t_0} = 0, \\
G_x(-1, t_0) = G_x(1, t_0) = 0, \\
D(t_0)G_x(t_0^-, t_0) - D(t_0)G_x(t_0^+, t_0) = 1, \\
G(t_0^-, t_0) - G(t_0^+, t_0) = 0, \\
D(x_b^-)G_x(x_b^-, t_0) - D(x_b^+)G_x(x_b^+, t_0) = 0, \\
G(x_b^-, t_0) - G(x_b^+, t_0) = 0,
\end{cases}
\tag{7.37}
$$

where δ_{t_0} is the Dirac delta distribution located at t_0.

Setting

$$
G(x, t_0) = \begin{cases}
A \frac{\cosh \theta_1 (x+1)}{\cosh \theta_1 (t_0+1)}, & -1 < x < t_0, \\
B \frac{\sinh \theta_1 (x-t_0)}{\sinh \theta_1 (x_b-t_0)} + A \frac{\sinh \theta_1 (x-x_b)}{\sinh \theta_1 (t_0-x_b)}, & t_0 < x < x_b, \\
B \frac{\cosh \theta_2 (x-1)}{\cosh \theta_2 (x_b-1)}, & x_b < x < 1
\end{cases}
\tag{7.38}
$$

then $G(x, t_0)$ is continuous at both $x = t_0$ and $x = x_b$. Next $D(x)G_x(x, t_0)$ jumps by -1 at $x = t_0$ and is continuous at $x = x_b$ and so we have

$$
\begin{aligned}
\frac{A}{\theta_1} \left(\tanh \theta_1 (t_0 + 1) + \coth \theta_1 (x_b - t_0) \right) - \frac{B}{\theta_1} \frac{1}{\sinh \theta_1 (x_b - t_0)} &= 1, \\
B \left(\frac{1}{\theta_1} \coth \theta_1 (x_b - t_0) + \frac{1}{\theta_2} \tanh \theta_2 (1 - x_b) \right) - \frac{A}{\theta_1} \frac{1}{\sinh \theta_1 (x_b - t_0)} &= 0.
\end{aligned}
\tag{7.39}
$$

From (7.39) we compute

$$
\begin{aligned}
G(t_0, t_0)^{-1} &= A^{-1} \\
&= \theta_1^{-1} \Bigg[\tanh \theta_1 (t_0 + 1) + \coth \theta_1 (x_b - t_0) \\
&\quad - \bigg(\sinh \theta_1 (x_b - t_0) \cosh \theta_1 (x_b - t_0)
\end{aligned}
$$

$$\left. + \frac{\theta_1}{\theta_2} \sinh^2 \theta_1 (x_b - t_0) \tanh \theta_2 (1 - x_b) \right)^{-1} \right]$$

$$= \theta_1^{-1} \left[\tanh \theta_1 (t_0 + 1) \right.$$

$$\left. + \frac{\theta_2 \sinh \theta_1 (x_b - t_0) + \theta_1 \cosh \theta_1 (x_b - t_0) \tanh \theta_2 (1 - x_b)}{\theta_2 \cosh \theta_1 (x_b - t_0) + \theta_1 \sinh \theta_1 (x_b - t_0) \tanh \theta_2 (1 - x_b)} \right]$$

$$=: \theta_1^{-1} u(t_0). \tag{7.40}$$

Setting $v(t_0) = \theta_2 \cosh \theta_1 (x_b - t_0) + \theta_1 \sinh \theta_1 (x_b - t_0) \tanh \theta_2 (1 - x_b)$, we have

$$u(t_0) = \tanh \theta_1 (t_0 + 1) - \theta_1^{-1} \frac{v'(t_0)}{v(t_0)}.$$

Note that $\theta_1^{-2} v''(t_0) = v(t_0)$. This implies for $u'(t_0) = \frac{d}{dt_0} u(t_0)$ that

$$\theta_1^{-1} u'(t_0) = 1 - \tanh^2 \theta_1 (t_0 + 1) - \theta_1^{-2} \frac{v''(t_0) v(t_0) - (v'(t_0))^2}{(v(t_0))^2}$$

$$= -\tanh^2 \theta_1 (t_0 + 1) + \theta_1^{-2} \frac{(v'(t_0))^2}{(v(t_0))^2}. \tag{7.41}$$

Note that $\frac{d}{dt_0} G(t_0, t_0) = 0$ if and only if $u'(t_0) = 0$ and $u(t_0) \neq 0$, since

$$\frac{d}{dt_0} G(t_0, t_0) = -\theta_1 \frac{u'(t_0)}{(u(t_0))^2}. \tag{7.42}$$

Further, we have

$$\theta_1^{-2} u''(t_0) = -2 \tanh \theta_1 (t_0 + 1) \left(1 - \tanh^2 \theta_1 (t_0 + 1) \right)$$

$$+ 2\theta_1^{-3} \frac{v(t_0) v'(t_0) v''(t_0) - v(t_0) (v'(t_0))^3}{(v(t_0))^3}$$

$$= -2 \tanh \theta_1 (t_0 + 1) \left(1 - \tanh^2 \theta_1 (t_0 + 1) \right)$$

$$+ 2\theta_1^{-3} \frac{v'(t_0) [\theta_1^2 (v(t_0))^2 - (v'(t_0))^2]}{(v(t_0))^3}.$$

Using the relations

$$\theta_1^2 (v(t_0))^2 - (v'(t_0))^2 = \theta_1^2 \theta_2^2 \left[1 - \left(\frac{\theta_1}{\theta_2} \tanh \theta_2 (1 - x_b) \right)^2 \right] > 0$$

(by (7.34)) and $v'(t_0) < 0$ it follows that $u''(t_0) < 0$. Next we compute the sign of $\frac{d^2}{dt_0^2} G(t_0, t_0)$, first calculating

$$\frac{d^2}{dt_0^2} G(t_0, t_0) = \theta_1 \frac{-u''(t_0)u(t_0) + 2(u'(t_0))^2}{(u(t_0))^3}.$$

Since $u'(t_0) = 0$ and $u''(t_0) < 0$, we have

$$\frac{d^2}{dt_0^2} G(t_0, t_0) > 0. \tag{7.43}$$

We need to investigate some other properties of the Green's function G. Let the regular part of the Green's function be given by

$$H(x, y) := \frac{\theta_i}{2} e^{-\theta_i |x-y|} - G(x, y).$$

Then we have

$$\frac{d}{dt_0} G(t_0, t_0) = \frac{d}{dt_0} \frac{\theta_i}{2} - \frac{d}{dt_0} H(t_0, t_0)$$

$$= -2\nabla_x|_{x=t_0} H(x, t_0) =: -2\nabla_{t_0} H(t_0, t_0), \tag{7.44}$$

where $i = 1$ if $t_0 < x_b$ and $i = 2$ if $t_0 > x_b$. Here we have used the notation

$$\nabla_x|_{x=t_0} H(x, t_0) := \left. \frac{\partial}{\partial x} \right|_{x=t_0} H(x, t_0)$$

and the fact that $H(x, y) = H(y, x)$. We also have

$$\frac{d^2}{dt_0^2} G(t_0, t_0) = -2(\nabla_x \nabla_y)_{x=y=t_0} H(x, y) - 2(\nabla_x^2)_{x=t_0} H(x, t_0)$$

$$=: -2\nabla_{t_0}^2 H(t_0, t_0). \tag{7.45}$$

For $t \in (-1, 1)$, we set

$$\hat{\xi}_0(t) = \frac{1}{G(t, t)}, \tag{7.46}$$

where $G(x, y)$ is given in (7.37).

Note that $\hat{\xi}(t)$ is in $C^1(-1, 1)$. Then we have

$$\nabla_t \hat{\xi}(t) = \frac{d}{dt} (G(t, t))^{-1} = 2\nabla_t (G(t, t))^{-1} = -2(\nabla_t G(t, t)) \hat{\xi}(t)^2.$$

The previous computations enable us to compute the derivative of

$$F(t) := (-2\nabla_t G(t, t)) \hat{\xi}^2(t) = \nabla_t \hat{\xi}(t)$$

as

$$\nabla_t F(t) = \nabla_t \frac{-2\nabla_t G(t,t)}{G^2(t,t)} = \frac{-2G(t,t)\nabla_t^2 G(t,t) + 4(\nabla_t G(t,t))^2}{G^3(t,t)}$$

and for $t = t_0$ we have

$$\nabla_{t_0} F(t_0) = \frac{-2G(t_0,t_0)\nabla_{t_0}^2 G(t_0,t_0)}{G^3(t_0,t_0)} \tag{7.47}$$

if $\nabla_{t_0} G(t_0,t_0) = 0$.

Similarly to Chap. 2, the previous computations will lead to the reduced problem which is given by

$$W_\epsilon(t) = -d_{00}\big[\nabla_x H(x,t_0)|_{x=t_0}\big] + O(\epsilon), \tag{7.48}$$

where

$$d_{00} = 2.4\hat{\xi}_0^2.$$

Then, following Chap. 4, we derive the following stability result:

Theorem 7.6 (Stability of an interior-spike solution) *The interior spike established in Theorem 7.4 is linearly stable.*

7.2.2 A Spike near the Jump Discontinuity of the Inhibitor Diffusivity

Our first result concerns the existence of a spike near the jump discontinuity of the inhibitor diffusivity.

Theorem 7.7 (Existence of spike near x_b) *Let*

$$I(L) := \int_L^\infty w^3(y)\big(\rho(y) - \beta\big)dy, \tag{7.49}$$

where

$$\beta = \alpha \frac{\theta_2 \tanh\theta_1(1 + x_b) - \theta_1 \tanh\theta_2(1 - x_b)}{\theta_2 \tanh\theta_1(x_b + 1) + \theta_1 \tanh\theta_2(1 - x_b)}, \tag{7.50}$$

and $\rho(y)$ is given in (7.32). Let L_0 be (uniquely) determined by $\rho(L_0) = \beta$, and set $\alpha = 3$.

(i) *If*

$$\begin{cases} \theta_1 < \theta_2 \quad and \\ 0 < \frac{\theta_2 \tanh\theta_1(1+x_b) - \theta_1 \tanh\theta_2(1-x_b)}{\theta_2 \tanh\theta_1(1+x_b) + \theta_1 \tanh\theta_2(1-x_b)} < \frac{\theta_2^2 - \theta_1^2}{2\theta_1^2} \frac{I(L_0)}{10.8}, \end{cases} \tag{7.51}$$

*there are precisely two spikes near x_b whose positions are given by (7.35) with
$t^\epsilon = x_b - \epsilon L + o(\epsilon)$ and L is one of the two solutions of (7.60).*

(ii) *Suppose that condition (7.34) is valid and $\theta_1 > \theta_2$, or that*

$$\begin{cases} \theta_1 < \theta_2 \quad \text{and} \\ \frac{\theta_2 \tanh\theta_1(1+x_b)-\theta_1 \tanh\theta_2(1-x_b)}{\theta_2 \tanh\theta_1(1+x_b)+\theta_1 \tanh\theta_2(1-x_b)} > \frac{\theta_2^2-\theta_1^2}{2\theta_1^2}\frac{I(L_0)}{10.8} > 0. \end{cases} \tag{7.52}$$

*There are no spikes which are located near x_b. That is, it is impossible to have
a steady state which satisfies (7.35) and $|t^\epsilon - x_b| = O(\epsilon)$.*

Remark 7.8

(i) By an inner-outer expansion for the eigenfunction (similar to the expansion of
the solution in the existence proof) it is shown in [230] that for the spike near
the jump there is exactly one small eigenvalue which satisfies

$$\lambda_\epsilon = -\frac{1}{1.2}\epsilon\hat{\xi}_0\big(\theta_2^2 - \theta_1^2\big)I'(L) + o(\epsilon). \tag{7.53}$$

All other eigenvalues are stable. Thus the spike further to the right corresponding to smaller L is stable, while the spikes further to the left corresponding to
larger L is unstable.

(ii) Further, in [230] the case of an arbitrary number of jumps (instead of just one
jump) is considered.

Next we make some remarks about possible positions of the interior spike which
follow from analysing (7.36).

We have $t_0 \to 0$ in both of the following limits:

(i) $\theta_2 \to \theta_1$ (for $\theta_1 = $ const.) (the system approaches the Gierer-Meinhardt system
with constant diffusivities).

(ii) $\theta_1 \to 0$ and $\theta_2 \to 0$ (the system approaches the shadow Gierer-Meinhardt system with $D = \infty$).

The spike moves to the centre of the domain.

(iii) In the limit $\theta_2 \to \infty$ (for $\theta_1 = $ const.) we have $t_0 \to (x_b - 1)/2 \in (-1, x_b)$.
Note that $-1 < (x_b - 1)/2 < 0$. The spike moves to the centre of the subin-
terval with finite inhibitor diffusivity if the inhibitor diffusivity in the other
subinterval tends to zero.

(iv) If the inequality in (7.34) approaches equality, the spike moves to x_b.

It follows that for a suitable choice of diffusion constants D_1 and D_2, we can
have a single spike located at any point of the open interval $((x_b - 1)/2, x_b)$. This
illustrates that for discontinuous inhibitor diffusivity an interior spike is typically
located away from the centre. This stands in marked contrast to the constant coefficient case for which an interior spike is always located in the centre of the domain.
Important implications of this phenomenon for biological applications have been
described, e.g. for understanding limb development (see Sect. 7.3).

Next we give a proof.

Proof of Theorem 7.7 Let

$$a_\epsilon(x) = \xi_0 w\left(\frac{x - t^\epsilon}{\epsilon}\right) \chi\left(\frac{x - t^\epsilon}{\epsilon}\right) + O(\epsilon) \quad \text{in } H^2(\Omega_\epsilon),$$

where $x_b - t^\epsilon = \epsilon L$. We compute an inner-outer expansion:

$$h_\epsilon(x) = \xi_0\left(\epsilon h_1\left(\frac{x - t^\epsilon}{\epsilon}\right) + h_2(x)\right) + O(\epsilon) \quad \text{in } H^{2,*}(\Omega_\epsilon), \qquad (7.54)$$

where $h_1(y)$ for $y = (x - t^\epsilon)/\epsilon$ is given by

$$\begin{cases} (D(t^\epsilon + \epsilon y)h_{1,y}(y))_y + w^2(y) = 0, \\ h_1(0) = 0, \qquad h_{1,y}(0) = 0 \end{cases} \qquad (7.55)$$

and $h_2(x)$ satisfies

$$\begin{cases} (D(x)h_{2,x}(x))_x - h_2(x) - \epsilon h_1(x) = 0, \\ h_{2,x}(\pm 1) = -h_{1,y}(\pm\infty). \end{cases} \qquad (7.56)$$

We integrate (7.55) and get

$$h_{1,y}(y) = \begin{cases} -\theta_1^2 \rho(y), & -\infty < y < L, \\ -\theta_2^2 \rho(y), & L < y < \infty, \end{cases} \qquad (7.57)$$

where $\theta_i = D_i^{-1/2}$ and $\rho(y)$ is given in (7.32).

Recalling from (7.33) that

$$\alpha = \int_0^\infty w^2(z)dz = 3$$

we have

$$h_{1,y}(-\infty) = \theta_1^2 \alpha, \qquad h_{1,y}(\infty) = -\theta_2^2 \alpha.$$

Integrating (7.57) once more gives in leading order

$$\epsilon h_1\left(\frac{x - t^\epsilon}{\epsilon}\right) = \begin{cases} \theta_1^2 \alpha(x - x_b), & -1 < x < x_b, \\ -\theta_2^2 \alpha(x - x_b), & x_b < x < 1. \end{cases} \qquad (7.58)$$

Hence h_2 satisfies (up to order $O(\epsilon)$, which is included in the error term in (7.54))

$$\begin{cases} (D(x)h_{2,x}(x))_x - h_2(x) - \epsilon h_1(x) = 0, \\ h_{2,x}(-1) = -\theta_1^2 \alpha, \qquad h_{2,x}(1) = \theta_2^2 \alpha. \end{cases} \qquad (7.59)$$

Using (7.58) and (7.59) we get

$$
h_2(x) = \begin{cases} -\theta_1^2 \alpha (x - x_b) + A\theta_1 \frac{\cosh \theta_1 (x+1)}{\cosh \theta_1 (x_b+1)}, & -1 < x < x_b, \\ \theta_2^2 \alpha (x - x_b) + B\theta_1 \frac{\cosh \theta_2 (x-1)}{\cosh \theta_2 (x_b-1)}, & x_b < x < 1. \end{cases}
$$

Since $h_2(x)$ is continuous at $x = x_b$ we get $A = B$. Further, since $D(x)h_{2,x}(x)$ is continuous at $x = x_b$, we have

$$
0 = D_1 h_{2,x}\left(x_b^-\right) - D_2 h_{2,x}\left(x_b^+\right) = A\left(\tanh \theta_1 (x_b + 1) + \frac{\theta_1}{\theta_2} \tanh \theta_2 (1 - x_b) \right) - 2\alpha
$$

which implies

$$
A = \frac{2\alpha\theta_2}{\theta_2 \tanh \theta_1 (x_b + 1) + \theta_1 \tanh \theta_2 (1 - x_b)}.
$$

Note that by (7.34) we have

$$
A > \frac{\alpha}{\tanh \theta_1 (x_b + 1)}.
$$

Thus we have

$$
D_1 h_{2,x}\left(x_b^-\right) = D_2 h_{2,x}\left(x_b^+\right) = A \tanh \theta_1 (x_b + 1) - \alpha
$$
$$
= \alpha \frac{\theta_2 \tanh \theta_1 (x_b + 1) - \theta_1 \tanh \theta_2 (1 - x_b)}{\theta_2 \tanh \theta_1 (x_b + 1) + \theta_1 \tanh \theta_2 (1 - x_b)}
$$

which gives

$$
h_{2,x}\left(x_b^-\right) = \theta_1^2 \beta, \qquad h_{2,x}\left(x_b^+\right) = \theta_2^2 \beta,
$$

where

$$
\beta = \alpha \frac{\theta_2 \tanh \theta_1 (x_b + 1) - \theta_1 \tanh \theta_2 (1 - x_b)}{\theta_2 \tanh \theta_1 (x_b + 1) + \theta_1 \tanh \theta_2 (1 - x_b)}.
$$

Solving the reduced problem as in Chap. 2, we derive the following condition which determines the position of the spike.

$$
0 = \xi_\epsilon^{-1} \int_{-\infty}^{\infty} w^3(y) h_x \left(t^\epsilon + \epsilon y\right) dy + O(\epsilon)
$$
$$
= \int_{-\infty}^{\infty} w^3(y) \left(h_{1,y}(y) + h_{2,x}\left(t^\epsilon + \epsilon y\right)\right) dy + O(\epsilon)
$$
$$
= \int_{-\infty}^{L} w^3(y) \left(-\theta_1^2 \rho(y) + h_{2,x}\left(x_b^-\right)\right) dy
$$
$$
+ \int_{L}^{\infty} w^3(y) \left(-\theta_2^2 \rho(y) + h_{2,x}\left(x_b^+\right)\right) dy + O(\epsilon)
$$

$$= \theta_1^2 \left(\int_{-\infty}^{L} w^3(y) \big(-\rho(y) + \beta \big) dy \right) + \theta_2^2 \left(\int_{L}^{\infty} w^3(y) \big(-\rho(y) + \beta \big) dy \right) + O(\epsilon)$$

$$= \theta_1^2 \left(\int_{-\infty}^{\infty} w^3(y) \big(-\rho(y) + \beta \big) dy - \int_{L}^{\infty} w^3(y) \big(-\rho(y) + \beta \big) dy \right)$$

$$+ \theta_2^2 \left(\int_{L}^{\infty} w^3(y) \big(-\rho(y) + \beta \big) dy \right) + O(\epsilon)$$

$$= \beta \theta_1^2 \int_{-\infty}^{\infty} w^3(y) dy + (\theta_2^2 - \theta_1^2) \int_{L}^{\infty} w^3(y) \big(-\rho(y) + \beta \big) dy + O(\epsilon).$$

Here we have used the fact that the function $\rho(y)$ is odd.

Thus we have to determine L such that

$$\beta \theta_1^2 \int_{-\infty}^{\infty} w^3(y) dy + (\theta_2^2 - \theta_1^2) \int_{L}^{\infty} w^3(y) \big(-\rho(y) + \beta \big) dy = 0, \qquad (7.60)$$

where $\theta_1, \theta_2, \beta$ are fixed and

$$\beta = \alpha \frac{\theta_2 \tanh \theta_1 (1 + x_b) - \theta_1 \tanh \theta_2 (1 - x_b)}{\theta_2 \tanh \theta_1 (1 + x_b) + \theta_1 \tanh \theta_2 (1 - x_b)}, \qquad \alpha = 3,$$

$$\rho(y) = \int_{0}^{y} w^2(z) dz = \frac{9}{2} \tanh \frac{y}{2} - \frac{3}{2} \tanh^3 \frac{y}{2}.$$

Next we study (7.60) in detail. We first observe that the integrand of

$$\int_{L}^{\infty} w^3(y) \big(-\rho(y) + \beta \big) dy$$

changes sign at $\rho(y) = \beta$. Further, the function ρ satisfies

$$\rho(0) = 0, \qquad \rho'(y) = w^2(y) > 0, \qquad \rho(-y) = -\rho(y),$$

$$\rho(y) \to \int_{0}^{\infty} w^2 dy = \alpha \ (= 3) \quad \text{as } y \to \infty$$

(7.61)

and for β we have

$$0 < \beta = \alpha \frac{\theta_2 \tanh \theta_1 (1 + x_b) - \theta_1 \tanh \theta_2 (1 - x_b)}{\theta_2 \tanh \theta_1 (x_b + 1) + \theta_1 \tanh \theta_2 (1 - x_b)} < \alpha.$$

Thus for any $0 < \beta < \alpha$ there exists exactly one $y =: L_0 > 0$ such that $\rho(L_0) - \beta = 0$. Further, we have $\rho(y) - \beta < 0$ for $0 < y < L_0$ and $\rho(y) - \beta > 0$ for $y > L_0$. Then L_0 can be computed explicitly from

$$\rho(L_0) = \frac{9}{2} \tanh \frac{L_0}{2} - \frac{3}{2} \tanh^3 \frac{L_0}{2} = \beta$$

(see (7.33)). It is easy to see that L_0 is unique.

Recalling from (7.49) that

$$I(L) := \int_L^\infty w^3(y)\big(\rho(y) - \beta\big)dy, \quad L \in \mathbb{R},$$

we have

(i) $I(L) \to 0$ as $L \to \infty$, $I(L) \to -7.2\beta < 0$ as $L \to -\infty$,
(ii) $I(L)$ achieves its unique maximum among all real L at $L = L_0 > 0$, where $I(L_0) > 0$,
(iii) $I(L)$ is monotone increasing on $(-\infty, L_0)$,
(iv) $I(L)$ is monotone decreasing on (L_0, ∞),
(v) $I(L) = 0$ for a unique $L = L_1 < 0$.

Thus the equation $I(L) = c$ has

$$\begin{cases} \text{two solutions} & \text{if } 0 < c < I(L_0), \\ \text{one solution} & \text{if } c = I(L_0) \text{ or } -7.2\beta < c \leq 0, \\ \text{no solution} & \text{if } c > I(L_0) \text{ or } c \leq -7.2\beta. \end{cases} \quad (7.62)$$

Combining (7.60) and (7.62), we get for

$$c = \frac{\theta_2 \tanh \theta_1(1 + x_b) - \theta_1 \tanh \theta_2(1 - x_b)}{\theta_2 \tanh \theta_1(1 + x_b) + \theta_1 \tanh \theta_2(1 - x_b)} \frac{10.8 \cdot 2\theta_1^2}{\theta_2^2 - \theta_1^2}.$$

Equation (7.60) has two solutions if

$$0 < \frac{\theta_2 \tanh \theta_1(1 + x_b) - \theta_1 \tanh \theta_2(1 - x_b)}{\theta_2 \tanh \theta_1(1 + x_b) + \theta_1 \tanh \theta_2(1 - x_b)} < \frac{\theta_2^2 - \theta_1^2}{2\theta_1^2} \frac{I(L_0)}{10.8}.$$

Equation (7.60) has no solution if

$$\frac{\theta_2 \tanh \theta_1(1 + x_b) - \theta_1 \tanh \theta_2(1 - x_b)}{\theta_2 \tanh \theta_1(1 + x_b) + \theta_1 \tanh \theta_2(1 - x_b)} > \frac{\theta_2^2 - \theta_1^2}{2\theta_1^2} \frac{I(L_0)}{10.8}.$$

This concludes the proof of Theorem 7.7.   □

Finally, we consider the existence versus non-existence and the locations of spikes near the jump discontinuity in typical limiting situations.

(i) Taking the limit $\theta_2 \to \infty$ (and $\theta_1 = \text{const.}$), we get $\beta \to 3$. Then (7.60) gives

$$7.2\beta \frac{\theta_1^2}{\theta_2^2 - \theta_1^2} = I(L)$$

and so (7.60) has solutions if and only if

$$7.2\beta \frac{\theta_1^2}{\theta_2^2 - \theta_1^2} < I(L_0). \quad (7.63)$$

We compute

$$e^{-2L_0} \sim c(3 - \beta), \qquad I(L_0) \sim c(3 - \beta)^{5/2}$$

and

$$\frac{\theta_1}{\theta_2} \sim c(3 - \beta)$$

as $\beta \to 3$, where $c > 0$ denotes constants which may vary one term to the other. Thus in (7.63) we get

$$\text{l.h.s.} \sim c(3 - \beta)^2 \quad \text{and} \quad \text{r.h.s.} \sim c(3 - \beta)^{5/2}.$$

This implies that (7.63) has no solutions if β is sufficiently close to 3. Considering β as a bifurcation parameter is can easily be seen that there is some point β_0 at which a saddle-node bifurcation occurs: For $0 < \beta < \beta_0$ there are two spike locations which approach the same point as β comes close to β_0. For $\beta_0 < \beta < 3$ there are no spikes near the jump.

(ii) In the limit $x_b \to 1$ we have $\tanh \theta_2(1 - x_b) \to 0$ which gives $\beta \to 3$. Further, we compute

$$e^{-2L_0} \sim c(3 - \beta), \qquad I(L_0) \sim c(3 - \beta)^{5/2}$$

and

$$\frac{\theta_1}{\theta_2} \sim c$$

as $\beta \to 3$. Then, following the analysis in Case (i), we conclude that spikes near the jump cease to exist as β increases.

(iii) In the limit $\beta \to 0$, we have from (7.60) that

$$(\theta_2^2 - \theta_1^2) \int_L^\infty w^3(y)\rho(y)dy \to 0.$$

Supposing $\theta_2^2 - \theta_1^2 \neq 0$, we get $L \to -\infty$ or $L \to \infty$. Here the spikes at the jump (moving to infinity on the $O(\epsilon)$ scale) connect to the interior spikes (moving to the jump on the $O(1)$ scale).

7.2.3 Numerical Simulations

We present some numerical simulations for the time-dependent behaviour system (7.29). We make the following selections: Set domain $\Omega = (-1, 1)$, time-relaxation constant $\tau = 0.1$ and activator diffusivity $\epsilon^2 = 0.0001$. Choose the jump of the inhibitor diffusivity at $x_b = 0$ or $x_b = 0.5$ and consider different constants for $D(x)$ on either side of x_b.

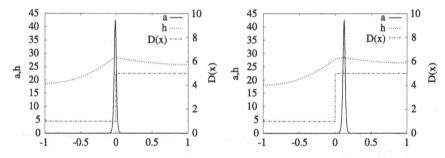

Fig. 7.4 Long-time limit of the solution to (7.29)–(7.31) with $\epsilon^2 = 0.0001$ and $D(x) = 1$ for $-1 < x < 0$, $D(x) = 5$ for $0 < x < 1$. We get a spike near the jump discontinuity of the inhibitor diffusivity and a spike in the right subinterval, respectively. The conditions (7.34) and (7.51), respectively, are valid. Equation (7.36) gives $t_0 \approx 0.10336$ in the second example, which agrees well with the figure

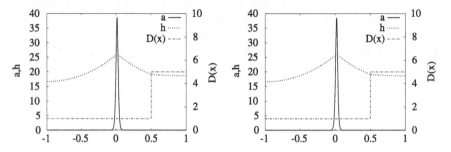

Fig. 7.5 Long-time limit of the solution to (7.29)–(7.31) with $\epsilon^2 = 0.0001$ and $D(x) = 1$ for $-1 < x < 0.5$, $D(x) = 5$ for $0.5 < x < 1$. We observe an interior spike in the left subinterval. The condition (7.34) is satisfied. Equation (7.36) implies $t_0 \approx 0.01556$ for the position of the interior spike, which is in good agreement with the figure

The figures show the solution for time $t = 10^5$. By this time we have observed that the computation has come to a standstill and the solution has approached a steady state with high precision.

We choose initial conditions as follows: For the activator, we take consider two cases (i) $a(x, 0) = 2 - \sin(x\pi/2)$, i.e. the maximum is located at the left boundary, or (ii) $a = 2 + \sin(x\pi/2)$, i.e. the maximum is located at the right boundary. In the following figures, we plot case (i) on the left-hand side and case (ii) on the right-hand side. For the inhibitor, we select $h(x, 0) \equiv 1$.

First we have Fig. 7.4.

Next we keep the same initial conditions but change the location of the jump discontinuity from $x_b = 0$ to $x_b = 0.5$. Then in both examples the spike moves to the same interior spike which is located close to (in fact, slightly to the right of) the centre $x = 0$ (Fig. 7.5).

Finally, we decrease the inhibitor diffusivity to $D(x) = 0.1$ for $-1 < x < x_b$ and $D(x) = 0.5$ for $x_b < x < 1$ for $x_b = 0$ or $x_b = 0.5$. As expected we get multiple spikes. Some examples are displayed in Figs. 7.6 and 7.7.

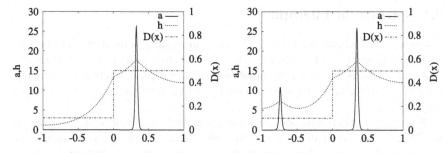

Fig. 7.6 Long-time limit of the solution to (7.29)–(7.31) with $\epsilon^2 = 0.0001$ and $D(x) = 0.1$ for $-1 < x < 0$, $D(x) = 0.5$ for $0 < x < 1$. We have an interior spike in the right subinterval or two interior spikes in different subintervals. Equation (7.36) gives $t_0 \approx 0.33057$ for the first example, which agrees well with the figure

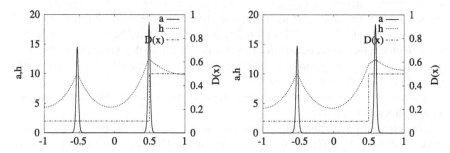

Fig. 7.7 Long-time limit of the solution to (7.29)–(7.31) with $\epsilon^2 = 0.0001$ and $D(x) = 0.1$ for $-1 < x < 0.5$, $D(x) = 0.5$ for $0.5 < x < 1$. We get an interior spike in the left subinterval combined with a spike near the jump discontinuity or two interior spikes in different subintervals

We summarise and discuss the observations from these numerical simulations: Dynamically, we have seen a spike moving from the left or right boundary toward the centre of the interval driven by the gradient of the regular part of the Green's function. It stops at the first (locally) stable position which could be near the jump in the interior of the subdomains away from the jump. If there is no stable position near the jump the spike will cross the jump and move from one subdomain into the other.

For small enough inhibitor diffusivity, steady states with two spikes have been computed which show that there are many possible two-spike patterns such as two interior spikes in the same subinterval, two interior spikes in different subintervals or a spike near the jump discontinuity in combination with an interior spike. Here the two spikes in general have different amplitudes unless they are both located in the same subinterval. Similar remarks apply to multi-spike solutions consisting of more than two spikes.

7.3 Notes on the Literature

For precursor gradients we follow [277]. It is interesting to recall that the Gierer-Meinhardt model [73, 145, 152] was originally introduced with precursor gradients. They played a crucial role in the initial biological example, namely in determining the position of the head structure in (re)generation experiments for the fresh-water polyp *Hydra*.

Precursor gradients have also been studied in the Brusselator model, introduced in [206], to constrain pattern formation within some subdomain. For example, in [90] the gradient causes the system to move in or out of the pattern-forming region according to linear parameter space, namely the Turing bifurcation is crossed. This results in a confinement of the region where peak formation is possible. A similar localisation effect occurs in the modelling of segmentation patterns for the fruit fly *Drosophila melanogaster* [89] and [137].

Precursor gradients are extremely popular in ecological modelling, where they represent the interaction between eco-systems and their heterogeneous environments. Typical variables include the flows of heat, gas or fluid, the movement of soil and chemical reactions. Reaction-diffusion systems have been successfully applied to model important pattern-forming phenomena and (de)stabilitation effects in ecosystems. The interaction of diverse length scales is frequently of crucial importance. For a survey we refer to [193].

Precursors have been shown to cause striped patterns for the Brusselator model in two dimensions [127].

Numerical investigations of precursor gradients are given in, among other references, [12, 84, 196, 197]. In Sect. 7.2.3, we have performed some numerical simulations to confirm our analytical results. We also refer to recent progress on pulse dynamics in heterogeneous media [187, 290].

The case of an inhibitor diffusivity with a jump discontinuity has been considered in [276]. For systems with jump discontinuities of diffusivities, Turing instabilities have been simulated numerically and studied analytically by Benson, Maini and Sherratt [10, 11, 141]. Implications have been drawn on dispersion relations and typical solution profiles. In particular, isolated patterns and asymmetric spatially oscillating patterns have been established which do not occur for spatially homogeneous Turing systems. These mathematical effects have been applied in biology to explain the anterior-posterior asymmetry of skeletal elements in vertebrate limbs and to theoretically confirm experimental results on double anterior limbs [12, 141, 289].

This dynamical behaviour of a spike for a jump in the inhibitor diffusivity is simpler than in [187], where travelling pulses close to bifurcation points such as drift instability or splitting instability are possible as well as pinning, splitting, rebound, penetration or oscillation of pulses.

The boundaries causing jumps in diffusivities are often formed due to spatially varying genetic expressions. The following typical examples are particularly well understood [17, 18, 101]. The A-P and D-V boundaries in the *Drosophila* wing imaginal disc and the compartments in the vertebrate hindbrain are examples of lineage boundaries. They are established because the gene expressions vary in different compartments. Then because of differing strengths of adhesion only a small proportion of the cells are able to cross the boundary and move into the neighbouring compartment. The imaginal disc is composed of cells which form adult structures during metamorphosis. It is crucial to unravel the mechanisms underlying their interaction in order to understand how a larva develops into a fly. After the compartments have been established, special border cells are frequently created which play an important role in morphogenesis due to their role as a signalling centre determining the later stages of pattern formation.

A model explaining the roles of boundaries as organising regions for secondary embryonic fields has been suggested by Meinhardt [149] following the concept of positional information of Wolpert [287]. It is postulated that near a sharp border a morphogen concentration peak is formed. Our results in this chapter support this postulate since they imply that even in a simple activator-inhibitor system it is possible to have a stable spike near a jump of the inhibitor diffusivity representing a sharp border. Using this postulate, it was possible to explain various behaviours in developmental biological such as the formation of duplicated and triplicated insect legs or the regeneration-duplication phenomenon of imaginal disc fragments.

An effect related to the results derived in this chapter is the well known distinction between so-called body modes (states localised away from boundaries) and wall modes (states localised near boundaries) which play a role in various branches of physics such as fluid mechanics [156] or plasma physics [21].

To conclude, the results derived in this chapter and their implications such as irregular positions or unequal amplitudes of spikes as well as localisation of patterns within a subdomain play an important role in many diverse areas of biological modelling such as the development of skeletal patterns in growing vertebrate limbs or the compartment boundaries of differentiated cells.

Chapter 8
Other Aspects of the Gierer-Meinhardt System

In this chapter we will consider the Gierer-Meinhardt system in new contexts. Previously, we always assumed that the diffusivity of the activator is small enough. In Sect. 8.1 we remove this assumption and prove that monotone patterned solutions are stable using an approach based on nonlocal eigenvalue problems. In Sect. 8.2 we consider the system for finite diffusivities in the limit that the reaction rate of the activator tends to infinity. We will show that there are concentration phenomena similar to those seen in the previous chapters. In Sect. 8.3 we will study Robin type boundary conditions (instead of Neumann type). We will see that this introduces new instabilities for spiky states. In Sect. 8.4 we will consider the Gierer-Meinhardt system on a manifold (instead of a bounded domain). We will unravel a coupling phenomenon between the Green's function (representing reaction-diffusion behaviour of the system) and the Gaussian curvature (representing the geometry of the manifold).

8.1 The Gierer-Meinhardt System with Finite Diffusivity

We consider monotone solutions for the shadow Gierer-Meinhardt system

$$
\begin{cases}
A_t = \epsilon^2 \Delta A - A + \frac{A^2}{\xi}, & x \in \Omega, t > 0, \\
\tau \xi_t = -\xi + \frac{1}{|\Omega|} \int_\Omega A^2 dx, \\
A > 0, \quad \frac{\partial A}{\partial \nu} = 0 \text{ on } \partial\Omega,
\end{cases}
\tag{8.1}
$$

where $\epsilon > 0$, $\tau > 0$ are positive constants, $\Delta := \sum_{i=1}^N \frac{\partial^2}{\partial x^2}$ is the usual Laplace operator and $\Omega \subset \mathbb{R}^n$ is a bounded and smooth domain. Note that here $\epsilon > 0$ is a fixed positive number and we do not assume that ϵ is small which stands in marked contrast to all the previous chapters.

Problem (8.1) is derived, at least formally, by taking the limit $D \to +\infty$ in the Gierer-Meinhardt system (1.5). For further details concerning the derivation of (8.1) from (1.5), we refer to [168, 178, 179, 249].

J. Wei, M. Winter, *Mathematical Aspects of Pattern Formation in Biological Systems*, Applied Mathematical Sciences 189, DOI 10.1007/978-1-4471-5526-3_8, © Springer-Verlag London 2014

We first consider the one-dimensional case $N = 1$. In Sect. 8.1.3, we will study some extensions to higher dimensions. Due to rescaling and translation with respect to the spatial variable, we may assume that $\Omega = (0, 1)$. Thus we have

$$\begin{cases} A_t = \epsilon^2 A_{xx} - A + \frac{A^2}{\xi}, & 0 < x < 1, t > 0, \\ \tau \xi_t = -\xi + \int_0^1 A^2 dx, \\ A > 0, \quad A_x(0, t) = A_x(1, t) = 0. \end{cases} \tag{8.2}$$

Setting $u(x) = \xi^{-1} A(x)$, then (A, ξ) is a monotone decreasing steady-state of (8.2) if and only if:

$$\xi^{-1} = \int_0^1 u^2(x) dx$$

and

$$\epsilon^2 u_{xx} - u + u^2 = 0, \qquad u_x(x) < 0, \quad 0 < x < 1, u_x(0) = u_x(1) = 0. \tag{8.3}$$

We let

$$L := \frac{1}{\epsilon} \tag{8.4}$$

and rescale $u(x) = w_L(y)$, where $y = Lx$. Then w_L solves

$$w_L'' - w_L + w_L^2 = 0, \qquad w_L'(y) < 0, \quad 0 < y < L, w_L'(0) = w_L'(L) = 0. \tag{8.5}$$

Now (8.5) has a nontrivial solution if and only if

$$\epsilon < \frac{1}{\pi} \quad \text{(which is equivalent to } L > \pi\text{).} \tag{8.6}$$

On the other hand, if $\epsilon \geq \frac{1}{\pi}$ (or $L \leq \pi$), then $w_L = 1$. This follows for example from (8.34) below.

By Theorem 1.1 of [177] we know that any stable solution to (8.2) is asymptotically monotone. More precisely, if $(A(x, t), \xi(t))$, $t \geq 0$ is a linearly neutrally stable solution to (8.2), then there exists a $t_0 > 0$ such that

$$A_x(x, t_0) \neq 0 \quad \text{for all } (x, t) \in (0, 1) \times [t_0, +\infty). \tag{8.7}$$

This implies that all non-monotone steady-state solutions are linearly unstable. Hence we will concentrate on monotone solutions. Obliviously there are two monotone solutions, the monotone increasing and the monotone decreasing one, and they are related by reflection. Without loss of generality, we will study the monotone decreasing solution which we denote by u_ϵ. By [178] it has the least energy among all positive solutions of (8.3). If $L \leq \pi$, then $w_L = 1$. For the solutions to (8.2) we set

$$A_L(x) = \xi_L w_L(Lx), \qquad \xi_L^{-1} = \int_0^1 w_L^2(Lx) dx. \tag{8.8}$$

In [179] and [178], under the assumption that L is sufficiently large, it has been shown that that (A_L, ξ_L) is linearly stable for τ small enough by the SLEP (singular limit eigenvalue problem) approach. In [249], it has been proved that, for ϵ sufficiently small, u_ϵ is linearly stable for τ small enough, using the NLEP (nonlocal eigenvalue problem) method.

Then the question arises if these stability results can be extended to the case of finite ϵ (corresponding to finite L). This is of huge practical relevance since in real-life experiments the physical constants are fixed and it is often hard to justify that they are small in a suitable sense. Therefore the results in this chapter will be useful for experimentalists and inform the setting up of models, testing of hypotheses and prediction of results. In fact, we will derive results on the stability of steady states for all finite ϵ (or L).

We begin our analysis by introducing some notation. For $I = (0, L)$ and $\phi \in H^2(I)$ we set

$$\mathcal{L}[\phi] = \phi'' - \phi + 2w_L\phi. \tag{8.9}$$

In Sect. 8.1.1, we will show that the spectrum of $\mathcal{L}$ is given by

$$\lambda_1 > 0, \qquad \lambda_j < 0, \quad j = 2, 3, \dots. \tag{8.10}$$

This implies for $\mathcal{L} : H^2(I) \to L^2(I)$ that its inverse

(H1c) $\mathcal{L}^{-1}$ exists.

Next we state

Theorem 8.1 *Assume that $L > \pi$ and*

(H2c) $\int_0^L w_L \mathcal{L}^{-1} w_L \, dy > 0.$

Then the steady state (A_L, ξ_L) of (8.2) given in (8.8) is linearly stable for τ small enough.

Thus to determine the stability we only have to compute the integral $\int_0^L w_L \times \mathcal{L}^{-1} w_L \, dy$. Whereas for general L this is quite hard, in the limiting cases $L \to +\infty$ or $L \to \pi$ this can be achieved by asymptotic analysis (see Lemma 8.5 below). If L is sufficiently large, we will see that (H2c) is valid. In particular, Theorem 8.1 recovers results of [179] and [249]. On the other hand, if L is near π, then $w_L \sim 1$, $\mathcal{L}^{-1} w_L \sim 1$, and thus $\int_0^L w_L \mathcal{L}^{-1} w_L \, dy > 0$.

For finite τ, we have the following result.

Theorem 8.2 *Assume that (H2c) holds and let $L > \pi$. Then there is a unique $\tau_c > 0$ such that for $\tau < \tau_c$, (A_L, ξ_L) is stable and for $\tau > \tau_c$ it is unstable. At $\tau = \tau_c$ there exists a unique Hopf bifurcation. The Hopf bifurcation is transversal, i.e.*

$$\left. \frac{d\lambda_R}{d\tau} \right|_{\tau=\tau_c} > 0, \tag{8.11}$$

where λ_R is the real part of the eigenvalue.

By the results of Sect. 8.1.1, we calculate that $\int_0^L w_L \mathcal{L}^{-1} w_L dy > 0$ for all $L > \pi$ using Weierstrass $p(z)$ functions and Jacobi elliptic integrals. Then we have

Theorem 8.3 *Assume that $L > \pi$. Then there exists a unique $\tau_c > 0$ such that for $\tau < \tau_c$, (A_L, ξ_L) is stable and for $\tau > \tau_c$, (A_L, ξ_L) is unstable. At $\tau = \tau_c$, there exists a Hopf bifurcation. Furthermore, the Hopf bifurcation is transversal.*

Thus for the shadow Gierer-Meinhardt system we have given a complete picture of the stability of nontrivial monotone solutions for all $\tau > 0$ and $L > 0$. Note that in case $L \leq \pi$ we necessarily have $w_L \equiv 1$ and there are only trivial monotone solutions. We remark that singular perturbation, which can be applied when ϵ is small enough, cannot be used here.

In the previous chapters, we have considered the existence and stability of multiple spikes for small activator diffusivity ϵ^2 and finite inhibitor diffusivity D. Now we study the complementary case of finite ϵ^2 and infinite D.

8.1.1   Some Properties of the Function w_L

In this subsection, we consider the unique solution of the boundary value problem

$$w_L'' - w_L + w_L^2 = 0, \qquad w_L'(0) = w_L'(L) = 0, \qquad w_L'(y) < 0 \quad \text{for } 0 < y < L.$$

$$(8.12)$$

Using Weierstrass functions and elliptic integrals we will derive some properties of w_L.

Recall that

$$\mathcal{L}[\phi] = \phi'' - \phi + 2w_L \phi.$$

Our first result is

Lemma 8.4 *For the eigenvalue problem*

$$\begin{cases} \mathcal{L}\phi = \lambda\phi, & 0 < y < L, \\ \phi'(0) = \phi'(L) = 0, \end{cases} \qquad (8.13)$$

the eigenvalues satisfy

$$\lambda_1 > 0, \qquad \lambda_j < 0, \quad j = 2, 3, \ldots. \qquad (8.14)$$

The eigenfunction Φ_1 for the eigenvalue λ_1 can be chosen to be positive.

Proof Let $\lambda_1 \geq \lambda_2 \geq \cdots$ be the eigenvalues of $\mathcal{L}$. It is well-known that $\lambda_1 > \lambda_2$ and that the eigenfunction Φ_1 of λ_1 can be made positive. Further, we have

$$-\lambda_1 = \min_{\int_0^L \phi^2 dy = 1} \left(\int_0^L (|\phi'|^2 + \phi^2 - 2w_L\phi^2) dy \right)$$

$$\leq \left(\int_0^L w_L^2 dy \right)^{-1} \left(\int_0^L (|w_L'|^2 + w_L^2 - 2w_L w_L^2) dy \right) < 0. \quad (8.15)$$

By a standard argument (see Theorem 2.11 of [133]) it follows that $\lambda_2 \leq 0$. We include a proof for the convenience of the reader. Using the variational characterisation of λ_2, we get

$$-\lambda_2 = \sup_{v \in H^1(I)} \inf_{\phi \in H^1(I), \phi \not\equiv 0} \left[\frac{\int_0^L (|\phi'|^2 + \phi^2 - 2w_L\phi^2) dy}{\int_0^L \phi^2 dy} : v \not\equiv 0, \int_0^L \phi v\, dy = 0 \right].$$

$$(8.16)$$

Since w_L has least energy, namely

$$E[w_L] = \inf_{u \not\equiv 0, u \in H^1(I)} E[u],$$

where

$$E[u] = \frac{\int_0^L (|u'|^2 + u^2) dy}{(\int_0^L u^3 dy)^{2/3}}$$

and so for

$$h(t) = E[w_L + t\phi], \quad \phi \in H^1(I),$$

we know that $h(t)$ attains its minimum at $t = 0$. Thus we get

$$h''(0) = 2 \left[\int_0^L (|\phi'|^2 + \phi^2) dy - 2 \int_0^L w_L\phi^2 dy + \frac{(\int_0^L w_L^2\phi\, dy)^2}{\int_0^L w_L^3 dy} \right]$$

$$\times \frac{1}{(\int_0^L w_L^3 dy)^{2/3}}$$

$$\geq 0.$$

By (8.16), we see that

$$-\lambda_2 \geq \inf_{\int_0^L \phi w\, dy = 0} \left[\int_0^L (|\phi'|^2 + \phi^2) dy - 2 \int_0^L w_L\phi^2 dy + \frac{(\int_0^L w_L^2\phi\, dy)^2}{\int_0^L w_L^3 dy} \right]$$

$$\times \frac{1}{(\int_0^L w_L^3 dy)^{2/3}}$$

$$\geq 0.$$

Now as in the proof of Lemma 13.4, Step 2, we can conclude that $\lambda_2 < 0$. $\qquad \square$

By Lemma 8.4, we know that $\mathcal{L}^{-1}$ exists. In the next step we calculate the integral $\int_0^L w_L \mathcal{L}^{-1} w_L dy$. Using a perturbation argument, we get

Lemma 8.5 *We have*

$$\lim_{L \to \pi} \int_0^L w_L \mathcal{L}^{-1} w_L dy = \pi, \tag{8.17}$$

$$\lim_{L \to +\infty} \int_0^L w_L \mathcal{L}^{-1} w_L dy = \frac{3}{4} \int_0^\infty w_\infty^2 dy, \tag{8.18}$$

where $w_\infty(y)$ is given by

$$w'' - w + w^2 = 0, \qquad w'(0) = 0, \qquad w'(y) < 0, \qquad w(y) > 0, \quad 0 < y < +\infty. \tag{8.19}$$

We will compute $\int_0^L w_L \mathcal{L}^{-1} w_L dy$ by using elliptic integrals and derive the following result.

Lemma 8.6 *We have*

$$\int_0^L w_L \mathcal{L}^{-1} w_L dy > 0$$

for all $L > \pi$.

Before proving Lemma 8.6, we rewrite w_L using Weierstrass functions. An introduction to Weierstrass functions can be found in [1].

Let $w_L(0) = M$, $w_L(L) = m$.

From (8.12), we have

$$\left(w_L'\right)^2 = w_L^2 - \frac{2}{3} w_L^3 - M^2 + \frac{2}{3} M^3 \tag{8.20}$$

and

$$-m^2 + \frac{2}{3} m^3 = -M^2 + \frac{2}{3} M^3. \tag{8.21}$$

From (8.21), we deduce that

$$\frac{Mm}{M+m} = M + m - \frac{3}{2}. \tag{8.22}$$

Now let

$$\hat{w} = -\frac{1}{6} w_L + \frac{1}{12}. \tag{8.23}$$

Elementary calculations give

$$\left(\hat{w}'\right)^2 = 4\hat{w}^3 - g_2 \hat{w} - g_3 = 4(\hat{w} - e_1)(\hat{w} - e_2)(\hat{w} - e_3), \tag{8.24}$$

where

$$g_2 = \frac{1}{12}, \qquad g_3 = -\frac{1}{216} - \frac{1}{36}\left(-M^2 + \frac{2}{3}M^3\right), \tag{8.25}$$

$$e_1 = \frac{1}{6}(M + m) - \frac{1}{6}, \qquad e_2 = -\frac{1}{6}m + \frac{1}{12}, \qquad e_3 = -\frac{1}{6}M + \frac{1}{12}. \tag{8.26}$$

For the Weierstrass function $p(z)$ we have [1]:

$$\hat{w}(x) = p(x + \alpha; g_2, g_3) \tag{8.27}$$

for some constant α. From now on, we will avoid the arguments g_2 and g_3 of p.
  We get

$$p(f_i) = e_i, \qquad p'(f_i) = 0, \quad i = 1, 2, 3, \ f_1 + f_2 + f_3 = 0 \tag{8.28}$$

which implies that

$$\hat{w}(x) = p(f_3 + x), \qquad L = f_1. \tag{8.29}$$

The Weierstrass function $\zeta(z)$ satisfies

$$\zeta(z) = \frac{1}{z} - \int_0^z \left(p(u) - \frac{1}{u^2}\right) du$$

and so we get

$$\zeta'(u) = -p(u), \qquad \zeta(f_i) = \eta_i, \quad i = 1, 2, 3, \ \eta_1 + \eta_2 + \eta_3 = 0. \tag{8.30}$$

We calculate

$$\int_0^L \hat{w}(x)dx = \int_0^{f_1} p(f_3 + x)dx = -\zeta(u)\big|_{f_3}^{-f_2}$$

$$= \zeta(f_3) + \zeta(f_2)$$

$$= -\zeta(f_1) = -\zeta(L). \tag{8.31}$$

This implies that

$$\int_0^L w_L^2 dy = \int_0^L w_L dy = \int_0^L \left(-6\hat{w} + \frac{1}{2}\right) dy = 6\zeta(L) + \frac{L}{2}. \tag{8.32}$$

By the formulas on p. 649 of [1], we get

$$\zeta(L) = \frac{K(k)}{3L}\big[3E(k) + (k-2)K(k)\big],$$

$$e_1 = \frac{(2-k)K^2(k)}{3L^2},$$

$$e_2 = \frac{(2k-1)K^2(k)}{3L^2},$$ (8.33)

$$e_3 = \frac{-(k+1)K^2(k)}{3L^2},$$

where e_1, e_2 and e_3 are given in (8.26) and

$$e_1 e_2 + e_2 e_3 + e_1 e_3 = -\frac{1}{4}g_2 = -\frac{1}{48}.$$

Here $E(k)$ and $K(k)$ denote Jacobi elliptic integrals defined as

$$E(k) = \int_0^{\pi/2} \sqrt{1 - k^2 \sin^2 \varphi}\, d\varphi, \qquad K(k) = \int_0^{\pi/2} \frac{1}{\sqrt{1 - k^2 \sin^2 \varphi}}\, d\varphi.$$

We get

$$L = 2\big(k^2 - k + 1\big)^{1/4} K(k).$$ (8.34)

Now (8.34) implies

$$\frac{dL}{dk} = \frac{4K^2((2k-1)K^2 + 4KK'(k^2 - k + 1))}{L^3},$$ (8.35)

where the argument k of K has been omitted. By (8.34), for every $L > \pi$ there is a unique k. Further, we have $\frac{dk}{dL} > 0$ and

$$(2k-1)K + 4K'\big(k^2 - k + 1\big) > 0.$$ (8.36)

Now we come to

Proof of Lemma 8.6 We set $\phi_L = \mathcal{L}^{-1}w_L$ and so ϕ_L solves

$$\phi_L'' - \phi_L + 2w_L\phi_L = w_L, \qquad \phi_L'(0) = \phi_L'(L) = 0.$$

Set

$$\phi_L = w_L + \frac{1}{2}yw_L'(y) + \Psi.$$ (8.37)

Then $\Psi(y)$ satisfies

$$\Psi'' - \Psi + 2w_L\Psi = 0,$$

$$\Psi'(0) = 0, \qquad \Psi'(L) = -\frac{1}{2}Lw_L''(L).$$

$$(8.38)$$

Next we set $\Psi_0 = \frac{\partial w_L}{\partial M}$. Then Ψ_0 solves

$$\Psi_0'' - \Psi_0 + 2w_L\Psi_0 = 0,$$

$$\Psi_0(0) = 1, \qquad \Psi_0'(0) = 0.$$

$$(8.39)$$

Integration of (8.39) gives

$$\Psi_0'(L) = \int_0^L \frac{\partial w_L}{\partial M}dy - 2\int_0^L w_L\frac{\partial w_L}{\partial M}dy$$

$$= \frac{d}{dM}\left(\int_0^L (w_L - w_L^2)dy\right) - (w_L(L) - w_L^2(L))\frac{dL}{dM}.$$

Using the equation for w_L, we have $\int_0^L (w_L - w_L^2)dy = 0$. Thus we obtain

$$\Psi_0'(L) = -\left(w_L(L) - w_L^2(L)\right)\frac{dL}{dM}.$$

$$(8.40)$$

Comparing (8.38) and (8.40), we have

$$\Psi(x) = \frac{L}{2}\left(\frac{dL}{dM}\right)^{-1}\Psi_0(x).$$

$$(8.41)$$

Thus we get

$$\int_0^L w_L\phi_L dy = \int_0^L \left(w_L + \frac{1}{2}yw_L' + \Psi\right)w_L dy$$

$$= \frac{3}{4}\int_0^L w_L^2 dy + \frac{1}{4}Lw_L^2(L) + \frac{L}{2}\left(\frac{dL}{dM}\right)^{-1}\int_0^L w_L\Psi_0 dy.$$

$$(8.42)$$

Further, we have

$$\int_0^L w_L\Psi_0 dy = \int_0^L w_L\frac{\partial w_L}{\partial M}dy$$

$$= \frac{1}{2}\frac{d}{dM}\int_0^L w_L^2 dy - \frac{1}{2}w_L^2(L)\frac{dL}{dM}$$

$$= \frac{1}{2}\left[\frac{d}{dL}\int_0^L w_L^2 dy - w_L^2(L)\right]\frac{dL}{dM}.$$

$$(8.43)$$

Substituting (8.43) into (8.42), we obtain

$$\int_0^L w_L \phi_L dy = \frac{3}{4} \int_0^L w_L^2 dy + \frac{1}{4} L \frac{d}{dL} \int_0^L w_L^2 dy$$

$$= \frac{L^{-2}}{4} \frac{d}{dL} \left(L^3 \int_0^L w_L^2 dy \right). \tag{8.44}$$

By (8.32) and (8.34), we derive

$$L^3 \int_0^L w_L^2 dy = L^3 \int_0^L w_L dy$$

$$= 2L^2 K \big[3E + (k-2)K \big] + \frac{L^4}{2}$$

$$= 8\sqrt{k^2 - k + 1} K^3 \big[3E + \big(k - 2 + \sqrt{k^2 - k + 1} \big) K \big]. \tag{8.45}$$

If $2k - 1 \geq 0$, we compute

$$\frac{1}{8} \frac{d}{dk} \left(L^3 \int_0^L w_L^2 \right) > 0.$$

If $2k - 1 < 0$, using (8.36) and

$$\frac{dK}{dk} = \frac{E - (k')^2 K}{k(k')^2}, \qquad \frac{dE}{dk} = \frac{E - K}{k},$$

where $k' = \sqrt{1 - k^2}$, we have

$$\frac{1}{8} \frac{d}{dk} \left(L^3 \int_0^L w_L^2 \right)$$

$$= \frac{d}{dk} \big[\sqrt{k^2 - k + 1} K^3 [3E + \rho_k K] \big]$$

$$= \sqrt{k^2 - k + 1} K^2$$

$$\times \left[9 \frac{dK}{dk} E + 3K \frac{dE}{dk} + \frac{d\rho_k}{dk} K^2 + 4\rho_k K \frac{dK}{dk} \right.$$

$$\left. + \frac{2k - 1}{2(k^2 - k + 1)} K[3E + \rho_k K] \right]$$

$$= \sqrt{k^2 - k + 1} K^2$$

$$\times \left[3 \frac{d(EK)}{dk} + 2E \left(\frac{dK}{dk} + \frac{2k - 1}{4(k^2 - k + 1)} K \right) + 4 \frac{dK}{dk} (E + \rho_k K) \right]$$

$$+ \sqrt{k^2 - k + 1} K^2 \left[K \left(\frac{d\rho_k}{dk} K + \frac{2k - 1}{2(k^2 - k + 1)} (2E + \rho_k K) \right) \right],$$

where $\rho_k = k - 2 + \sqrt{k^2 - k + 1}$. In the previous expression each term is positive which follows from basic calculations.

This completes the proof.   □

8.1.2 Nonlocal Eigenvalue Problems

Since the nonlocal eigenvalue problem in this problem is defined in a finite interval, in contrast to all previous studies in the book, we have to derive and study it afresh.

Linearising (8.2) around the steady state

$$A_L = \xi w_L(Lx), \qquad \xi_L^{-1} = \int_0^1 w_L^2(Lx)dx, \tag{8.46}$$

we get the eigenvalue problem

$$\epsilon^2 \phi_{xx} - \phi + 2w_L\phi - \eta w_L^2 = \lambda\phi,$$
$$-\eta + 2\xi_L \int_0^1 w_L\phi dx = \tau\lambda\eta. \tag{8.47}$$

We also rescale:

$$y = Lx. \tag{8.48}$$

Solving the second equation for η and putting it into the first equation, we derive the following NLEP:

$$\phi'' - \phi + 2w_L\phi - \frac{2}{1+\tau\lambda}\frac{\int_0^L w_L\phi dy}{\int_0^L w_L^2 dy}w_L^2 = \lambda\phi, \quad y \in (0, L), \tag{8.49}$$

with

$$\phi'(0) = \phi'(L) = 0$$

and

$$\lambda = \lambda_R + \sqrt{-1}\lambda_I \in \mathbb{C}. \tag{8.50}$$

In this subsection, we assume that $\tau = 0$. Thus (8.49) can be written as

$$L_\gamma[\phi] := \mathcal{L}[\phi] - \gamma\frac{\int_0^L w_L\phi dy}{\int_0^L w_L^2 dy}w_L^2 = \lambda\phi, \qquad \phi'(0) = \phi'(L) = 0. \tag{8.51}$$

Then we have

Lemma 8.7 *Suppose that $\gamma \neq 1$. Then $\lambda = 0$ is not an eigenvalue of* (8.49).

Proof Supposing $\lambda = 0$, we get

$$0 = \mathcal{L}[\phi] - \gamma \frac{\int_0^L w_L \phi dy}{\int_0^L w_L^2 dy} w_L^2 = \mathcal{L}\left(\phi - \gamma \frac{\int_0^L w_L \phi dy}{\int_0^L w_L^2 dy} w_L\right).$$

By Lemma 8.4,

$$\phi - \gamma \frac{\int_0^L w_L \phi dy}{\int_0^L w_L^2 dy} w_L = 0.$$

Multiplying this equation by w_L and integrating, we get

$$(1 - \gamma) \int_0^L w_L \phi dy = 0.$$

Hence, since $\gamma \neq 1$, we have

$$\int_0^L w_L \phi dy = 0.$$

This implies

$$\mathcal{L}[\phi] = 0$$

and by Lemma 8.4 we get

$$\phi = 0. \qquad \qquad \square$$

Next we prove that the unstable eigenvalues are bounded uniformly in τ.

Lemma 8.8 *Let λ be an eigenvalue of (8.49) with $\mathrm{Re}(\lambda) \geq 0$. Then there is a constant C independent of $\tau > 0$ which satisfies*

$$|\lambda| \leq C. \tag{8.52}$$

Proof We multiply (8.49) by the complex conjugate $\bar{\phi}$ of ϕ and integrate. Then we get

$$\lambda \int_0^L |\phi|^2 dy = -\int_0^L \left(|\phi'|^2 + |\phi|^2 - 2w_L |\phi|^2\right) dy$$
$$- \frac{2}{1 + \tau\lambda} \frac{(\int_0^L w_L \phi dy)(\int_0^L w_L^2 \bar{\phi} dy)}{\int_0^L w_L^2 dy}, \tag{8.53}$$

where $|\phi|^2 = \phi\bar{\phi}$. Using

$$\left|\frac{1}{1 + \tau\lambda}\right| \leq 1 \quad \text{for } \mathrm{Re}(\lambda) \geq 0, \tag{8.54}$$

we have

$$\left| \frac{2}{1+\tau\lambda} \frac{(\int_0^L w_L \phi \, dy)(\int_0^L w_L^2 \bar{\phi} \, dy)}{\int_0^L w_L^2 \, dy} \right| \leq C \int_0^L |\phi|^2 \, dy, \tag{8.55}$$

where C is independent of τ.

Now (8.52) follows from (8.53) and (8.55). $\square$

Next we study the eigenvalue problem (8.49) and complete the proof of Theorem 8.1. We remark that the operator L_γ is not self-adjoint.

Assuming that $\tau = 0$, we have

Lemma 8.9 *Assume that* (H2c) *holds, i.e.*

$$\int_0^L w_L \mathcal{L}^{-1} w_L \, dy > 0. \tag{8.56}$$

Let λ be an eigenvalue of (8.51). *Then*

$$\mathrm{Re}(\lambda) < 0.$$

The proof of Lemma 8.9 requires the following result:

Lemma 8.10 *Assuming that* (H2c) *is valid, there is an $a_1 > 0$ such that*

$$Q[\phi, \phi] := \int_0^L \left(|\phi'|^2 + \phi^2 - 2w_L \phi^2 \right) dy + \frac{2 \int_0^L w_L^2 \phi \, dy \int_0^L w_L \phi \, dy}{\int_0^L w_L^2 \, dy}$$

$$- \frac{\int_0^L w_L^3 \, dy}{(\int_0^L w_L^2 \, dy)^2} \left(\int_0^L w_L \phi \, dy \right)^2$$

$$\geq a_1 d_{L^2}^2(\phi, X_1) \quad \text{for all } \phi \in H^1(0, L). \tag{8.57}$$

Here $X_1 = \mathrm{span}\{w\}$ and d_{L^2} denotes distance with respect to the L^2-norm.

Using Lemma 8.10, we have

Lemma 8.11 *Let (λ, ϕ) satisfy* (8.49) *with $\mathrm{Re}(\lambda) \geq 0$. Assuming that* (H2c) *is valid, we get*

$$\mathrm{Re}\left[\bar{\lambda}\chi(\tau\lambda) - \lambda\right] + |\chi(\tau\lambda) - 1|^2 \left(\frac{\int_0^L w_L^3 \, dy}{\int_0^L w_L^2 \, dy} \right) \leq 0, \tag{8.58}$$

where

$$\chi(\tau\lambda) = \frac{2}{1+\tau\lambda} \tag{8.59}$$

and $\bar{\lambda}$ denotes the conjugate of λ.

Proof Let (λ, ϕ) solve (8.49) and set $\lambda = \lambda_R + \sqrt{-1}\lambda_I$ and $\phi = \phi_R + \sqrt{-1}\phi_I$. Let $\chi(\tau\lambda)$ be given in (8.59). By (8.49) and its complex conjugate, we have

$$\mathcal{L}\phi - \chi(\tau\lambda)\frac{\int_0^L w_L\phi dy}{\int_0^L w_L^2 dy}w_L^2 = \lambda\phi, \tag{8.60}$$

$$\mathcal{L}\bar{\phi} - \bar{\chi}(\tau\lambda)\frac{\int_0^L w_L\bar{\phi} dy}{\int_0^L w_L^2 dy}w_L^2 = \bar{\lambda}\bar{\phi}. \tag{8.61}$$

We multiply (8.60) by $\bar{\phi}$ and integrate by parts to get

$$-\lambda\int_0^L |\phi|^2 dy - \chi(\tau\lambda)\frac{(\int_0^L w_L\phi dy)(\int_0^L w_L^2\bar{\phi} dy)}{\int_0^L w_L^2 dy}$$

$$= \int_0^L \left(|\phi'|^2 + |\phi|^2\right)dy - 2\int_0^L w_L|\phi|^2 dy. \tag{8.62}$$

Multiplication of (8.61) by w_L gives

$$\int_0^L w_L^2\bar{\phi} dy - \bar{\chi}(\tau\lambda)\frac{\int_0^L w_L\bar{\phi} dy}{\int_0^L w_L^2 dy}\int_0^L w_L^3 dy = \bar{\lambda}\int_0^L w_L\bar{\phi} dy. \tag{8.63}$$

Multiplying (8.63) by $\int_0^L w_L\phi dy$ and substituting the result into (8.62), we have

$$\int_0^L \left(|\phi'|^2 + |\phi|^2 - 2w_L|\phi|^2\right)dy + \lambda\int_0^L |\phi|^2 dy$$

$$= -\chi(\tau\lambda)\left[\bar{\lambda} + \overline{\chi}(\tau\lambda)\left(\frac{\int_0^L w_L^3 dy}{\int_0^L w_L^2 dy}\right)\right]\frac{|\int_0^L w_L\phi dy|^2}{\int_0^L w_L^2 dy}. \tag{8.64}$$

We express (8.64) by the quadratic functional Q defined in Lemma 8.10. Using (8.63), we have

$$\left[\text{Re}[\bar{\lambda}\chi(\tau\lambda) - \lambda] + |\chi(\tau\lambda) - 1|^2\left(\frac{\int_0^L w_L^3 dy}{\int_0^L w_L^2 dy}\right)\right]\frac{|\int_0^L w_L\phi dy|^2}{\int_0^L w_L^2 dy}$$

$$= -Q[\phi_R, \phi_R] - Q[\phi_I, \phi_I] - \text{Re}(\lambda)\left[\int_0^L |\phi|^2 dy - \frac{|\int_0^L w_L\phi dy|^2}{\int_0^L w_L^2 dy}\right]$$

$$\leq 0. \tag{8.65}$$

The lemma follows.   □

We give

Proof of Lemma 8.9 Assuming $\tau = 0$, from (8.58) we get

$$\mathrm{Re}\big[\bar{\lambda}\chi(\tau\lambda) - \lambda\big] + |\chi(\tau\lambda) - 1|^2\left(\frac{\int_0^L w_L^3 dy}{\int_0^L w_L^2 dy}\right)$$

$$= (\gamma - 1)\,\mathrm{Re}(\lambda) + |\gamma - 1|^2\left(\frac{\int_0^L w_L^3 dy}{\int_0^L w_L^2 dy}\right)$$

$$\leq 0$$

which implies

$$\mathrm{Re}(\lambda) \leq -(\gamma - 1)\left(\frac{\int_0^L w_L^3 dy}{\int_0^L w_L^2 dy}\right) < 0$$

since $\gamma > 1$. $\square$

Finally, we provide

Proof of Lemma 8.10 The operator

$$\mathcal{L}_1\phi := \mathcal{L}\phi - \frac{\int_0^L w_L\phi dy}{\int_0^L w_L^2 dy}w_L^2$$

$$- \frac{\int_0^L w_L^2\phi dy}{\int_0^L w_L^2 dy}w_L + \frac{\int_0^L w_L^3 dy \int_0^L w_L\phi dy}{(\int_0^L w_L^2 dy)^2}w_L \qquad (8.66)$$

is self-adjoint and

$$Q[\phi, \phi] \geq 0 \quad \Longleftrightarrow \quad \mathcal{L}_1 \text{ has no positive eigenvalues.}$$

Simple computations give

$$\mathcal{L}_1 w_L = 0.$$

If $\mathcal{L}_1\phi = 0$, then we have

$$\mathcal{L}\phi = c_1(\phi)w_L + c_2(\phi)w_L^2,$$

where

$$c_1(\phi) = \frac{\int_0^L w_L^2\phi dy}{\int_0^L w_L^2 dy} - \frac{\int_0^L w_L^3 dy \int_0^L w_L\phi dy}{(\int_0^L w_L^2 dy)^2}, \qquad (8.67)$$

$$c_2(\phi) = \frac{\int_0^L w_L\phi dy}{\int_0^L w_L^2 dy}. \qquad (8.68)$$

Thus we get

$$\phi - c_1(\phi)(\mathcal{L}^{-1}w_L) - c_2(\phi)w_L = 0. \tag{8.69}$$

Substitution of (8.69) into (8.67) gives

$$c_1(\phi) = c_1(\phi)\frac{\int_0^L w_L^2 \mathcal{L}^{-1}w_L dy}{\int_0^L w_L^2 dy} - c_1(\phi)\frac{\int_0^L w_L^3 dy \int_0^L w_L \mathcal{L}^{-1}w_L dy}{(\int_0^L w_L^2 dy)^2}$$

$$= c_1(\phi) - c_1(\phi)\frac{\int_0^L w_L^3 dy \int_0^L w_L \mathcal{L}^{-1}w_L dy}{(\int_0^L w_L^2 dy)^2}.$$

Now (H2c) gives $c_1(\phi) = 0$. Thus we have $\phi = c_2(\phi)w_L$. This implies that w_L is the only eigenfunction of $\mathcal{L}_1$ with eigenvalue zero.

Next we assume that the operator $\mathcal{L}_1$ has a positive eigenvalue $\lambda_0 > 0$ with eigenfunction ϕ_0. Due to the self-adjointness of $\mathcal{L}_1$, we have

$$\int_0^L w_L \phi_0 dy = 0 \tag{8.70}$$

and so

$$(\mathcal{L} - \lambda_0)\phi_0 = \frac{\int_0^L w_L^2 \phi_0 dy}{\int_0^L w_L^2 dy} w_L. \tag{8.71}$$

Note that $\int_0^L w_L^2 \phi_0 dy \neq 0$. In fact, if $\int_0^L w_L^2 \phi_0 dy = 0$, then $\lambda_0 > 0$ is an eigenvalue of $\mathcal{L}$. By Lemma 8.4, $\lambda_0 = \lambda_1$ and ϕ_0 does not change sign. This contradicts $\phi_0 \perp w_L$ and so $\lambda_0 \neq \lambda_1$. Thus $\mathcal{L} - \lambda_0$ is invertible. From (8.71), we get

$$\phi_0 = \frac{\int_0^L w_L^2 \phi_0 dy}{\int_0^L w_L^2 dy}(\mathcal{L} - \lambda_0)^{-1}w_L.$$

Thus

$$\int_0^L w_L^2 \phi_0 dy = \frac{\int_0^L w_L^2 \phi_0 dy}{\int_0^L w_L^2 dy}\int_0^L ((\mathcal{L} - \lambda_0)^{-1}w_L)w_L^2 dy.$$

Since $\int_0^L w_L^2 \phi_0 dy \neq 0$, we have

$$\int_0^L w_L^2 dy = \int_0^L ((\mathcal{L} - \lambda_0)^{-1}w_L)w_L^2 dy$$

and therefore

$$\int_0^L w_L^2 dy = \int_0^L ((\mathcal{L} - \lambda_0)^{-1}w_L)((\mathcal{L} - \lambda_0)w_L + \lambda_0 w_L)dy.$$

Using $\lambda_0 > 0$, we get

$$0 = \int_0^L \left((\mathcal{L} - \lambda_0)^{-1} w_L\right) w_L \, dy. \tag{8.72}$$

For $\beta(t) = \int_0^L ((\mathcal{L} - t)^{-1} w_L) w_L \, dy$ with $t > 0$, $t \neq \lambda_1$, we compute

$$\beta(0) = \int_0^L \left(\mathcal{L}^{-1} w_L\right) w_L \, dy > 0$$

using assumption (H2c) and

$$\beta'(t) = \int_0^L \left((\mathcal{L} - t)^{-2} w_L\right) w_L \, dy > 0.$$

Thus we have $\beta(t) > 0$ for all $t \in (0, \lambda_1)$. Further, we get

$$\beta(t) \to 0 \quad \text{as } t \to +\infty$$

which implies $\beta(t) < 0$ for $t > \lambda_1$.

To summarise, we have $\beta(t) \neq 0$ for $t > 0$, $t \neq \lambda_1$. Therefore (8.72) must be false and so $\mathcal{L}_1$ cannot have any positive eigenvalues.

Since

$$Q[\phi, \phi] = -\int_0^L (\mathcal{L}_1 \phi) \phi \, dy,$$

we get $Q[\phi, \phi] \geq 0$ for all ϕ with equality if and only if $\phi = c w_L$ for some constant c.

This completes the proof. $\square$

For the uniqueness and transversality of the Hopf bifurcation for some positive $\tau = \tau_0$ we refer to [266].

8.1.3 Extensions to Higher Dimensions

In the previous subsections, we have studied the one-dimensional case.

We now consider the case of general domains in $\mathbb{R}^n$, $N \geq 2$, namely the problem

$$\begin{cases} A_t = \Delta A - A + \frac{A^2}{\xi}, & x \in \Omega_L, t > 0, \\ \tau \xi_t = -\xi + \frac{1}{|\Omega_L|} \int_{\Omega_L} A^2 dx, \\ A > 0, & \frac{\partial A}{\partial \nu} = 0 \text{ on } \partial \Omega_L. \end{cases} \tag{8.73}$$

$\Omega_L = \frac{1}{\epsilon} \Omega \subset \mathbb{R}^n$ with $L = \frac{1}{\epsilon}$ denotes the rescaled domain and we assume it is a smooth and bounded. Letting the dimension satisfy $N \leq 5$, then the exponent 2 is subcritical. A steady state (8.73) is given by

$$A = \xi u, \qquad \xi^{-1} = \frac{1}{|\Omega_L|} \int_{\Omega_L} u^2 dx, \qquad (8.74)$$

where u solves

$$\begin{cases} \Delta u - u + u^2 = 0, & u > 0 \text{ in } \Omega_L, \\ \frac{\partial u}{\partial v} = 0 & \text{on } \partial \Omega_L. \end{cases} \qquad (8.75)$$

The energy minimising solution $w_L(x)$ of (8.75) is defined by

$$E[w_L] = \inf_{u \in H^1(\Omega_L), u \not\equiv 0} E[u], \qquad (8.76)$$

where

$$E[u] = \frac{\int_{\Omega_L} (|\nabla u|^2 + u^2) dy}{(\int_{\Omega_L} u^3 dy)^{2/3}}.$$

Then

$$A_L = \xi_L w_L, \qquad \xi_L^{-1} = \frac{1}{|\Omega_L|} \int_{\Omega_L} w_L^2 dx \qquad (8.77)$$

is a steady-state of the shadow system (8.73). Letting

$$\mathcal{L}[\phi] = \Delta \phi - \phi + 2w_L \phi,$$

then we have

Lemma 8.12 *Consider the following eigenvalue problem:*

$$\begin{cases} \mathcal{L}\phi = \lambda \phi & \text{in } \Omega_L, \\ \frac{\partial \phi}{\partial v} = 0 & \text{on } \partial \Omega_L. \end{cases} \qquad (8.78)$$

Then $\lambda_1 > 0$ and $\lambda_2 \leq 0$.

The proof of this lemma follows that of Lemma 8.4.

Again we make two key assumptions:

(H1c) $\mathcal{L}^{-1}$ exists.
(H2c) $\int_{\Omega_L} w_L(\mathcal{L}^{-1} w_L) dy > 0$.

Then we have the following result:

Theorem 8.13 *Assume that (H1c) and (H2c) hold. Then, for τ small enough, the steady state (A_L, ξ_L) is linearly stable. There is a unique $\tau = \tau_c$ such that (A_L, ξ_L) is stable for $\tau < \tau_c$, unstable for $\tau > \tau_c$, and there is a Hopf bifurcation at $\tau = \tau_c$. This Hopf bifurcation is transversal.*

The proof of Theorem 8.13 follows the same lines as the one dimensional case.

If L is large, by [8] and [253] we know that (H1) holds and (H2) holds for $N \leq 3$. This recovers the results of [249].

For general ϵ, it is difficult to verify (H1c) and (H2c). We expect that (H1c) holds for general domains.

8.2   The Gierer-Meinhardt System with Large Reaction Rates

Previously it was assumed that the diffusivity for the activator is much smaller than for the inhibitor. We considered (1.1), (1.2), (1.3) under the condition

$$\epsilon^2 = D_a \ll D_h = D. \tag{8.79}$$

This relation is related to the condition for Turing instability that the ratio of inhibitor and activator diffusivities must exceed a certain threshold [232].

In this section we do not assume the large diffusivity ratio (8.79) but we study the regime of large reaction rates for the activator.

Further, we make the assumption that

$$\tau = 0 \tag{8.80}$$

which will be important for the stability analysis. Arguing as in Chap. 3, the analysis can be extended to the case $\tau > 0$ for τ small enough.

We continue to use the notation $D_a = \epsilon^2$ and $D_h = h$ for the diffusion constants, as usual. After rescaling the spatial variable by $\hat{x} = x\frac{x}{\epsilon}$ and defining $\hat{D} = \frac{D}{\epsilon^2}, \hat{L} = \frac{1}{\epsilon}$, the Gierer-Meinhardt system for $q = 1$, $s = 0$ can be stated as

$$A_t = A_{\hat{x}\hat{x}} - A + \frac{A^p}{H}; \qquad \tau H_t = \hat{D} H_{\hat{x}\hat{x}} - H + A^{(p-1)r}, \quad x \in (-\hat{L}, \hat{L}),$$

$$A_{\hat{x}}(\pm\hat{L}) = 0 = H_{\hat{x}}(\pm L), \tag{8.81}$$

$$\hat{L}, \hat{D} = O(1); \qquad p \gg 1; \qquad r > 1 \quad \text{and} \quad r = O(1).$$

Note that here we consider exponents of the activator (reaction rates) which are given by p and $(p-1)r$ with p very large and $r = O(1)$. From now on, we simplify the notation by omitting hats.

We now present the main results. The first concerns the existence of one-spike or two-spike steady states.

Theorem 8.14

(a) *We consider the system*

$$0 = A_{xx} - A + \frac{A^p}{H}; \qquad 0 = DH_{xx} - H + A^{(p-1)r}, \tag{8.82}$$

where

$$x \in (-L, L), \qquad A_x(\pm L) = 0 = H_x(\pm L) \tag{8.83}$$

and the constants D, r, L are positive and fixed. We define the parameter

$$\alpha := \frac{1}{p-1}.$$

Suppose that p is large enough (i.e. α is small enough). Then (8.82) has a solution $(A, H) \in (H_N^2(-L, L))^2$ which satisfies

$$A(x) = \begin{cases} (\frac{H_0\eta}{3\alpha})^\alpha w^\alpha(\frac{\sqrt{\eta}}{\alpha}x)(1 + O(\alpha)), & |x| \ll 1, \\ \frac{1}{\alpha^\alpha}\frac{\cosh(|x|-L)}{\cosh(L)}(1 + O(\alpha)), & |x| \gg O(\alpha), \end{cases} \tag{8.84}$$

$$H(x) = H_0\frac{\cosh((|x| - L)/\sqrt{D})}{\cosh((L)/\sqrt{D})}(1 + O(\alpha)), \tag{8.85}$$

where

$$H_0 = \alpha\eta^{-(r-1/2)/(r-1)}\left[2\beta^{-1}D^{1/2}\tanh\left(\frac{L}{\sqrt{D}}\right)\right]^{1/(r-1)}(1 + O(\alpha)),$$

$$\eta = \tanh^2(L),$$

$$\beta = \int_{-\infty}^{\infty}\left(\frac{1}{2}\right)^r \operatorname{sech}^{2r}\left(\frac{y}{2}\right)dy$$

and w is given by (2.3). Further, we have the error estimate

$$\left\|\alpha A^{1/\alpha}(\alpha z) - \frac{H_0\eta}{3}w(\sqrt{\eta}z)\right\|_{H^2(-L/\alpha, L/\alpha)} = O(\alpha).$$

(b) *If we restrict the solution of Part (a) to the interval $(0, L)$, we obtain a solution $(A, H) \in (H_N^2(0, L))^2$ of (8.82) with*

$$A_x(0) = A_x(L) = 0 = H_x(0) = H_x(L). \tag{8.86}$$

(c) *If we extend the solution of Part (b) to the interval $(0, 2L)$ by even reflection at $x = L$, we get a solution $(A, H) \in (H_N^2(0, 2L))^2$ of (8.82) with*

$$A_x(0) = A_x(2L) = 0 = H_x(0) = H_x(2L). \tag{8.87}$$

Remark 8.15 The solution of Part (a) possesses an interior spike for A^p which is located at the centre $x = 0$ of the interval. The solution of Part (b) has a boundary spike for A^p located at the left boundary $x = 0$. The solution of Part (c) has two boundary spikes for A^p located at the boundaries $x = 0$ and $x = 2L$ (see Fig. 8.1).

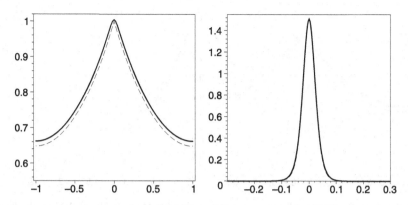

Fig. 8.1 *Left*: The steady state $A(x)$ (*solid line*) and its asymptotic approximation in the outer region (*dashed line*) defined by $\cosh(|x| - L)/\cosh(L)$. *Right*: The function $A(x)^p$ near the origin. Its asymptotic approximation given in (8.84) is also shown, but is indistinguishable by the naked eye from the full solution. We have selected the parameters $p = 90$, $r = 2$, $D = 1$, $L = 1$. Note that A^p is localised near the centre, in contrast to A

Next we consider the stability of spiky steady states.

Theorem 8.16 *Suppose that p is large enough. Then for the steady states given in Theorem 8.14 we have:*

(a) *The interior spike is unstable. The eigenvalue problem has an eigenvalue with positive real part of exact order $O(1)$ which is given in (8.143) and an odd eigenfunction.*
(b) *The boundary spike is stable.*
(c) *The two boundary spike steady state is stable for $D < D_c$ and unstable for $D > D_c$. Here D_c is given by*

$$D_c = \left(\frac{L}{\operatorname{artanh}(1/\sqrt{r})}\right)^2. \tag{8.88}$$

For $D > D_c$ there is an eigenvalue λ with $\operatorname{Re}(\lambda) = O(p^2)$ and an eigenfunction which is odd about $x = L$.

In Fig. 8.2 we show the instabilities given in Theorem 8.16. Due to the instability of the interior spike, it moves towards the boundary on a slow timescale of exact order $O(1)$ related to a small eigenvalue. The instability of the boundary spike implies spike annihilation which happens on a faster timescale of exact order $O(\frac{1}{p^2})$ related to a large eigenvalue.

As described in Chap. 2, a multi interior spike steady state can be constructed from a single interior spike solution by even reflection. Since a single interior spike is unstable, this multi-spike configuration is also unstable since the unstable eigenfunction can be extended in the same way.

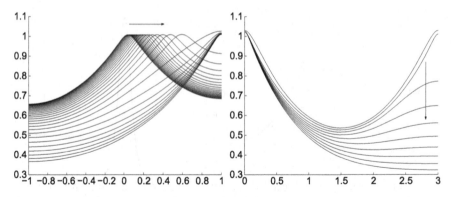

Fig. 8.2 *Left*: The interior spike moves towards the right boundary. Profiles of $A(x)$ are displayed with increments of 0.1 time steps. The parameters were selected as $p = 50$, $r = 2$, $D = 1$, $L = 1$. The initial condition was chosen slightly to the right of the centre. *Right*: Competition instability of two boundary spikes. The profiles of $A(x)$ are plotted with increments of 0.1 time steps. The spike at the right boundary eventually vanishes. Here, we have used $p = 50$, $r = 2$, $D = 4$, $L = 1.5$ and $x \in (0, 2L)$

8.2.1 Construction of the Steady State

In this subsection we give a construction of the steady state using asymptotic matching and prove Theorem 8.14. First, we observe that the problem

$$v_{xx} - v + v^p = 0, \quad x \in \mathbb{R}, v > 0, \max_{x \in \mathbb{R}} v(x) = v(0) \tag{8.89}$$

has the unique ground state

$$v(x) = \left[\left(\frac{p+1}{2} \right) \operatorname{sech}^2 \left(\frac{p-1}{2} x \right) \right]^{1/(p-1)}. \tag{8.90}$$

Motivated by the scalings of spatial variable and amplitude for $1 \ll p$, we set

$$A(x) = \left(\frac{u(z)}{\alpha} \right)^\alpha, \quad z = \frac{x}{\alpha}, \tag{8.91}$$

where

$$\alpha = \frac{1}{p-1} \ll 1.$$

Using this rescaling, we expect that in leading order u will be independent of p. Then we get the following inner problem:

$$\begin{cases} 0 = u_{zz} - \frac{u_z^2}{u} + \frac{u^2}{H} + \alpha(-u + \frac{u_z^2}{u}) + O(\alpha^2), & |z| \ll \frac{1}{\alpha}, \\ 0 = \frac{1}{\alpha^2} D H_{zz} - H + u^r \alpha^{-r}. \end{cases} \tag{8.92}$$

For $|z| \ll \frac{1}{\alpha}$, we calculate

$$u(z) = U_0(z)\big(1 + O(\alpha)\big),$$
$$H(z) = H_0\big(1 + O(\alpha)\big).$$

We will later show that $u = O(\alpha)$ and $H = O(\alpha)$ implying that

$$U_{0zz} - \frac{U_{0z}^2}{U_0} + \frac{U_0^2}{H_0} = 0, \qquad H_{0,zz} = 0. \qquad (8.93)$$

Using boundary conditions, we have that H_0 is a constant. By direct calculation, (8.93) possesses the one-parameter family of solutions

$$U_0(z) = \frac{H_0}{3} \eta w(\sqrt{\eta} z),$$

where w is given by (2.3) and η is a parameter that is related to the scaling symmetry $U_0 = \eta \hat{U}_0$, $z = \eta^{-1/2}\hat{z}$ of (8.93). The values for η and H_0 remain to be computed.

In the inner region, we use

$$w(y) = 6\exp\big(-|y|\big) + O\big(\exp(-2|y|)\big) \quad \text{as } |y| \to \infty$$

and get

$$\begin{aligned}
A(z) &= \frac{1}{\alpha^\alpha} \exp\big(\alpha \ln u(z)\big) \\
&= \frac{1}{\alpha^\alpha} \exp\big(\alpha \ln(U_0(z)) + O(\alpha^2)\big) \\
&= \frac{1}{\alpha^\alpha}\big(1 - \alpha\sqrt{\eta}|z| + O(\alpha^2)\big) \quad \text{for } |z| \ll \frac{1}{\alpha}.
\end{aligned} \qquad (8.94)$$

Thus we have $A(z) \to \frac{1}{\alpha^\alpha}$ as $|z| \to \infty$.

In the outer region, $|x| \gg \alpha$, corresponding to $|z| \gg 1$, we compute

$$A_{xx} - A \sim 0; \qquad A_x(\pm L) = 0.$$

Matching with the inner solution implies $A(0) = \frac{1}{\alpha^\alpha}$. Thus we have

$$A(x) = \frac{1}{\alpha^\alpha}\left(\frac{\cosh(L - |x|)}{\cosh L} + O(|x|^2)\right), \qquad \alpha \ll |x|.$$

Next we match inner and outer solutions to second order. For $\alpha \ll |x| \ll 1$ we compute, using Taylor series,

$$\begin{aligned}
A(x) &= \frac{1}{\alpha^\alpha}\big(1 - \tanh(L)|x| + O(|x|^2)\big) \\
&= \frac{1}{\alpha^\alpha}\big(1 - \alpha\tanh(L)|z| + O(\alpha^2|z|^2)\big).
\end{aligned} \qquad (8.95)$$

The $O(\alpha)$ terms in (8.94) and (8.95) give

$$\eta = \tanh^2 L.$$

In the outer region we have $|A^{(p-1)r}| = o(1)$ and thus

$$DH_{xx} - H \sim -C_0\delta(x); \qquad H_x(\pm L) = 0,$$

where

$$C_0 = \int_{-\infty}^{\infty} A^{(p-1)r} dx$$

$$= \frac{\alpha}{\sqrt{\eta}} \left(\frac{H_0\eta}{3\alpha}\right)^r \int_{-\infty}^{\infty} w^r dy \left(1 + O(\alpha)\right)$$

$$= \frac{\alpha}{\sqrt{\eta}} \left(\frac{H_0\eta}{\alpha}\right)^r \beta \left(1 + O(\alpha)\right) \qquad (8.96)$$

and

$$\beta = \int_{-\infty}^{\infty} \left(\frac{1}{2}\right)^r \mathrm{sech}^{2r}\left(\frac{y}{2}\right) dy.$$

Thus we get

$$H(x) = B \cosh\left(\frac{L - |x|}{\sqrt{D}}\right)\left(1 + O(\alpha)\right),$$

where B satisfies

$$\sqrt{D}B \sinh\left(\frac{L}{\sqrt{D}}\right) = \frac{1}{2}C_0.$$

It follows that

$$H(0) = H_0\left(1 + O(\alpha)\right) = \frac{1}{2}\frac{C_0}{\sqrt{D}} \coth\left(\frac{L}{\sqrt{D}}\right)\left(1 + O(\alpha)\right),$$

$$(8.97)$$

$$H_0 = \alpha\eta^{-(r-1/2)/(r-1)} \left[2\beta^{-1}D^{1/2}\tanh\left(\frac{L}{\sqrt{D}}\right)\right]^{1/(r-1)}\left(1 + O(\alpha)\right).$$

We remark that the consistency assumptions $U_0 = O(\alpha)$ and $H_0 = O(\alpha)$ required in (8.93) are valid. The construction of a boundary spike is complete.

Using extension by reflection an interior spike on $(-L, L)$ or two boundary spikes on $(0, 2L)$ are constructed from the boundary spike on $(0, L)$. This concludes the derivation of Theorem 8.14.

A rigorous proof based on Liapunov-Schmidt reduction and following the approach of Sect. 2.3.3 can be sketched as follows:

For given $u \in H_N^2(-L/\alpha, L/\alpha)$, we express the solution $(A, H) \in (H_N^2(-L, L))^2$ by using a suitable Green's function. Then we construct an approximate solution with

$$u = \frac{H_0\eta}{3\alpha}w\left(\frac{\sqrt{\eta}}{\alpha}x\right)(1 + O(\alpha))$$

for which the boundary conditions $u_z(-L/\alpha) = u_z(L/\alpha) = 0$ hold exactly by using a cutoff function argument. We show that this approximation solves the Gierer-Meinhardt system (8.82) up to an error of order $O(\alpha)$ with respect to the norm in $(L^2(-L/\alpha, L/\alpha))^2$. Then we complete the proof by the Liapunov-Schmidt reduction method, calculating η when solving the reduced problem. Thus we have rigorously proved the existence of a solution.

8.2.2 Stability

Next we study the (linear) stability of an inhomogeneous steady state $(A(x), H(x))$ given by Theorem 8.14. We linearise around the steady state by

$$A(x, t) = A(x) + e^{\lambda t}\phi(x),$$

$$H(x, t) = H(x) + e^{\lambda t}\psi(x)$$

and for $\tau = 0$ we obtain

$$\begin{cases} \lambda\phi = \phi_{xx} - \phi + p\frac{A^{p-1}\phi}{H} - \frac{A^p}{H^2}, \\ 0 = D\psi_{xx} - \psi + r(p-1)A^{(p-1)r-1}\phi. \end{cases} \tag{8.98}$$

The change of variables introduced in (8.91) gives

$$A^p = \frac{u}{\alpha}A \sim \frac{u}{\alpha}$$

since $A \sim 1$ near $x \sim 0$. We obtain

$$\alpha^2(\lambda + 1)\phi \sim \alpha^2\phi_{xx} + \frac{u}{H}\phi + \alpha\frac{u}{H}\phi - \alpha\frac{u}{H^2}\psi, \tag{8.99}$$

$$0 \sim D\psi_{xx} - \psi + r\alpha^{-r-1}u^r\phi. \tag{8.100}$$

Around an interior spike which is even about the origin, there are eigenfunctions of two types which are either odd or even about the origin. An even eigenfunction can be restricted to $(0, L)$ and coincides with the eigenfunction for a single boundary spike. The double boundary spike on $(0, 2L)$ allows another type of eigenfunction which is odd about $x = L$.

Altogether, there are three different types of eigenfunctions on $(0, L)$, which are given as follows:

(1) An eigenfunction on $(0, L)$ for a boundary spike or an interior spike on $(-L, L)$:

$$\phi_x(0) = 0, \qquad \phi_x(L) = 0; \qquad \psi_x(0) = 0, \qquad \psi_x(L) = 0. \qquad (8.101)$$

(2) An odd eigenfunction on $(-L, L)$ for an interior spike:

$$\phi(0) = 0, \qquad \phi_x(L) = 0; \qquad \psi(0) = 0, \qquad \psi_x(L) = 0. \qquad (8.102)$$

(3) An eigenfunction on $(0, 2L)$ which is odd about $x = L$ for a double boundary spike:

$$\phi_x(0) = 0, \qquad \phi(L) = 0; \qquad \psi_x(0) = 0, \qquad \psi(L) = 0. \qquad (8.103)$$

We will show that problems (8.101) and (8.103) allow eigenvalues of exact order $O(p^2)$ which we call large eigenvalues. They will be studied in Sect. 8.2.3.

Problem (8.102) possesses an eigenvalue of order $O(1)$ which we refer to as a small eigenvalue. It will be analysed in Sect. 8.2.4.

8.2.3 Large Eigenvalues

We study the large eigenvalues. For inner variables

$$x = \frac{\alpha}{\sqrt{\eta}} y; \qquad u \sim \frac{H_0}{3} \eta w(y),$$

we have

$$\frac{\alpha^2}{\eta}(\lambda + 1)\phi \sim \phi_{yy} + \frac{1}{3} w\phi - \frac{\alpha}{3} \frac{\psi_0}{H_0} w,$$

where

$$\psi_0 = \psi(0)$$

and $\psi(0)$ is given by

$$D\psi_{xx} - \psi \sim C_1 \delta(x); \qquad \psi_x(\pm L) = 0. \qquad (8.104)$$

This implies

$$C_1 = \int_{-\infty}^{\infty} \left(\frac{u}{\alpha}\right)^r \frac{r}{\alpha} \phi \, dx = \frac{r}{\sqrt{\eta}} \left(\frac{H_0 \eta}{3\alpha}\right)^r \int_{-\infty}^{\infty} w^r \phi \, dy$$

and

$$\psi(x) \sim -C_1 G(0),$$

where the Green's function

$$G(x) = \frac{\cosh((L - |x|)/\sqrt{D})}{2\sqrt{D}\sinh(L/\sqrt{D})}$$

is given by

$$DG_{xx} - G = -\delta(x), \qquad G_x(\pm L) = 0.$$

Using (8.97), we get

$$H_0 = \frac{\alpha}{\sqrt{\eta}} \left(\frac{H_0\eta}{3\alpha}\right)^r \int_{-\infty}^{\infty} w^r \, dy \, G(0).$$

Using the boundary conditions (8.101), we get the NLEP

$$\lambda_0 \phi = \phi_{yy} + \frac{1}{3}w\phi - \frac{r}{3}w\frac{\int_{-\infty}^{\infty} w^r \phi \, dy}{\int_{-\infty}^{\infty} w^r \, dy}, \qquad \lambda_0 \sim \frac{\alpha^2}{\eta}\lambda. \tag{8.105}$$

Switching to boundary conditions (8.103), we have to use the Green's function for Dirichlet boundary conditions:

$$G_d(x) = \frac{\sinh((L - |x|)/\sqrt{D})}{2\sqrt{D}\cosh(L/\sqrt{D})}.$$

We compute

$$\lambda_0 \phi = \phi_{yy} + \frac{1}{3}w\phi - \frac{r}{3}\tanh^2\left(\frac{L}{\sqrt{D}}\right)w\frac{\int_{-\infty}^{\infty} w^r \phi \, dy}{\int_{-\infty}^{\infty} w^r \, dy}, \qquad \lambda_0 \sim \frac{\alpha^2}{\eta}\lambda. \tag{8.106}$$

For both (8.105) and (8.106) the following two results are crucial:

Theorem 8.17 *Let*

$$L_0\phi = \phi_{yy} + \frac{1}{3}w\phi \tag{8.107}$$

and consider the following NLEP on the real line:

$$L_0\phi - \gamma w \int_{-\infty}^{\infty} w^r \phi \, dy = \lambda\phi, \qquad \phi \in L^{\infty}(\mathbb{R}), r \geq 1 \tag{8.108}$$

where w is given by (2.3). Let

$$\gamma_0 = \frac{1}{3}\frac{1}{\int_{-\infty}^{\infty} w^r \, dy}. \tag{8.109}$$

Then we have:

(a) If $\gamma < \gamma_0$, then (8.108) admits a positive eigenvalue $\lambda > 0$.

(b) *If $\gamma > \gamma_0$ and $r = 2$, then we have $\text{Re}(\lambda) \le 0$ for all λ. The only eigenfunction for $\lambda = 0$ is $\phi = \frac{w_y}{w}$ which is an odd function; all other eigenvalues have strictly negative real part.*

Remark 8.18 Note that the nonlocal eigenvalue problem (8.107) is of a new type which is different from those in Chap. 3.

To be precise, the inner problem (8.108) is posed on a finite but large interval $(-R, R)$ with $R = L/\alpha$. The following theorem shows that this does not influence the sign of the real part of the large eigenvalues:

Theorem 8.19 *For $R > 0$ we study the NLEP*

$$L_0\phi - \gamma w \int_{-R}^{R} w^r \phi dy = \lambda_R \phi, \quad \phi \in L^\infty(-R, R), \phi_y(\pm R) = 0, r \ge 1. \quad (8.110)$$

There exists an $R_0 > 0$ such that for $R > R_0$, we get the following:

(a) *If $\gamma < \gamma_0$, then (8.110) possesses a positive eigenvalue $\lambda_R > 0$.*
(b) *If $\gamma > \gamma_0$ and $r = 2$, then either $\lim_{R \to \infty} \lambda_R = 0$ or $\text{Re}(\lambda_R) \le -c_0$ for some $c_0 > 0$.*

Combining Theorems 8.17 and 8.19 it follows that there is a stability threshold given by (8.88).

We first study the operator L_0 and have

$$L_0 w = w - \frac{2}{3}w^2; \qquad L_0^{-1} w = 3; \qquad (8.111)$$

$$\int_{-\infty}^{\infty} w^2 dy = 6; \qquad \int_{-\infty}^{\infty} w dy = 6. \qquad (8.112)$$

The spectrum of L_0 can be described as follows:

Lemma 8.20 *The eigenvalue problem*

$$L_0\phi = \phi_{yy} + \frac{1}{3}w\phi = \lambda\phi, \quad \phi \in L^\infty(\mathbb{R}) \qquad (8.113)$$

possesses two nonnegative eigenvalues. The first eigenvalue

$$\lambda_1 = \frac{1}{4}$$

has the even eigenfunction

$$\phi_1 = w^{1/2}.$$

The second eigenvalue

$$\lambda_2 = 0$$

has the odd eigenfunction

$$\phi_2 = \frac{w_y}{w}.$$

All other eigenvalues satisfy $\lambda < 0$ and are embedded in the continuous spectrum covering the negative real axis.

Finally, we state the following key lemma.

Lemma 8.21 *The NLEP*

$$L_0\phi - \frac{1}{18}w \int_{-\infty}^{\infty} w\phi \, dy = \lambda\phi, \quad \phi \in L^{\infty}(\mathbb{R}) \tag{8.114}$$

has an eigenvalue $\lambda = 0$ with two-dimensional eigenspace spanned by the even eigenfunction $\phi = 1$ and the odd eigenfunction $\phi = \frac{w_y}{w}$. All other eigenvalues satisfy $\lambda < 0$. Thus we get

$$\int_{-\infty}^{\infty} \left((\phi_y)^2 - \frac{1}{3}w\phi^2 \right) dy + \frac{1}{18} \left(\int_{-\infty}^{\infty} w\phi \, dy \right)^2 \geq 0, \quad \text{for all } \phi \in H^1(\mathbb{R}). \tag{8.115}$$

Proof of Lemma 8.20 We use hypergeometric functions as in Sect. 3.4. Let $\gamma = \sqrt{\lambda}$ with the principal branch of the square root. We substitute $\phi(y) = w^{\gamma}(y)F(y)$ and get

$$F_{yy} + 2\gamma \frac{w_y}{w} F_y + \left(\frac{1}{3} - \left(\gamma + \frac{2}{3}\gamma(\gamma - 1) \right) \right) wF = 0. \tag{8.116}$$

Introducing

$$z = \frac{1}{2}\left(1 - \frac{w_y}{w} \right), \tag{8.117}$$

we have

$$\frac{w_y}{w} = 1 - 2z, \qquad w = 6z(1-z), \qquad \frac{dz}{dx} = z(1-z).$$

For $F(z)$ we get

$$z(1-z)F'' + \left(c - (a+b+1)z \right)F' - abF = 0, \tag{8.118}$$

where

$$a+b+1 = 2+4\gamma, \qquad ab = 2\left(2\gamma(\gamma-1) - 3\left(\frac{1}{3} - \gamma \right) \right), \qquad c = 1+2\gamma. \tag{8.119}$$

The solutions to (8.118) are standard hypergeometric functions. We refer to [217] for more details. Note that (8.118) has two solutions:

$$F(a, b; c; z) \quad \text{and} \quad z^{1-c}F(a-c+1, b-c+1; 2-c; z).$$

Then F is regular at $z = 0$ but at $z = 1$ it has a singularity since

$$\lim_{z \to 1} (1 - z)^{-(c-a-b)} F(a, b; c; z) = \frac{\Gamma(c)\Gamma(a + b - c)}{\Gamma(a)\Gamma(b)},$$

where $c - a - b = -2\gamma < 0$. Since $\gamma = \sqrt{\lambda}$, the real part of γ is positive. In summary, a solution that is regular at both $z = 0$ and $z = 1$ can only exist if $\Gamma(x)$ has a pole at a or b, respectively. Written differently, we get

$$a = 0, -1, -2, \dots \quad \text{or} \quad b = 0, -1, -2, \dots . \tag{8.120}$$

By (8.119) we have

$$a = 2\gamma - 1, \qquad b = 2\gamma + 2 \quad \text{or} \quad b = 2\gamma - 1, \qquad a = 2\gamma + 2.$$

Since we assume $\text{Re}(\gamma) \geq 0$, we need to have $\gamma = 0$ or $\gamma = \frac{1}{2}$ in order to satisfy (8.120). This implies

$$\lambda = \frac{1}{4} \quad \text{or} \quad \lambda = 0.$$

If $\lambda = \frac{1}{4}$, we have $\gamma = 1/2$, $a = 0$, $b = 3$, $c = 2$, $F(0, 3; 2; z) = 1$ and the eigenfunction is $w^{1/2}$. Note that the choice $a = 3$, $b = 0$ gives the same eigenfunction. If $\lambda = 0$, we get $a = -1$, $b = 2$, $c = 1$ which implies $F(-1, 2, 1, z) = 1 - 2z$ and by (8.117) the eigenfunction is $\phi = \frac{w_y}{w}$. The first part of the lemma is shown. The second part follows from standard spectral properties of elliptic operators. $\qquad\square$

Proof of Lemma 8.21 We study two cases separately. Firstly, if

$$\int_{-\infty}^{\infty} w\phi \, dy = 0, \tag{8.121}$$

we get $L_0\phi = \lambda\phi$. Then by Lemma 8.20 we either have $\phi = Cw^{1/2}$, $\lambda = \frac{1}{4}$ or $\phi = C\frac{w_y}{w}$, $\lambda = 0$, where C is some nonzero constant. The former case contradicts (8.121) since $w > 0$. In the latter case, since w is even, any odd eigenfunction ϕ satisfies (8.121). Thus by Lemma 8.20 it is equal to a multiple of the eigenfunction $\phi = \frac{w_y}{w}$ with zero eigenvalue.

Suppose that (8.121) does not hold. We rescale ϕ so that

$$\int_{-\infty}^{\infty} w\phi \, dy = 18.$$

Then (8.114) becomes

$$(L_0 - \lambda)\phi = w.$$

We define

$$f(\lambda) \equiv \int_{-\infty}^{\infty} w(L_0 - \lambda)^{-1} w \, dy$$

and so λ solves

$$f(\lambda) = 18. \tag{8.122}$$

Since the operator L_0 defined in (8.114) is self-adjoint, all its eigenvalues are real. Thus it is enough to prove that $f(\lambda) \neq 18$ for $\lambda > 0$. Using

$$L_0 1 = \frac{1}{3}w, \tag{8.123}$$

we have

$$f(0) = 18 \tag{8.124}$$

and so $\lambda = 0$ is an eigenvalue of (8.114) with eigenfunction $\phi = 1$. Taking derivatives, we have

$$f'(\lambda) = \int_{-\infty}^{\infty} w(L_0 - \lambda)^{-2} w \, dy = \int_{-\infty}^{\infty} \left[(L_0 - \lambda)^{-1} w \right]^2 dy > 0.$$

Thus $f(\lambda)$ is an increasing function. Since L_0 admits a single positive eigenvalue $\lambda_0 = \frac{1}{4}$, we conclude that $f(\lambda)$ has a single pole at $\lambda = \frac{1}{4}$ and no other poles along the positive real axis $\lambda > 0$. For large values of λ, we have

$$f(\lambda) \sim -\frac{1}{\lambda} \int_{-\infty}^{\infty} w^2 \, dy \to 0^- \quad \text{as } \lambda \to +\infty.$$

The properties of f are collected as follows: $f(\lambda)$ has a vertical asymptote at $\lambda = \frac{1}{4}$; $f(0) = 18$, $f \to 0^-$ as $\lambda \to \infty$ and f is increasing for $\lambda \neq \frac{1}{4}$. It follows that $f(\lambda) \neq 18$ for all $\lambda > 0$ and the first part of the lemma is shown.

To prove (8.115), we proceed by contradiction. Suppose (8.115) is false. Then there is a function $\phi \in H^1(\mathbb{R})$ such that

$$\int_{\mathbb{R}} \left(|\phi_y|^2 - \frac{1}{3}w\phi^2 \right) dy + \frac{1}{18} \left(\int_{\mathbb{R}} w\phi \, dy \right)^2 < 0. \tag{8.125}$$

Thus for R sufficiently large, the first eigenvalue of the problem

$$L_0 \phi_R - \frac{1}{18} w \int_{-R}^{R} w\phi_R \, dy = \lambda_R \phi_R, \quad y \in (-R, R), \phi_R(\pm R) = 0 \tag{8.126}$$

is positive. Using the variational characterisation of eigenvalues, we have

$$-\lambda_R = \inf_{\phi \in H_0^1(-R,R)} \frac{\int_{-R}^{R} ((\phi_y)^2 - (1/3)w\phi^2) dy + (1/18)(\int_{-R}^{R} w\phi \, dy)^2}{\int_{-R}^{R} \phi^2 dy = 1}, \tag{8.127}$$

where $\lambda_R > 0$. Using (8.125), we conclude $\lambda_R \geq \lambda_0 > 0$ for R large. Without loss of generality, we have $\max_{y \in (-R,R)} \phi_R(y) = 1$. Taking the limit $R \to +\infty$, we have

$\lambda_R \to \lambda$ and $\phi_R \to \phi$, where ϕ satisfies (8.114) with eigenvalue $\lambda > 0$. This is impossible and the result follows by contradiction.   □

Proof of Theorem 8.17 (a) We assume that $\gamma < \gamma_0$ and define a function

$$f(\lambda) \equiv \int_{-\infty}^{\infty} w^r (L_0 - \lambda)^{-1} w \, dy. \tag{8.128}$$

Then λ is an eigenvalue if and only if

$$f(\lambda) = \frac{1}{\gamma}. \tag{8.129}$$

Using (8.111), we have

$$f(0) = \frac{1}{\gamma_0} < \frac{1}{\gamma}. \tag{8.130}$$

Next, to show that

$$f(\lambda) \to +\infty \quad \text{as } \lambda \to \lambda_0^-, \tag{8.131}$$

we let $\lambda = \lambda_0 + \delta$ with $\delta \ll 1$ and $\phi = (L_0 - \lambda)^{-1} w$. Let $\phi_0 = w^{1/2}$ be the eigenfunction of L_0 with eigenvalue λ_0. We project w onto ϕ_0, that is, we write

$$w = a\phi_0 + w_1; \qquad a = \frac{\int_{-\infty}^{\infty} w \phi_0 \, dy}{\int_{-\infty}^{\infty} \phi_0^2 \, dy}; \qquad \int_{-\infty}^{\infty} w_1 \phi_0 \, dy = 0.$$

Taking the limit $\delta \to 0$, we get

$$\phi \sim -\frac{a}{\delta} \phi_0 + O(1) \quad \text{as } \delta \to 0$$

and

$$f(\lambda_0 + \delta) \sim -\frac{1}{\delta} \frac{\int_{-\infty}^{\infty} w \phi_0 \, dy \int_{-\infty}^{\infty} w^r \phi_0 \, dy}{\int_{-\infty}^{\infty} \phi_0^2 \, dy} \quad \text{as } \delta \to 0.$$

Since the only positive eigenvalue of L_0 is λ_0, the function $f(\lambda)$ has no other vertical asymptotes. By (8.130), (8.131) and the continuity of $f(\lambda)$, the intermediate value theorem implies that (8.129) has a solution with $\lambda < 0 \le \lambda_0$ if $0 \le \gamma < \gamma_0$. Part (a) is proved.

Next we show Part (b). Assume that $\gamma > \gamma_0$. Since the operator (8.108) is not self-adjoint, the eigenvalues are in general complex and so we decompose

$$\lambda = \lambda^r + \sqrt{-1}\lambda^i$$

$$\phi = \phi^r + \sqrt{-1}\phi^i.$$

For $r = 2$ we have

$$L_0 \phi^r - \gamma w \int_{-\infty}^{\infty} w^2 \phi^r \, dy = \lambda^r \phi^r - \lambda^i \phi^i \tag{8.132}$$

$$L_0 \phi^i - \gamma w \int_{-\infty}^{\infty} w^2 \phi^i \, dy = \lambda^r \phi^i + \lambda^i \phi^r. \tag{8.133}$$

We multiply (8.132) by ϕ^r and (8.133) by ϕ^i, then integrate. This gives

$$\int_{-\infty}^{\infty} \left(\phi^r L_0 \phi^r + \phi^i L_0 \phi^i \right) dy - \gamma A = \lambda^r B, \tag{8.134}$$

where

$$A = \int_{-\infty}^{\infty} w \phi^r \, dy \int_{-\infty}^{\infty} w^2 \phi^r \, dy + \int_{-\infty}^{\infty} w \phi^i \, dy \int_{-\infty}^{\infty} w^2 \phi^i \, dy, \tag{8.135}$$

$$B = \int_{-\infty}^{\infty} \left((\phi^r)^2 + (\phi^i)^2 \right) dy. \tag{8.136}$$

Multiplication of (8.132) and (8.133) by w and integration gives

$$\int_{-\infty}^{\infty} \left(w - \frac{2}{3} w^2 \right) \phi^r \, dy - 6\gamma \int_{-\infty}^{\infty} w^2 \phi^r \, dy = \lambda^r \int_{-\infty}^{\infty} w \phi^r \, dy - \lambda^i \int_{-\infty}^{\infty} w \phi^i \, dy$$

$$\int_{-\infty}^{\infty} \left(w - \frac{2}{3} w^2 \right) \phi^i \, dy - 6\gamma \int_{-\infty}^{\infty} w^2 \phi^i \, dy = \lambda^r \int_{-\infty}^{\infty} w \phi^i \, dy + \lambda^i \int_{-\infty}^{\infty} w \phi^r \, dy.$$

After elimination of λ^i we have

$$(\lambda^r - 1) C + \left(\frac{2}{3} + 6\gamma \right) A = 0, \tag{8.137}$$

where A is given by (8.135) and

$$C = \left(\int_{-\infty}^{\infty} w \phi^r \, dy \right)^2 + \left(\int_{-\infty}^{\infty} w \phi^i \, dy \right)^2.$$

Using Lemma 8.21, we get from (8.134)

$$\gamma A + \lambda^r B \leq \frac{1}{18} C.$$

From (8.137), we have

$$\lambda^r B - \gamma \frac{\lambda^r - 1}{2/3 + 6\gamma} C - \frac{1}{18} C \leq 0$$

which implies

$$\lambda^r \left[B - \frac{\gamma}{2/3 + 6\gamma} C \right] \leq \left[\frac{1}{18} - \frac{\gamma}{2/3 + 6\gamma} \right] C. \qquad (8.138)$$

Since

$$\gamma_0 = \frac{1}{18} \qquad (8.139)$$

$$\frac{1}{18} \leq \frac{\gamma}{2/3 + 6\gamma} < \frac{1}{6} \quad \text{whenever } \gamma_0 \leq \gamma < \infty \qquad (8.140)$$

we have

$$\lambda^r \left[B - \frac{\gamma}{2/3 + 6\gamma} C \right] \leq 0, \quad \gamma \geq \gamma_0.$$

Using the Cauchy-Schwarz inequality, we conclude

$$C \leq 6B \quad \Longleftrightarrow \quad B - \frac{1}{6} C \geq 0. \qquad (8.141)$$

Combining (8.140) and (8.141), we get

$$B - \frac{\gamma}{2/3 + 6\gamma} C > 0 \quad \text{for } \gamma > \gamma_0.$$

Therefore $\lambda^r \leq 0$. Further, if $\lambda^r = 0$, from (8.138) and (8.140) we get

$$0 \leq \frac{1}{18} - \frac{\gamma}{2/3 + 6\gamma} \leq 0.$$

This is only possible if $\gamma = \frac{1}{18} = \gamma_0$. However, this case is excluded by the assumptions of Theorem 8.17(b). We conclude that $\lambda_r < 0$ for $\gamma > \gamma_0$. □

Proof of Theorem 8.19 Note that since L_0 is self-adjoint, it admits a single positive eigenvalue $\lambda_{0,R} = \frac{1}{4} + o(1)$ for R large. Now Part (a) follows from a simple perturbation argument.

We prove Part (b) by contradiction. Suppose that for R large (8.110) has an eigenfunction ϕ with eigenvalue λ_R such that $\text{Re}(\lambda_R) \geq -c_0$ for some sufficiently small $c_0 > 0$ independent of R. We write ϕ as a sum of an odd and even function.

Suppose that ϕ is odd. Then $\int_{-R}^{R} w^r \phi \, dy = 0$ and (8.110) gives

$$L_0 \phi = \lambda_R \phi, \quad \phi \in L^\infty(-R, R), \phi \text{ is odd}, \phi_y(\pm R) = 0.$$

By Lemma 8.20 we conclude that $\lambda_R \to 0$ as $R \to +\infty$, proving the first alternative in Theorem 8.19(b).

Suppose that ϕ is even. Following the proof of Theorem 8.17, we have

$$\int_{-R}^{R} \left(\phi^r L_0 \phi^r + \phi^i L_0 \phi^i \right) dy - \gamma A = \lambda_R^r B, \tag{8.142}$$

where λ_R^r is the real part of λ_R and $\phi = \phi^r + \sqrt{-1}\phi^i$. This implies that $|\lambda_R^r| \le C$ where C is independent of R for R large. Similarly, we also get $|\lambda_R^i| \le C$. Together, we have $|\lambda_R| \le C$. Thus we may assume that $\lambda_R \to \lambda$ as $R \to +\infty$. Without loss of generality we have $\|\phi\|_{L^\infty(-R,R)} \le 1$. Thus the limit of ϕ exists and it satisfies (8.108) with $\mathrm{Re}(\lambda) \ge -c_0$ for some $c_0 > 0$. Since ϕ is even, we have derived a contradiction to Theorem 8.17(b) if c_0 is chosen small enough. This concludes the proof of Theorem 8.19(b). □

8.2.4 Small Eigenvalues

Finally, we study the stability of the small eigenvalues. We have

Theorem 8.22 *If $p \gg 1$ the eigenvalue problem (8.98) with boundary conditions (8.102) admits a positive eigenvalue λ which satisfies*

$$\sqrt{\lambda + 1}\tanh L \tanh(L\sqrt{\lambda + 1}) = 1 + O\left(\frac{1}{p}\right). \tag{8.143}$$

Proof We expand the steady state and eigenfunction in the inner region to the two leading orders:

$$x = \alpha z;$$

$$u = U_0(z) + \alpha U_1(z) + \cdots \qquad H = H_0 + \alpha H_1(z) + \cdots$$

$$\phi = \Phi_0(z) + \alpha \Phi_1(z) + \cdots \qquad \psi = \Psi_0 + \cdots$$

The leading order equations are

$$\Phi_{0zz} + \frac{U_0}{H_0}\Phi_0 = 0; \qquad U_{0zz} - \frac{U_{0z}^2}{U_0} + \frac{U_0^2}{H_0} = 0; \qquad H_0 \equiv \text{const.} \tag{8.144}$$

Then Φ_0 is given by

$$\Phi_0(z) = \frac{U_{0z}}{U_0}. \tag{8.145}$$

Next we derive a solvability condition with Φ_0 as a test function. Multiplying (8.99) by $\frac{1}{\alpha}\Phi_0(\frac{x}{\alpha})$ and integrating on the half-interval $(0, L)$, we have,

$$\alpha^2(\lambda + 1)\int_0^L \phi(x)\Phi_0\left(\frac{x}{\alpha}\right)\frac{dx}{\alpha}$$

$$= \int_0^L \left(\alpha^2\phi_{xx} + \frac{u}{H}\phi + \alpha\frac{u}{H}\phi - \alpha\frac{u}{H^2}\psi \right)(x)\Phi_0\left(\frac{x}{\alpha}\right)\frac{dx}{\alpha}. \tag{8.146}$$

First we estimate the left-hand side of (8.146). Using $w(y) \sim C \exp(-|y|)$, $|y| \to \infty$ we have

$$\Phi_0 \sim -\sqrt{\eta}, \quad |z| \gg 1.$$

On the other hand, up to exponentially small terms we have

$$\phi_{xx} \sim (\lambda + 1)\phi, \quad |x| \gg \alpha; \qquad \phi_x(L) = 0$$

which gives

$$\phi \sim A_0 \frac{\cosh(\sqrt{\lambda + 1}(x - L))}{\cosh(\sqrt{\lambda + 1}L)},$$

where A_0 is obtained by matching ϕ as $x \to 0$ to Φ_0 as $z \to \infty$. We conclude

$$A_0 = -\sqrt{\eta}$$

and estimate

$$\int_0^L \phi(x)\Phi_0\left(\frac{x}{\alpha}\right)dx \sim \eta \int_0^L \frac{\cosh(\sqrt{\lambda + 1}(x - L))}{\cosh(\sqrt{\lambda + 1}L)}dx$$

$$\sim \frac{\eta}{\sqrt{\lambda + 1}}\tanh(\sqrt{\lambda + 1}L)$$

which implies

$$\text{l.h.s. } (8.146) = \alpha\eta\sqrt{\lambda + 1}\tanh(\sqrt{\lambda + 1}L). \tag{8.147}$$

Next we compute the right-hand side of (8.146). Since u decays exponentially as $|z| \to \infty$, the inner region gives the dominant contribution. Changing variables $x = \alpha z$ and expanding, we have,

$$\text{r.h.s. } (8.146) = \int_0^\infty \Phi_0\left(\Phi_{0zz} + \frac{U_0}{H_0}\Phi_0\right)dz + \alpha\int_0^\infty \Phi_0\left(\Phi_{1zz} + \frac{U_0}{H_0}\Phi_1\right)dz$$

$$+ \alpha\int_0^\infty \Phi_0^2\left(\frac{U_1}{H_0} - \frac{U_0 H_1}{H_0^2}\right)dz + \alpha\int_0^\infty \Phi_0^2\frac{U_0}{H_0}dz$$

$$- \alpha\int_0^\infty \frac{U_0\Phi_0}{H_0^2}\Psi_0 dz + O(\alpha^2).$$

By (8.144) the first term vanishes. The remaining terms are

$$\text{r.h.s. } (8.146) = \alpha(I_0 + I_1 + I_2 + I_3),$$

where

$$I_0 = \int_0^\infty \Phi_0\left(\Phi_{1zz} + \Phi_1\frac{U_0}{H_0}\right)dz,$$

$$I_1 = \int_0^\infty \Phi_0^2 \left(\frac{U_1}{H_0} - \frac{U_0 H_1}{H_0^2} \right) dz,$$

$$I_2 = \int \Phi_0^2 \frac{U_0}{H_0} dz,$$

$$I_3 = -\int_0^\infty \frac{U_0 \Phi_0}{H_0^2} \Psi_0 dz.$$

Defining

$$L_0 \Phi \equiv \Phi_{zz} + \frac{U_0}{H_0} \Phi,$$

then integration by parts gives

$$I_0 = \int_0^\infty \Phi_0 L_0 \Phi_1 dz = [\Phi_{1z}\Phi_0 - \Phi_1 \Phi_{0z}]_0^\infty = 0$$

by (8.144). Next, U_1 satisfies

$$U_{1zz} - \frac{2U_{0z}U_{1z}}{U_0} + \frac{U_{0z}^2}{U_0^2}U_1 + 2\frac{U_0 U_1}{H_0} - \frac{U_0^2}{H_0^2}H_1 - U_0 + \frac{U_{0z}^2}{U_0} = 0. \qquad (8.148)$$

Now define

$$\hat{U}_1 \equiv \frac{U_1}{U_0}.$$

Then $\hat{U}_1$ satisfies

$$\hat{U}_{1zz} + \frac{U_0}{H_0}\hat{U}_1 - \frac{U_0 H_1}{H_0^2} - 1 + \frac{U_{0z}^2}{U_0^2} = 0. \qquad (8.149)$$

Differentiating (8.149), we obtain

$$L_0 \hat{U}_{1z} = -\frac{U_{0z}\hat{U}_1}{H_0} + \frac{U_{0z}H_1}{H_0^2} + \frac{U_0 H_{1z}}{H_0^2} - \left(\frac{U_{0z}^2}{U_0^2} \right)_z$$

$$= -\Phi_0 \frac{U_1}{H_0} + \frac{\Phi_0 U_0 H_1}{H_0^2} + \frac{U_0 H_{1z}}{H_0^2} + 2\frac{U_{0z}}{H_0}$$

by (8.144). Therefore we have

$$I_1 = -\int_0^\infty \Phi_0 L_0 \hat{U}_{1z} dz + \int_0^\infty \frac{\Phi_0 U_0 H_{1z}}{H_0^2} dz + 2\int_0^\infty \frac{\Phi_0 U_{0z}}{H_0} dz.$$

Integrating by parts, we get

$$\int_0^\infty \Phi_0 L_0 \hat{U}_{1z} dz = \Phi_0(\infty)\hat{U}_{1zz}(\infty).$$

Note that $U_0 \to 0$, $\Phi_0 \to -\sqrt{\eta}$ as $z \to \infty$ and using (8.149) we obtain

$$\hat{U}_{1zz}(\infty) = 1 - \eta$$

so that

$$\int_0^\infty \Phi_0 L_0 \hat{U}_{1z} dz = -\sqrt{\eta} + \eta^{3/2}.$$

Next we compute

$$\int_0^\infty \frac{\Phi_0 U_0 H_{1z}}{H_0^2} dz = \frac{1}{H_0^2} \int_0^\infty U_{0z} H_{1z} dz = -\frac{1}{H_0^2} \int_0^\infty U_0 H_{1zz} dz.$$

Note that by (8.92) H satisfies

$$0 = DH_{xx} - H + u^r \alpha^{-r}$$

so that

$$DH_{1zz}(z) \sim -\alpha^{1-r} U_0^r(z)$$

and

$$\int_0^\infty \frac{\Phi_0 U_0 H_{1z}}{H_0^2} dz \sim \frac{\alpha^{1-r}}{DH_0^2} \int_0^\infty U_0^{r+1} dz.$$

Finally,

$$2 \int_0^\infty \frac{\Phi_0 U_{0z}}{H_0} dz = \frac{2}{H_0} \int_0^\infty \frac{U_{0z}^2}{U_0} dz = \frac{2}{3} \eta^{3/2} \int_0^\infty \frac{(w_y(y))^2}{w(y)} dy = \frac{2}{3} \eta^{3/2}.$$

In summary, we obtain

$$I_1 = \sqrt{\eta} - \frac{1}{3} \eta^{3/2} + \frac{\alpha^{1-r}}{DH_0^2} \int_0^\infty U_0^{r+1} dz.$$

Now

$$I_2 = \int_0^\infty \Phi_0^2 \frac{U_0}{H_0} dz = \frac{1}{3} \eta^{3/2}$$

and finally, we write

$$I_3 = \int_0^\infty U_0 \frac{\Psi_{0z}}{H_0^2} dz.$$

Now we have

$$\frac{D\Psi_{0zz}}{\alpha^2} - \Psi_0 + r\alpha^{-r-1} U_0^r \frac{U_{0z}}{U_0} = 0$$

so that

$$\Psi_{0z} \sim -\frac{\alpha^{1-r}}{D} U_0^r,$$

$$I_3 \sim -\frac{\alpha^{1-r}}{DH_0^2} \int_{-\infty}^{\infty} U_0^{r+1} dz.$$

Therefore, we finally obtain

$$\text{r.h.s. } (8.146) = \alpha \sqrt{\eta}.$$

We combine this result with (8.147) and recall by Theorem 8.14 that $\eta = \tanh^2 L$. Then we get (8.143). We note that l.h.s. $(8.143)|_{\lambda=0} = \tanh^2 L < 1$ and l.h.s. $(8.143) \to \infty$ as $\lambda \to \infty$ and so (8.143) has a positive eigenvalue. □

Finally, we complete the Proof of Theorem 8.16 by putting together the results on large and small eigenvalues. Firstly, in the case of a single boundary spike, an eigenfunction satisfies the boundary condition (8.101) and the eigenvalue problem given by (8.105) which is equivalent to (8.108) for $\gamma = \frac{r}{3} \frac{1}{\int_{-\infty}^{\infty} w^r dy}$. Since we assume $r > 1$, we have $\gamma > \gamma_0$, where γ_0 is given by (8.109) and thus $\text{Re}(\lambda) \leq -c_0 < 0$ by Theorem 8.19. This implies the stability of a single boundary spike.

Secondly, we consider an interior spike centred at the origin. There are two relevant eigenvalues whose eigenfunctions satisfy the boundary conditions given in (8.101) or (8.102), respectively. The former is stable as just shown. However the latter (small eigenvalue) always has negative real part, as shown in Theorem 8.22.

Thirdly, we consider the double boundary spike case, which admits two eigenvalues. The first corresponding eigenfunction satisfies boundary conditions (8.101), the second boundary conditions (8.103). The former is stable as just shown. The latter leads to the nonlocal eigenvalue problem (8.106) as derived in Sect. 8.2.3. Theorem 8.17 implies that the corresponding eigenvalue is unstable if $r \tanh^2(\frac{L}{\sqrt{D}}) < 1$ and stable if $r = 2$ and $r \tanh^2(\frac{L}{\sqrt{D}}) > 1$. The threshold is given by $D = D_c$ where D_c is given by (8.88). The proof of Theorem 8.16 is complete.

Finally, we verify Theorem 8.22 numerically, by computing the eigenvalue problem (8.98), (8.102). The numerical algorithm proceeds by reformulating the eigenvalue problem as a boundary value problem, where an extra equation $\frac{d}{dx}\lambda(x) = 0$ is adjoined along with an extra boundary condition $\phi_x(0) = 1$. We started by using (8.145) as a first approximation. Then we compared the resulting λ_{numeric} with $\lambda_{\text{asymptotic}}$, computed by numerically solving the algebraic equation (8.143). For parameters $r = 2$, $D = 1$, $L = 1$ and $p = 90$ or $p = 180$ we get:

$p = 90$: $\lambda_{\text{numeric}} = 1.21197,$ $\lambda_{\text{asymptotic}} = 1.13769;$ error $= 6.5\%$

$p = 180$: $\lambda_{\text{numeric}} = 1.1729,$ $\lambda_{\text{asymptotic}} = 1.13769;$ error $= 3.4\%.$

These data indicate that doubling p halves the error thus giving a reasonable numerical verification of Theorem 8.22.

8.3 The Gierer-Meinhardt System with Robin Boundary Conditions

We now consider the effect of the boundary conditions on spiky patterns. Previously we have always studied Neumann (also called no-flux) boundary conditions. In this section we will investigate Robin (also called third type) boundary conditions and encounter some striking phenomena.

We study the Gierer-Meinhardt system

$$
\begin{cases}
\frac{\partial A}{\partial t} = D_a \Delta A - A + \frac{A^p}{H^q}, \\
\tau \frac{\partial H}{\partial t} = D_H \Delta H - H + \frac{A^r}{H^s}, \\
p > 1, \qquad q > 0, \qquad r > 0, \qquad s \geq 0, \\
0 < \frac{p-1}{q} < \frac{r}{s+1}
\end{cases}
\tag{8.150}
$$

with Robin boundary conditions

$$
\epsilon \frac{\partial A}{\partial \nu} + a_A A = 0, \qquad \sqrt{D} \frac{\partial H}{\partial \nu} + a_H H = 0.
\tag{8.151}
$$

Taking the limit $D \to \infty$, at least formally, we obtain the shadow system

$$
\begin{cases}
A_t = \epsilon^2 \Delta A - A + \frac{A^p}{\xi^q}, & x \in \Omega, t > 0, \\
\tau |\Omega| \xi_t = -(|\Omega| + \sqrt{D} a_H |\partial \Omega|)\xi + \frac{1}{\xi^s} \int_\Omega A^r dx, & t > 0, \\
\epsilon \frac{\partial A}{\partial \nu} + a_A A = 0, & x \in \partial \Omega, t > 0,
\end{cases}
$$

where $\xi(t)$ ($t \geq 0$) is the spatial average of the inhibitor $H(x,t)$, namely

$$
\xi(t) = \frac{1}{|\Omega|} \int_\Omega H(x,t) dx \quad \text{for all } t \geq 0.
$$

Here we have used

$$
\int_\Omega D \Delta H dx = \int_{\partial \Omega} D \frac{\partial H}{\partial \nu} ds
$$

$$
= -\int_{\partial \Omega} \sqrt{D} a_H H ds \sim \sqrt{D} a_H |\partial \Omega| |\xi|
$$

for all $t \geq 0$. From now on we assume $a_H = 0$ to simplify the presentation. Then we get the shadow system

$$
\begin{cases}
A_t = \epsilon^2 \Delta A - A + \frac{A^p}{\xi^q}, & x \in \Omega, t > 0, \\
\tau |\Omega| \xi_t = -|\Omega| \xi + \frac{1}{\xi^s} \int_\Omega A^r dx, & t > 0, \\
\epsilon \frac{\partial A}{\partial \nu} + a_A A = 0, & x \in \partial \Omega, t > 0.
\end{cases}
\tag{8.152}
$$

To derive stationary solutions to the shadow system (8.152), we set $A(x) = \xi^{q/(p-1)}u(x)$, $a_A = a$, where u satisfies

$$\begin{cases} \epsilon^2 \Delta u - u + u^p = 0, & x \in \Omega, \\ u > 0, & x \in \Omega, \\ \epsilon \frac{\partial u}{\partial \nu} + au = 0, & x \in \partial\Omega \end{cases} \qquad (8.153)$$

and for ξ we have

$$0 = -|\Omega|\xi + \frac{\xi^{qr/(p-1)}}{\xi^s} \int_\Omega u^r dx$$

which gives

$$\xi = \left(\frac{1}{|\Omega|} \int_\Omega u^r dx \right)^{-(p-1)/(qr-(p-1)(s+1))}.$$

Problem (8.153) has been studied by Berestycki and Wei in [14] and the following result has been proved:

Theorem A   *Let* $1 < p < (\frac{N+2}{N-2})_+$. *Then there exists a number* $a(N, p)$, *where* $a(1, p) = 1$ *and* $a(N, p) > 1$ *for* $N \geq 2$, *such that problem* (8.153) *has a solution* $u_{\epsilon,a}$ *satisfying*

(1) $u_{\epsilon,a}$ *has the least energy among all solutions to* (8.153), *i.e.*

$$E_\epsilon[u_{\epsilon,a}] \leq E_\epsilon[u] \qquad (8.154)$$

for all solutions u *to* (8.153), *where* E_ϵ *is the energy functional defined by*

$$E_\epsilon[u] = \frac{\epsilon^2}{2} \int_\Omega |\nabla u|^2 dx + \frac{1}{2} \int_\Omega u^2 dx - \frac{1}{p+1} \int_\Omega u^{p+1} dx + \frac{\epsilon a}{2} \int_{\partial\Omega} u^2 ds \qquad (8.155)$$

and $u_+ = \max\{u, 0\}$.

(2) *If* $0 < a < a(N, p)$, *then* $u_{\epsilon,a}$ *has a local maximum point* $x_\epsilon \in \Omega$ *with*

$$\frac{d(x_\epsilon, \partial\Omega)}{\epsilon} \to d_0 > 0. \qquad (8.156)$$

(3) *If* $a > a(N, p)$, *then* $u_{\epsilon,a}$ *has a unique local maximum point* $x_\epsilon \in \Omega$ *with*

$$d(x_\epsilon, \partial\Omega) \to \max_{x \in \Omega} d(x, \partial\Omega). \qquad (8.157)$$

Remark 8.23

(1) The solution in part (2) of Theorem A is called a near-boundary spike (see Fig. 8.3).
(2) The solution in part (3) of Theorem A is called the interior spike (see Fig. 8.4).

(3) We remark that in part (2) of Theorem A, i.e. for $0 < a < a(N, p)$, there also exists an interior spike which is a solution of (8.153) but does not minimise (8.155) among the solutions of (8.153).

Next we study the stability of the steady state $(A_{\epsilon,a}, \xi_{\epsilon,a})$ of the shadow system (8.152), where

$$\begin{cases} A_{\epsilon,a} = \xi_{\epsilon,a}^{q/(p-1)} u_{\epsilon,a}, \\ \xi_{\epsilon,a} = (\frac{1}{|\Omega|} \int_\Omega u_{\epsilon,a}^r dx)^{-(p-1)/(qr-(p-1)(s+1))} \end{cases} \tag{8.158}$$

and $u_{\epsilon,a}$ is the minimal energy solution of (8.153) given in Theorem A.

In [249] a stability result has been obtained for Neumann boundary conditions in the cases

$$r = 2 \quad \text{and} \quad 1 < p < 1 + \frac{4}{N},$$

and

$$r = p + 1 \quad \text{and} \quad 1 < p < \left(\frac{N+2}{N-2}\right)_+. \tag{8.159}$$

We will now derive a result for Robin boundary conditions under similar assumptions.

The first result states that for $a > a(N, p)$ the interior spike is stable.

Theorem 8.24 (Stability of the interior spike) *Assume that $a > a(N, p)$. Further, assume that either*

$$r = 2 \quad \text{and} \quad 1 < p < 1 + \frac{4}{N}$$

or

$$r = p + 1 \quad \text{and} \quad 1 < p < \left(\frac{N+2}{N-2}\right)_+.$$

Then there exists a $\tau_0 > 0$ such that for all $0 < \epsilon \ll 1$ and $0 \leq \tau < \tau_0$ the interior spike $(A_{\epsilon,a}, \xi_{\epsilon,a})$ is a (linearly) stable steady state of the shadow system (8.152).

The second results states that for $N = 1$, i.e. in case Ω is an interval, for all $1 < p \leq 3$ and $0 < a < 1$ the near-boundary spike is stable.

Theorem 8.25 (Stability of the near-boundary spike) *Assume that*

$$N = 1 \quad \text{and} \quad 0 < a < 1. \tag{8.160}$$

Further, assume that either

$$r = 2 \quad \text{and} \quad 1 < p \leq 3$$

or

$$r = p + 1 \quad \text{and} \quad 1 < p < \infty.$$

Then there exists a $\tau_0 > 0$ such that for all $0 < \epsilon \ll 1$ and $0 \leq \tau < \tau_0$ the near-boundary spike $(A_{\epsilon,a}, \xi_{\epsilon,a})$ is a (linearly) stable steady state of the shadow system (8.152).

The last result states that for $p > 3$ the near-boundary spike may become unstable.

Theorem 8.26 (Instability of the near-boundary spike) *Assume that (8.160) is valid. Further, assume that $r = 2$ and $p > 3$. Then there exist $a_0 > 0$ and $\mu_0 > 1$ such that if*

$$a_0 < a < 1 \quad \text{and} \quad 1 < \mu := \frac{2q}{(p-1)(s+1)} < \mu_0 \qquad (8.161)$$

then for all $0 < \epsilon \ll 1$ and $\tau \geq 0$ the near-boundary spike $(A_{\epsilon,a}, \xi_{\epsilon,a})$ is an unstable steady state of the shadow system (8.152).

Remark 8.27

(1) By the proof of Theorem 8.26 the unstable eigenfunction has an eigenvalue of order $O(1)$ as $\epsilon \to 0$. It is important to note that for Neumann boundary conditions in case $N = 1$ the minimal energy solution, which is given by a boundary spike, is stable for all p, q, s such that

$$1 < \mu := \frac{2q}{(p-1)(s+1)} \quad \text{and} \quad 1 < p < 5,$$

see [253]. Therefore the instability of Theorem 8.26 occurs only for Robin and not for Neumann boundary conditions.
(2) The threshold $\mu = \mu_0$ is linked to a Hopf bifurcation.
(3) Note that in Theorem 8.26 we assume that both constants $a < 1$ and $\mu := \frac{qr}{(p-1)(s+1)} > 1$ are sufficiently close to 1.

We begin by highlighting the main points of the proofs of Theorems 8.24–8.26.
We linearise the shadow system (8.152) around the spiky steady state (8.158) and get

$$\begin{cases} \epsilon^2 \Delta \phi_\epsilon - \phi_\epsilon + p \frac{A_\epsilon^{p-1}}{\xi_\epsilon^q} \phi_\epsilon - q \frac{A_\epsilon^p}{\xi_\epsilon^{q+1}} \eta_\epsilon = \alpha_\epsilon \phi_\epsilon, \\ \frac{r}{\tau |\Omega|} \int_\Omega \frac{A_\epsilon^{r-1} \phi_\epsilon}{\xi_\epsilon^s} dx - \frac{s+1}{\tau} \eta_\epsilon = \alpha_\epsilon \eta_\epsilon, \end{cases} \qquad (8.162)$$

where $(\phi_\epsilon, \eta_\epsilon)$ in $H_{rob}^1(\Omega) \times \mathbb{R}$ and

$$H_{rob}^1(\Omega) = \left\{ \phi \in H^1(\Omega) : \epsilon \frac{\partial \phi}{\partial \nu} + a\phi = 0 \text{ on } \partial \Omega \right\}.$$

By (8.158), if $a_H = 0$ the eigenvalues of (8.162) in $H^1_{rob}(\Omega) \times \mathbb{R}$ coincide with those of

$$\epsilon^2 \Delta \phi_\epsilon - \phi_\epsilon + p u_\epsilon^{p-1} \phi_\epsilon - \frac{qr}{s+1+\tau\alpha_\epsilon} \frac{\int_\Omega u_\epsilon^{r-1} \phi_\epsilon dx}{\int_\Omega u_\epsilon^r dx} u_\epsilon^p$$

$$= \alpha_\epsilon \phi_\epsilon, \quad \phi_\epsilon \in H^1_{rob}(\Omega). \tag{8.163}$$

If $N = 1$ and $0 < a < 1$, we get $u_{\epsilon,a}(x) \sim w(\frac{x-x_\epsilon}{\epsilon}) = w(\frac{x}{\epsilon} - \frac{x_\epsilon}{\epsilon}) =: w_{x_\epsilon/\epsilon}(\frac{x}{\epsilon})$, where w is given by (8.89). Using the Robin boundary condition $w' = aw$, we have $\frac{x_\epsilon}{\epsilon} \to y_0$, where $y_0 > 0$ is satisfies

$$w'(-y_0) = aw(-y_0). \tag{8.164}$$

For the ground state of (8.89) we recall that

$$w(y) = \left(\frac{p+1}{2} \cosh^{-2} \frac{(p-1)y}{2} \right)^{1/(p-1)}$$

and compute

$$w'(y) = -\tanh \frac{(p-1)y}{2} w(y)$$

and finally

$$\frac{w'(-y_0)}{w(-y_0)} = \tanh \frac{(p-1)y_0}{2} = a.$$

Thus, if $0 < a < 1$, we have

$$y_0 = \frac{2}{p-1} \operatorname{artanh} a \tag{8.165}$$

which implies

$$w(y_0) = \left(\frac{(p+1)(1-a^2)}{2} \right)^{1/(p-1)}. \tag{8.166}$$

Then we have

Lemma A *Let α_ϵ be an eigenvalue of (8.163).*

(1) *For $a > a(N, p)$ we have $\alpha_\epsilon = o(1)$ as $\epsilon \to 0$ if and only if $\alpha_\epsilon = (1 + o(1))\tau^\epsilon_j$ for some $j = 1, \ldots, N$, where τ^ϵ_j will be given in Theorem 8.45 below (interior spike case).*

For $N = 1$ and $a < a(N, p)$, there are no eigenvalues $\alpha_\epsilon = o(1)$ (near-boundary spike case).

(2) *If* $\alpha_\epsilon \to \alpha_0 \neq 0$, *then*

$$\Delta\phi - \phi + pw_{y_0}^{p-1}\phi - \frac{qr}{s+1+\tau\alpha_0}\frac{\int_0^\infty w_{y_0}^{r-1}\phi}{\int_0^\infty w_{y_0}^r}w_{y_0}^p = \alpha_0\phi, \qquad (8.167)$$

where (i) *for* $a > a(N, P)$ *we have* $w_{y_0} = w$, $\phi \in H^1(\mathbb{R}^N)$ *(interior spike case) and* (ii) *for* $N = 1$ *and* $a < a(1, P) = 1$ *we have* $w_{y_0} = w(y - y_0)$, *where* $y_0 > 0$ *solves* $w'(y_0) + aw(y_0) = 0$ *and* $\phi \in H^1_{rob}(\mathbb{R}^+)$ *(near-boundary spike case).*

Proof In case $a > a(N, p)$ the proof of Part (1) in Lemma A for the Robin boundary condition is similar to that in Sect. 4.1.2 for the Neumann boundary condition. In both cases, we analyse interior spikes with exponential decay at infinity. However, for Neumann boundary conditions interior spikes are unstable, whereas for Robin boundary conditions they are stable. This comes from the fact that the terms $\varphi_{\epsilon,P_\epsilon}(P_\epsilon)$ which will be defined in (8.202) below, have different signs for Neumann and Robin boundary conditions, respectively.

If $N = 1$ and $a < 1$ the proof of Part (1) in Lemma A for Robin boundary conditions resembles that of Sect. 4.1.2 for a boundary spike with Neumann boundary condition: In both cases there are no small eigenvalues $\alpha_\epsilon = o(1)$.

The proof of Part (2) follows by a convergence result and a compactness result of Dancer [37] similarly to Chap. 4. □

We remark that the eigenvalue problem in Part (2) of Lemma A (near-boundary spike case) is a half-line nonlocal eigenvalue problem NLEP with Robin boundary condition. In the next section, we will prove results on its spectral and stability properties.

From now on we assume that $\tau = 0$. By a regular perturbation argument as in Chap. 3 the results also hold when τ is sufficiently small.

8.3.1 Study of the NLEP

In this subsection, we analyse the NLEP

$$\phi'' - \phi + pw_{y_0}^{p-1}\phi - \frac{qr}{s+1}\frac{\int_0^\infty w_{y_0}^{r-1}\phi dy}{\int_0^\infty w_{y_0}^r dy}w_{y_0}^p = \lambda\phi, \qquad \phi \in H^1_{rob}(\mathbb{R}^+), \quad (8.168)$$

where $w_{y_0}(y) = w(y - y_0)$ for some $y_0 > 0$ and w satisfies (8.89). Letting

$$L_0\phi := \phi'' - \phi + pw_{y_0}^{p-1}\phi, \qquad \phi \in H^1_{rob}(\mathbb{R}^+),$$

we set

$$L\phi := L_0\phi - \mu(p-1)\frac{\int_0^\infty w_{y_0}^{r-1}\phi dy}{\int_0^\infty w_{y_0}^r dy}w_{y_0}^p, \qquad \phi \in H^1_{rob}(\mathbb{R}^+),$$

where

$$\mu = \frac{qr}{(s+1)(p-1)} > 1.$$

First we prove

Lemma 8.28 *Let $\phi \in H_{rob}^1(\mathbb{R}^+)$ satisfy*

$$\phi'' - \phi + pw_{y_0}^{p-1}\phi = 0, \qquad \|\phi\|_{H^1(\mathbb{R}^+)} = 1. \tag{8.169}$$

Then $\phi \equiv 0$.

Proof By the Robin boundary condition for w_{y_0} we have

$$a = \frac{w_{y_0}'(0)}{w_{y_0}(0)}. \tag{8.170}$$

Using (8.89), w_{y_0} satisfies

$$w_{y_0}'' = w_{y_0} - w_{y_0}^p, \qquad (w_{y_0}')^2 = w_{y_0}^2 - \frac{2}{p+1}w_{y_0}^{p+1}. \tag{8.171}$$

Multiplying (8.169) by w_{y_0}', integration by parts, using (8.171) and the Robin boundary condition for ϕ, we get

$$0 = \phi'(0)w_{y_0}'(0) - \phi(0)w_{y_0}''(0) = \phi(0)\left[aw_{y_0}'(0) - w_{y_0}''(0)\right]. \tag{8.172}$$

By (8.170) and (8.171), we have

$$\begin{aligned}
aw_{y_0}'(0) - w_{y_0}''(0) &= \frac{(w_{y_0}'(0))^2 - w_{y_0}(0)w_{y_0}''(0)}{w_{y_0}(0)} \\
&= \frac{(w_{y_0}'(0))^2 - (w_{y_0}(0))^2 + (w_{y_0}(0))^{p+1}}{w_{y_0}(0)} \\
&= \frac{p-1}{p+1}w_{y_0}^p(0) > 0. \tag{8.173}
\end{aligned}$$

Now (8.172) implies

$$\phi(0) = 0 \tag{8.174}$$

and by the Robin boundary condition we get $\phi'(0) = 0$. Using the uniqueness of initial value problems for ODEs, we conclude that $\phi(y) \equiv 0$ on $\mathbb{R}^+$. The lemma follows. $\square$

By the Fredholm Alternative (see Theorem 13.2), we conclude from Lemma 8.28 that the operator L_0, defined on $H_{rob}^1(\mathbb{R}^+)$, is invertible.

Using

$$L_0 w_{y_0} = (p-1)w_{y_0}^p, \qquad w'_{y_0}(0) - a w_{y_0}(0) = 0,$$

we have

$$L_0^{-1}\left(w_{y_0}^p\right) = \frac{1}{p-1} w_{y_0}. \qquad (8.175)$$

By elementary calculations we have that

$$L_0\left(\frac{1}{p-1} w_{y_0} + \frac{1}{2} y w'_{y_0}\right) = w_{y_0}. \qquad (8.176)$$

However, it is important to note that $\frac{1}{p-1} w_{y_0} + \frac{1}{2} y w'_{y_0}$ does not satisfy the Robin boundary condition and so $\frac{1}{p-1} w_{y_0} + \frac{1}{2} y w'_{y_0} \notin H^1_{rob}(\mathbb{R}^+)$. This implies $L_0^{-1}(w_{y_0}) \neq \frac{1}{p-1} w_{y_0} + \frac{1}{2} y w'_{y_0}$ and $L_0^{-1}(w_{y_0})$ is given as follows.

Lemma 8.29 *For $a \neq 1$ we have*

$$L_0^{-1}(w_{y_0}) = \frac{1}{p-1} w_{y_0} + \frac{1}{2} y w'_{y_0} + A w'_{y_0},$$

where

$$A = \frac{a}{(p-1)(1-a^2)}.$$

Proof To satisfy the Robin boundary condition, we need to select A such that

$$A\left(w''_{y_0}(0) - a w'_{y_0}(0)\right) + \frac{1}{2} w'_{y_0}(0) = 0.$$

By (8.173) we compute

$$A = \frac{p+1}{2(p-1)} w'_{y_0}(0) w_{y_0}^{-p}(0). \qquad (8.177)$$

Substituting (8.165), (8.166) into (8.177), we get

$$A = \frac{p+1}{2(p-1)}\left(\frac{p+1}{2}\right)^{-1} \frac{a}{1-a^2} = \frac{a}{(p-1)(1-a^2)}.$$

The lemma follows.   □

Remark 8.30

(1) A difficulty in the multidimensional case is to find a term corresponding to $A w'_{y_0}$ in Lemma 8.29.

(2) The term $A w'_{y_0}$ in Lemma 8.29 can lead to an instability of the near-boundary spike.

(3) Asymptotically, we have $A \to \infty$ as $a \to 1$ and $A \to 0$ as $a \to 0$. The first limit will play a key role in the analysis. The second limit occurs when the near-boundary spike in the Robin boundary condition case approaches the boundary spike in the Neumann boundary condition case.

As an essential step, we next compute the sign of the integral

$$\rho(y_0) := \int_0^\infty w_{y_0} L_0^{-1}(w_{y_0}) dy.$$

Using Lemma 8.29, we get

$$
\begin{aligned}
\rho(y_0) &= \int_0^\infty w_{y_0} L_0^{-1}(w_{y_0}) dy \\
&= \frac{1}{p-1} \int_0^\infty w_{y_0}^2 dy + \frac{1}{2} \int_0^\infty y w_{y_0} w'_{y_0} dy + A \int_0^\infty w_{y_0} w'_{y_0} dy \\
&= \left(\frac{1}{p-1} - \frac{1}{4} \right) \int_0^\infty w_{y_0}^2 dy - \frac{A}{2} w_{y_0}^2(0) \\
&= \left(\frac{1}{p-1} - \frac{1}{4} \right) \int_0^\infty w_{y_0}^2 dy - \frac{a}{2(p-1)(1-a^2)} \left(\frac{(p+1)(1-a^2)}{2} \right)^{2/(p-1)} \\
&= \left(\frac{1}{p-1} - \frac{1}{4} \right) \int_{-y_0}^\infty w^2 dy - \frac{(p+1)^{2/(p-1)} a}{2^{(p+1)/(p-1)}(p-1)(1-a^2)^{(p-3)/(p-1)}}.
\end{aligned}
$$

$$(8.178)$$

Differentiating $\rho(y_0)$ and using (8.177), we have

$$
\begin{aligned}
\rho'(y_0) &= \frac{5-p}{4(p-1)} w_{y_0}^2(0) + \frac{p+1}{4(p-1)} \big[w_{y_0}^{2-p}(0) w''_{y_0}(0) \\
&\quad + (2-p) w_{y_0}^{1-p}(0) \left(w'_{y_0}(0) \right)^2 \big] \\
&= \frac{5-p}{4(p-1)} w_{y_0}^2(0) + \frac{p+1}{4(p-1)} \Big[(w_{y_0} - w_{y_0}^p) w_{y_0}^{2-p} \\
&\quad + (2-p) w_{y_0}^{1-p} \left(w_{y_0}^2 - \frac{2}{p+1} w_{y_0}^{p+1} \right) \Big](0) \\
&= \frac{(p+1)(3-p)}{4(p-1)} w_{y_0}^{3-p}(0)
\end{aligned}
$$

$$(8.179)$$

by (8.171). Using these computations, we have

Proposition 8.31 *For $1 < p \leq 3$, we have*

$$\int_0^\infty w_{y_0} L_0^{-1}(w_{y_0}) dy > 0. \tag{8.180}$$

Proof If $1 < p \leq 3$, for $y_0 = 0$ (corresponding to $a = 0$), (8.178) implies

$$\rho(0) = \left(\frac{1}{p-1} - \frac{1}{4}\right) \int_0^\infty w_{y_0}^2 dy > 0.$$

By (8.179), we have $\rho'(y_0) \geq 0$ for all $y_0 \in (0, \infty)$ and so $\rho(y_0) \geq 0$ for all $y_0 \in [0, \infty)$. □

Remark 8.32 It is interesting to note that for $p = 3$ and $p = 5$ the integral in (8.180) can be calculated explicitly. For $p = 3$, we get $\rho'(y_0) = 0$ and

$$\int_0^\infty w_{y_0} L_0^{-1}(w_{y_0}) dy = \rho(0) = \frac{1}{4} \int_0^\infty w^2 dy = \frac{1}{2} \int_0^\infty \cosh^{-2}(y) dy$$

$$= [\tanh y]_0^\infty = \frac{1}{2} > 0 \quad \text{for all } y_0 > 0.$$

For $p = 5$, we have

$$\rho(y_0) = \frac{\sqrt{3}}{2} \frac{a}{\sqrt{1-a^2}}.$$

Unlike in Proposition 8.31, for $p > 3$, the integral $\int_0^\infty w_{y_0} L_0^{-1}(w_{y_0}) dy$ can be negative and we have

Proposition 8.33 *Suppose that $p > 3$ and*

$$\frac{5-p}{p-1} \int_{-(2/(p-1))\operatorname{artanh} a}^\infty w^2 dy < \frac{2^{(p-3)/(p-1)}(p+1)^{2/(p-1)}a}{(p-1)(1-a^2)^{(p-3)/(p-1)}}. \tag{8.181}$$

Then we have $\int_0^\infty w_{y_0} L_0^{-1}(w_{y_0}) dy < 0$. Further, there is some constant $a_0(p) < 1$ such that for $a_0(p) < a < 1$ condition (8.181) holds.

Before proving Proposition 8.33, we discuss the implications of condition (8.181) in a series of remarks.

Remark 8.34

(1) If $p = 5$, (8.181) is satisfied for all $0 < a < 1$ since l.h.s. $= 0$ and r.h.s. > 0. By continuity, this implies that $a_0(p) \to 0$ as $p \to 5^-$.
(2) If $p = 3$, we have r.h.s. $= 2a$ and l.h.s. $= 2 + 2a$. Thus there is some $p_0 > 3$ such that for $3 < p < p_0$ the solution set of (8.181) is empty for $0 < a < 1$.

Proof of Proposition 8.33 From (8.178) we get (8.181). The left-hand side of (8.181) is positive and bounded for $a \in (0, 1)$. The right-hand side of (8.181) tends to 0 as $a \to 0^+$ and to $+\infty$ as $a \to 1^-$. By continuity, there is an $a_0(p) \in (0, 1)$ such that (8.181) holds for $a_0(p) < a < 1$.   □

Next we need

Lemma 8.35 *The first eigenvalue of L_0, denoted by μ_1, is positive. The second eigenvalue of L_0 is negative.*

Proof Let

$$Q[u] = \frac{\int_0^\infty [(u')^2 + u^2]dy + au^2(0)}{(\int_0^\infty u^{p+1}dy)^{2/(p+1)}}.$$

Then up to a scaling factor the unique minimiser of $Q[u]$ in $H^1_{rob}(\mathbb{R}^+)$ is given by w_{y_0}.

As in the proof of Lemma 13.5, if follows that the second eigenvalue of L_0 is non-positive. Further, by Lemma 8.28, the kernel is trivial and so the second eigenvalue is negative.   □

If $r = 2$ we define the operator

$$L_1\phi := L_0\phi - (p - 1)\frac{\int_0^\infty w_{y_0}\phi dy}{\int_0^\infty w_{y_0}^2 dy}w_{y_0}^p - (p - 1)\frac{\int_0^\infty w_{y_0}^p\phi dy}{\int_0^\infty w_{y_0}^2 dy}w_{y_0}$$

$$+ (p - 1)\frac{\int_0^\infty w_{y_0}^{p+1}dy \int_0^\infty w_{y_0}\phi dy}{(\int_0^\infty w_{y_0}^2 dy)^2}w_{y_0}, \quad \phi \in H^1_{rob}(\mathbb{R}^+). \quad (8.182)$$

Then we have

Lemma 8.36

(1) *The operator L_1 is self-adjoint and its kernel X_1 is given by* span$\{w_{y_0}\}$.
(2) *There is a constant $c_0 > 0$ such that*

$$L_1(\phi, \phi) := \int_0^\infty [(\phi')^2 + \phi^2 - pw_{y_0}^{p-1}\phi^2]dy$$

$$+ \frac{2(p - 1)\int_0^\infty w_{y_0}\phi dy \int_0^\infty w_{y_0}^p\phi dy}{\int_0^\infty w_{y_0}^2 dy}$$

$$- (p - 1)\frac{\int_0^\infty w_{y_0}^{p+1}dy}{(\int_0^\infty w_{y_0}^2 dy)^2}\left(\int_0^\infty w_{y_0}\phi dy\right)^2$$

$$\geq c_0 d^2_{L^2(\mathbb{R}^+)}(\phi, X_1)$$

for all $\phi \in H^1_{rob}(\mathbb{R}^+)$, where $d_{L^2(0,\infty)}$ denotes the distance in L^2-norm.

Proof By definition (8.182), it is easy to compute that $(L_1\phi, \psi)_{L^2(0,\infty)} = (L_1\psi, \phi)_{L^2(0,\infty)}$ for all $\phi, \psi \in H^1_{rob}(\mathbb{R}^+)$, and thus the operator L_1 is self-adjoint.

To compute the kernel of L_1, we note that $w_{y_0} \in \text{kernel}(L_1)$. For $\phi \in \text{kernel}(L_1)$, Lemma 8.29 implies,

$$L_0\phi = c_1(\phi)w_{y_0} + c_2(\phi)w^p_{y_0}$$
$$= c_1(\phi)L_0\left(\frac{1}{p-1}w_{y_0} + \frac{1}{2}yw'_{y_0} + Aw'_{y_0}\right) + c_2(\phi)L_0\left(\frac{1}{p-1}w_{y_0}\right),$$

where

$$c_1(\phi) = (p-1)\frac{\int_0^\infty w^p_{y_0}\phi\, dy}{\int_0^\infty w^2_{y_0}\, dy} - (p-1)\frac{\int_0^\infty w^{p+1}_{y_0}\, dy \int_0^\infty w_{y_0}\phi\, dy}{(\int_0^\infty w^2_{y_0}\, dy)^2},$$

$$c_2(\phi) = (p-1)\frac{\int_0^\infty w_{y_0}\phi\, dy}{\int_0^\infty w^2_{y_0}\, dy}.$$

This implies

$$\phi = c_1(\phi)L_0^{-1}(w_{y_0}) + c_2(\phi)L_0^{-1}(w^p_{y_0})$$
$$= c_1(\phi)L_0^{-1}(w_{y_0}) + \frac{1}{p-1}c_2(\phi)w_{y_0}. \tag{8.183}$$

By (8.183), we have

$$c_1(\phi) = c_1(\phi)\left[(p-1)\frac{\int_0^\infty w^p_{y_0}L_0^{-1}(w_{y_0})\, dy}{\int_0^\infty w^2_{y_0}\, dy}\right.$$
$$\left. - (p-1)\frac{\int_0^\infty w^{p+1}_{y_0}\, dy \int_0^\infty w_{y_0}L_0^{-1}(w_{y_0})\, dy}{(\int_0^\infty w^2_{y_0}\, dy)^2}\right]$$
$$= c_1(\phi)\left[1 - (p-1)\frac{\int_0^\infty w^{p+1}_{y_0}\, dy \int_0^\infty w_{y_0}L_0^{-1}(w_{y_0})\, dy}{(\int_0^\infty w^2_{y_0}\, dy)^2}\right].$$

Therefore, $c_1(\phi) = 0$. Using (8.183) and Lemma 8.28, the proof of Part (1) is finished.

The proof of Part (2) follows along the same lines as in Chap. 3 for the case of Neumann boundary conditions. $\qquad\square$

The following result on the stability or instability of a near-boundary spike for a Robin boundary condition is similar to the Neumann boundary condition case.

Theorem 8.37 *Assume* $0 < a < 1$. *Further, assume*

$$r = 2 \quad and \quad 1 < p \le 3$$

or

$$r = p + 1 \quad and \quad 1 < p < \infty.$$

Then the NLEP

$$\phi'' - \phi + p w_{y_0}^{p-1} \phi - \mu(p-1) \frac{\int_0^\infty w_{y_0}^{r-1} \phi \, dy}{\int_0^\infty w_{y_0}^r \, dy} w_{y_0}^p$$

$$= \lambda \phi, \quad \phi \in H_{rob}^1(\mathbb{R}^+), \tag{8.184}$$

for $\mu > 1$ *admits only stable eigenvalues but for* $\mu < 1$ *allows unstable eigenvalues.*

For the exponents $r = 2$, $p > 3$, there is a new instability for the near-boundary spike in the case of Robin boundary conditions (not present for Neumann boundary conditions).

Theorem 8.38 *Assume*

$$r = 2 \quad and \quad p > 3.$$

Then there are $a_0(p) \in (0, 1)$ *and* $\mu_0(a, p) > 1$ *such that for*

$$a_0 < a < 1 \tag{8.185}$$

and

$$\mu < \mu_0(a, p) \tag{8.186}$$

the NLEP (8.184) *possesses a positive eigenvalue.*

Remark 8.39

(1) The number $a_0(p)$ can be chosen by (8.181) and Remark 8.34.
(2) By continuity, we have $\mu_0(a, p) \to 1$ as $a \to a_0(p)$ for fixed $p > 3$.

As a preparation for proving the instability part of Theorem 8.37, we prove the following

Theorem 8.40

(1) *Suppose that* $\mu < 1$ *and* $r = 2$, $1 < p \le 3$ *or* $r = p + 1$, $1 < p < \infty$. *Then the NLEP* (8.184) *has a positive eigenvalue.*
(2) *Suppose that* $r = 2$ *and*

$$\int_0^\infty w_{y_0} L_0^{-1} w_{y_0} \, dy < 0.$$

If (8.186) *holds, then the NLEP* (8.184) *has a positive eigenvalue.*

Proof

(1) Suppose that $\mu < 1$. We will show that there is an eigenvalue $\alpha > 0$ for (8.184), or equivalently,

$$\begin{cases} \phi = \mu(p-1)\dfrac{\int_0^\infty w_{y_0}^{r-1}\phi\, dy}{\int_0^\infty w_{y_0}^r dy}(L_0-\alpha)^{-1}w_{y_0}^p, & 0 < y < +\infty, \\ \phi'(0) - a\phi(0) = 0. \end{cases}$$

Multiplication by $w_{y_0}^{r-1}$ and integration gives

$$\int_0^\infty w_{y_0}^r dy = \mu(p-1)\int_0^\infty \left[(L_0-\alpha)^{-1}w_{y_0}^p\right]w_{y_0}^{r-1}dy.$$

By the identity

$$(p-1)(L_0-\alpha)^{-1}w_{y_0}^p = w_{y_0} + \alpha(L_0-\alpha)^{-1}w_{y_0},$$

we have

$$\int_0^\infty w_{y_0}^r dy = \mu\left(\int_0^\infty w_{y_0}^r dy + \alpha\int_0^\infty \left[(L_0-\alpha)^{-1}w_{y_0}\right]w_{y_0}^{r-1}dy\right),$$

or equivalently,

$$\frac{1}{\alpha}\left(\frac{1}{\mu}-1\right)\int_0^\infty w_{y_0}^r dy = \int_0^\infty \left[(L_0-\alpha)^{-1}w_{y_0}\right]w_{y_0}^{r-1}dy. \tag{8.187}$$

If $r=2$ and $1 < p \le 3$, Proposition 8.31 implies that the right-hand side of (8.187) is positive for $\alpha = 0$.

If $r = p+1$ and $1 < p < \infty$, the right-hand side of (8.187) is positive for $\alpha = 0$ due to

$$\int_0^\infty \left[L_0^{-1}w_{y_0}\right]w_{y_0}^p dy = \frac{1}{p-1}\int_0^\infty w_{y_0}^2 dy > 0.$$

Thus, as $\alpha \to 0^+$, the left-hand side of (8.187) tends to $+\infty$ and the right-hand side approaches some positive number. In the limit $\alpha \to \mu_1^-$, the left-hand side tends to some positive number and the right-hand side approaches $+\infty$. By continuity, (8.187) has a solution α.

(2) If $r = 2$, (8.187) can be written as

$$\left(\frac{1}{\mu}-1\right)\int_0^\infty w_{y_0}^2 dy = \alpha\int_0^\infty \left[(L_0-\alpha)^{-1}w_{y_0}\right]w_{y_0} dy. \tag{8.188}$$

Then the left-hand side of (8.188) does not depend on α. Denoting the right-hand side of (8.188) by $g(\alpha)$, we have

$$g(0) = 0, \qquad g'(0) = \int_0^\infty w_{y_0} L_0^{-1} w_{y_0} dy < 0,$$

$$g(\alpha) \to \infty \quad \text{as } \alpha \to \mu_1^-, \qquad g(\alpha) \to -\infty \quad \text{as } \alpha \to \mu_1^+,$$

$$g(\alpha) \to -\int_0^\infty w_{y_0}^2 dy \quad \text{as } \alpha \to \mu_1^+.$$

Further, we calculate

$$g''(\alpha) = 2 \int_0^\infty w_{y_0} \big[(L_0 - \alpha)^{-2} w_{y_0} + \alpha (L_0 - \alpha)^{-3} w_{y_0} \big] dy$$

$$= \frac{1}{\alpha} \frac{d}{d\alpha} \int_0^\infty \big[(L_0/\alpha - 1)^{-1} w_{y_0} \big]^2 dy$$

which gives

$$g''(\alpha) > 0 \quad \text{if } 0 < \alpha > \mu_1 \quad \text{and} \quad g''(\alpha) < 0 \quad \text{if } \mu_1 < \alpha < \infty.$$

Letting

$$\mu_0(a, p) = \frac{1}{1 + \min_{0 < \alpha < \mu_1} g(\alpha)},$$

Proposition 8.33 implies that $1 < \mu_0(a, p)$ and $\mu_0(a, p) \to 1$ as $a \to a_0(p)$.

By the properties of $g(\alpha)$, problem (8.188) has exactly two eigenvalues $0 < \alpha_1 < \alpha_2 < \mu_1$ if $1 < \mu < \mu_0(a, p)$ and exactly one eigenvalue if $\mu \le 1$.   □

Proof of Theorem 8.38 Theorem 8.38 follows from Proposition 8.33 and Part (2) of Theorem 8.40.   □

Proof of Theorem 8.37 The instability part of Theorem 8.37 follows from Part (1) of Theorem 8.40.

To prove the stability part of Theorem 8.37, we consider two cases separately.

Case 1. $r = 2, 1 < p \le 3$.
Case 2. $r = p + 1, 1 < p < \infty$.

Let $\alpha_0 = \alpha_R + i\alpha_I$ be an eigenvalue of (8.184) with eigenfunction $\phi = \phi_R + i\phi_I$. Since $\alpha_0 \ne 0$, we can choose $\phi \perp \text{kernel}(L_0)$. Separating real and imaginary parts, we have

$$L_0 \phi_R - (p-1)\mu \frac{\int_0^\infty w_{y_0} \phi_R dy}{\int_0^\infty w_{y_0}^2 dy} w_{y_0}^p = \alpha_R \phi_R - \alpha_I \phi_I, \qquad (8.189)$$

$$L_0 \phi_I - (p-1)\mu \frac{\int_0^\infty w_{y_0} \phi_I dy}{\int_0^\infty w_{y_0}^2 dy} w_{y_0}^p = \alpha_R \phi_I + \alpha_I \phi_R. \qquad (8.190)$$

Multiplication of (8.189) by ϕ_R, of (8.190) by ϕ_I, integration and addition gives

$$-\alpha_R \int_0^\infty (\phi_R^2 + \phi_I^2) dy$$

$$= L_1(\phi_R, \phi_R) + L_1(\phi_I, \phi_I)$$

$$+ (p-1)(\mu - 2)$$

$$\times \frac{\int_0^\infty w_{y_0}^p \phi_R dy \int_0^\infty w_{y_0}^p \phi_R dy + \int_0^\infty w_{y_0} \phi_I dy \int_0^\infty w_{y_0}^p \phi_I dy}{\int_0^\infty w_{y_0}^2 dy}$$

$$+ (p-1)\frac{\int_0^\infty w_{y_0}^{p+1} dy}{(\int_0^\infty w_{y_0}^2)^2 dy}\left[\left(\int_0^\infty w_{y_0} \phi_R dy\right)^2 + \left(\int_0^\infty w_{y_0} \phi_I dy\right)^2\right].$$

Multiplication of both (8.189) and (8.190) by w_{y_0} and integration yield

$$(p-1)\int_0^\infty w_{y_0}^p \phi_R dy - (p-1)\mu \frac{\int_0^\infty w_{y_0} \phi_R dy}{\int_0^\infty w_{y_0}^2 dy} \int_0^\infty w_{y_0}^{p+1} dy$$

$$= \alpha_R \int_0^\infty w_{y_0} \phi_R dy - \alpha_I \int_0^\infty w_{y_0} \phi_I dy, \tag{8.191}$$

$$(p-1)\int_0^\infty w_{y_0}^p \phi_I dy - (p-1)\mu \frac{\int_0^\infty w_{y_0} \phi_I dy}{\int_0^\infty w_{y_0}^2 dy} \int_0^\infty w_{y_0}^{p+1} dy$$

$$= \alpha_R \int_0^\infty w_{y_0} \phi_I dy + \alpha_I \int_0^\infty w_{y_0} \phi_R dy. \tag{8.192}$$

Multiplication of (8.191) by $\int_0^\infty w_{y_0} \phi_R dy$, of (8.192) by $\int_0^\infty w_{y_0} \phi_I dy$, integration and addition gives

$$(p-1)\int_0^\infty w_{y_0} \phi_R dy \int_0^\infty w_{y_0}^p \phi_R dy + (p-1)\int_0^\infty w_{y_0} \phi_I dy \int_0^\infty w_{y_0}^p \phi_I dy$$

$$= \left(\alpha_R + (p-1)\mu \frac{\int_0^\infty w_{y_0}^{p+1} dy}{\int_0^\infty w_{y_0}^2 dy}\right)\left[\left(\int_0^\infty w_{y_0} \phi_R dy\right)^2 + \left(\int_0^\infty w_{y_0} \phi_I dy\right)^2\right].$$

This implies

$$-\alpha_R \int_0^\infty (\phi_R^2 + \phi_I^2) dy$$

$$= L_1(\phi_R, \phi_R) + L_1(\phi_I, \phi_I)$$

$$+ (p-1)(\mu - 2)\left(\frac{1}{p-1}\alpha_R + \mu \frac{\int_0^\infty w_{y_0}^{p+1} dy}{\int_0^\infty w_{y_0}^2 dy}\right)$$

$$\times \frac{(\int_0^\infty w_{y_0}\phi_R dy)^2 + (\int_0^\infty w_{y_0}\phi_I dy)^2}{\int_0^\infty w_{y_0}^2 dy}$$

$$+ (p-1)\frac{\int_0^\infty w_{y_0}^{p+1} dy}{(\int_0^\infty w_{y_0}^2)^2}\left[\left(\int_0^\infty w_{y_0}\phi_R dy\right)^2 + \left(\int_0^\infty w_{y_0}\phi_I dy\right)^2\right].$$

Setting

$$\phi_R = c_R w_{y_0} + \phi_R^\perp, \quad \phi_R^\perp \perp X_1, \quad \phi_I = c_I w_{y_0} + \phi_I^\perp, \quad \phi_I^\perp \perp X_1,$$

where the kernel X_1 of L_1 has been introduced in Lemma 8.36, we have

$$\int_0^\infty w_{y_0}\phi_R dy = c_R \int_0^\infty w_{y_0}^2 dy, \quad \int_0^\infty w_{y_0}\phi_I dy = c_I \int_0^\infty w_{y_0}^2 dy,$$

$$d_{L^2(\mathbb{R}_+)}^2(\phi_R, X_1) = \|\phi_R^\perp\|_{L^2}^2, \quad d_{L^2(\mathbb{R}_+)}^2(\phi_I, X_1) = \|\phi_I^\perp\|_{L^2}^2.$$

Elementary computations imply

$$L_1(\phi_R, \phi_R) + L_1(\phi_I, \phi_I)$$

$$+ (\mu - 1)\alpha_R (c_R^2 + c_I^2)\int_0^\infty w_{y_0}^2 dy + (p-1)(\mu-1)^2(c_R^2 + c_I^2)\int_0^\infty w_{y_0}^{p+1} dy$$

$$+ \alpha_R\left(\|\phi_R^\perp\|_{L^2}^2 + \|\phi_I^\perp\|_{L^2}^2\right) = 0.$$

By Lemma 8.36(2), we have

$$(\mu - 1)\alpha_R (c_R^2 + c_I^2)\int_0^\infty w_{y_0}^2 dy$$

$$+ (p-1)(\mu-1)^2(c_R^2 + c_I^2)\int_0^\infty w_{y_0}^{p+1} dy$$

$$+ (\alpha_R + a_1)\left(\|\phi_R^\perp\|_{L^2}^2 + \|\phi_I^\perp\|_{L^2}^2\right) \leq 0$$

for some $a_1 > 0$. Now since $\mu > 1$, we have $\alpha_R < 0$. Theorem 8.37 follows in Case 1: $r = 2$, $1 < p \leq 3$.

Next we consider Case 2: $r = p + 1$, $1 < p < \infty$.

The nonlocal operator in (8.184) becomes

$$L\phi = L_0\phi - \mu(p-1)\frac{\int_0^\infty w_{y_0}^p \phi dy}{\int_0^\infty w_{y_0}^{p+1} dy} w_{y_0}^p$$

and we introduce

$$L_2\phi := L_0\phi - (p-1)\frac{\int_0^\infty w_{y_0}^p \phi dy}{\int_0^\infty w_{y_0}^{p+1} dy} w_{y_0}^p. \tag{8.193}$$

Then we have

Lemma 8.41

(1) L_2 is self-adjoint and its kernel X_2 is spanned by w_{y_0}.
(2) There is a constant $c_3 > 0$ such that

$$L_2(\phi, \phi) := \int_0^\infty [(\phi')^2 + \phi^2 - p w_{y_0}^{p-1} \phi^2] dy + \frac{(p-1)(\int_0^\infty w_{y_0}^p \phi \, dy)^2}{\int_0^\infty w_{y_0}^{p+1} dy}$$

$$\geq c_3 d_{L^2(\mathbb{R}^+)}^2(\phi, X_2), \quad \text{for all } \phi \in H_{rob}^1(\mathbb{R}^+).$$

Proof The proof of Part (1) follows along the same lines as that of Lemma 8.36. We prove (2) by contradiction. Suppose (2) is false, then by (1) there is (α, ϕ) such that (i) $\alpha > 0$, (ii) $\phi \perp w_{y_0}$, (iii) $L_2\phi = \alpha\phi$.

We show that this implies a contradiction. Using (ii) and (iii), we get

$$(L_0 - \alpha)\phi = \frac{(p-1)\int_0^\infty w_{y_0}^p \phi \, dy}{\int_0^\infty w_{y_0}^{p+1} dy} w_{y_0}^p. \tag{8.194}$$

Following the proof of Lemma 8.36, we compute that $\int_0^\infty w_{y_0}^p \phi \, dy \neq 0$ and $\alpha \neq \mu_1$ and so the operator $L_0 - \alpha$ is invertible. Thus (8.194) yields

$$\phi = \frac{(p-1)\int_0^\infty w_{y_0}^p \phi \, dy}{\int_0^\infty w_{y_0}^{p+1} dy} (L_0 - \alpha)^{-1} w_{y_0}^p.$$

Therefore, we have

$$\int_0^\infty w_{y_0}^p \phi \, dy = (p-1)\frac{\int_0^\infty w_{y_0}^p \phi \, dy}{\int_0^\infty w_{y_0}^{p+1} dy} \int_0^\infty ((L_0 - \alpha)^{-1} w_{y_0}^p) w_{y_0}^p dy$$

and

$$\int_0^\infty w_{y_0}^{p+1} dy = (p-1)\int_0^\infty ((L_0 - \alpha)^{-1} w_{y_0}^p) w_{y_0}^p dy. \tag{8.195}$$

Letting

$$h_2(\alpha) = (p-1)\int_0^\infty ((L_0 - \alpha)^{-1} w_{y_0}^p) w_{y_0}^p dy - \int_0^\infty w_{y_0}^{p+1} dy,$$

we compute

$$h_2(0) = (p-1)\int_0^\infty (L_0^{-1} w_{y_0}^p) w_{y_0}^p dy - \int_0^\infty w_{y_0}^{p+1} dy = 0$$

and

$$h_2'(\alpha) = (p-1) \int_0^\infty ((L_0 - \alpha)^{-2} w_{y_0}^p) w_{y_0}^p dy$$

$$= (p-1) \int_0^\infty ((L_0 - \alpha)^{-1} w_{y_0}^p)^2 dy > 0.$$

We conclude that $h_2(\alpha) > 0$ for all $\alpha \in (0, \mu_1)$. Using $h_2(\alpha) < 0$ for $\alpha \in (\mu_1, \infty)$, we have derived a contradiction to (8.195) and the lemma follows. $\square$

Finally, we complete the proof of Theorem 8.37 in Case 2.
In analogy to Case 1, we get

$$L_0 \phi_R - (p-1)\mu \frac{\int_0^\infty w_{y_0}^p \phi_R dy}{\int_0^\infty w_{y_0}^{p+1} dy} w_{y_0}^p = \alpha_R \phi_R - \alpha_I \phi_I, \qquad (8.196)$$

$$L_0 \phi_I - (p-1)\mu \frac{\int_0^\infty w_{y_0}^p \phi_I dy}{\int_0^\infty w_{y_0}^{p+1} dy} w_{y_0}^p = \alpha_R \phi_I + \alpha_I \phi_R. \qquad (8.197)$$

Multiplication of (8.196) by ϕ_R, of (8.197) by ϕ_I, integration and addition gives

$$-\alpha_R \int_0^\infty (\phi_R^2 + \phi_I^2) dy$$

$$= L_2(\phi_R, \phi_R) + L_2(\phi_I, \phi_I)$$

$$+ (p-1)(\mu-1) \frac{(\int_0^\infty w_{y_0}^p \phi_R dy)^2 + (\int_0^\infty w_{y_0}^p \phi_I dy)^2}{\int_0^\infty w_{y_0}^{p+1} dy}.$$

By Lemma 8.41(2), we have

$$\alpha_R \int_0^\infty (\phi_R^2 + \phi_I^2) + a_2 d_{L^2}^2(\phi, X_1)$$

$$+ (p-1)(\mu-1) \frac{(\int_0^\infty w_{y_0}^p \phi_R)^2 + (\int_0^\infty w_{y_0}^p \phi_I)^2}{\int_0^\infty w_{y_0}^{p+1}} \leq 0$$

for some $a_2 > 0$. Since $\mu > 1$ we conclude that $\alpha_R < 0$.
The proof is complete. $\square$

By Theorem 8.37 we have Theorem 8.25, and Theorem 8.38 gives Theorem 8.26.

8.3.2 Eigenvalue Estimates

In this subsection, we will study eigenvalue estimates for the linear operator $L_\epsilon :=$ $\epsilon^2 \Delta - 1 + p(u_\epsilon)^{p-1}$, defined on $H_{rob}^1(\Omega)$, for an interior spike. This will complete the proof of Theorem 8.24.

Before stating a result on the small (i.e. $o(1)$) eigenvalues, we introduce some important notation and mention a few crucial lemmas.

Let

$$d\mu_{P_0}(z) = \lim_{\epsilon \to 0} \frac{e^{-2|z-P_0|/\epsilon} dz}{\int_{\partial\Omega} e^{-2|z-P_0|/\epsilon} dz}. \tag{8.198}$$

Then the support of $d\mu_{P_0}(z)$ is a subset of $\bar{B}_{d(P_0,\partial\Omega)}(P_0) \cap \partial\Omega$.

A point P_0 is called a nondegenerate peak point if it satisfies (8.199) and (8.200):

(i)   There exists an $a \in \mathbb{R}^n$ such that

$$\int_{\partial\Omega} e^{\langle z-P_0, a\rangle}(z - P_0)d\mu_{P_0}(z) = 0. \tag{8.199}$$

(ii)   The matrix

$$\left(\int_{\partial\Omega} e^{\langle z-P_0, a\rangle}(z - P_0)_i (z - P_0)_j d\mu_{P_0}(z) \right) := G(P_0) \text{ is nonsingular.} \tag{8.200}$$

Note that the vector a in (8.200) is unique and the matrix $G(P_0)$ is positive definite. A geometric characterisation of a nondegenerate peak point P_0 is given by

$$P_0 \in \text{interior}\big(\text{convex hull of support}\big(d\mu_{P_0}(z)\big)\big).$$

These statements are stated in proved in Theorem 5.1 of [247].

For $P \in \Omega$, let $w_{\epsilon,P}$ be the unique solution of

$$\epsilon^2 \Delta u - u + w^P\left(\frac{x - P}{\epsilon}\right) = 0 \quad \text{in } \Omega, \qquad \epsilon \frac{\partial u}{\partial v} + au = 0 \quad \text{on } \partial\Omega. \tag{8.201}$$

Letting $\varphi_{\epsilon,P}(x) = w(\frac{x-P}{\epsilon}) - w_{\epsilon,P}(x)$, then $\varphi_{\epsilon,P}$ satisfies

$$\begin{cases} \epsilon^2 \Delta\varphi_{\epsilon,P} - \varphi_{\epsilon,P} = 0 \quad \text{in } \Omega, \\ a\varphi_{\epsilon,P} + \epsilon \frac{\partial\varphi_{\epsilon,P}}{\partial v} = aw(\frac{x-P}{\epsilon}) + \epsilon \frac{\partial w((x-P)/\epsilon)}{\partial v} \quad \text{on } \partial\Omega \end{cases} \tag{8.202}$$

and, for $x \in \partial\Omega$, we have

$$aw\left(\frac{x - P}{\epsilon}\right) + \epsilon \frac{\partial w((x - P)/\epsilon)}{\partial v}$$
$$= aw\left(\frac{x - P}{\epsilon}\right) + w'\left(\frac{x - P}{\epsilon}\right)\frac{\langle x - P, v\rangle}{|x - P|}$$
$$= w\left(\frac{x - P}{\epsilon}\right)\left(a - \frac{\langle x - P, v\rangle}{|x - P|} + o\left(\frac{\epsilon}{d(P, \partial\Omega)}\right)\right)$$
$$\geq (a - 1 - \delta)w\left(\frac{x - P}{\epsilon}\right),$$

where $w'(y) = \frac{dw(r)}{dr}$ for $r = |y|$ and $a - 1 - \delta > 0$. Thus there are two positive constants C_1 and C_2 such that

$$C_1 \varphi_{\epsilon,P,1} \le \varphi_{\epsilon,P} \le C_2 \varphi_{\epsilon,P,1}, \qquad (8.203)$$

where $\varphi_{\epsilon,P,1}$ satisfies

$$\begin{cases} \epsilon^2 \Delta \varphi_{\epsilon,P,1} - \varphi_{\epsilon,P,1} = 0 & \text{in } \Omega, \\ \varphi_{\epsilon,P,1} + a^{-1} \epsilon \frac{\partial \varphi_{\epsilon,P,1}}{\partial \nu} = w(\frac{x-P}{\epsilon}) & \text{on } \partial\Omega. \end{cases} \qquad (8.204)$$

To estimate $\varphi_{\epsilon,P,1}$, we use

Lemma 8.42 *Suppose that $d(P, \partial\Omega) > d_0 > 0$. Let $\varphi_{\epsilon,P}^D$ be the unique solution of*

$$\begin{cases} \epsilon^2 \Delta \varphi_{\epsilon,P}^D - \varphi_{\epsilon,P}^D = 0 & \text{in } \Omega, \\ \varphi_{\epsilon,P}^D = w(\frac{x-P}{\epsilon}) & \text{on } \partial\Omega. \end{cases} \qquad (8.205)$$

Then for any $\delta > 0$ and ϵ sufficiently small, we have

$$\left| \epsilon \frac{\partial \varphi_{\epsilon,P}^D}{\partial \nu} \right| \le (1+\delta) \varphi_{\epsilon,P}^D. \qquad (8.206)$$

Lemma 8.42 implies that on $\partial\Omega$ we have

$$\varphi_{\epsilon,P}^D + a^{-1} \epsilon \frac{\varphi_{\epsilon,P}^D}{\partial \nu} \le \varphi_{\epsilon,P}^D \left(1 + a^{-1}(1+\delta)\right) \le \left(1 + a^{-1}(1+\delta)\right) w\left(\frac{x-P}{\epsilon}\right)$$

and

$$\varphi_{\epsilon,P}^D + a^{-1} \epsilon \frac{\varphi_{\epsilon,P}^D}{\partial \nu} \ge \varphi_{\epsilon,P}^D \left(1 - a^{-1}(1-\delta)\right) \ge \left(1 - a^{-1}(1-\delta)\right) w\left(\frac{x-P}{\epsilon}\right).$$

By a standard elliptic comparison principle (see [74]), we get

Lemma 8.43 *There are two positive constants C_1 and C_2 such that*

$$C_1 \varphi_{\epsilon,P}^D \le \varphi_{\epsilon,P,1} \le C_2 \varphi_{\epsilon,P}^D.$$

The asymptotic behaviour of (8.205) for small ϵ has been studied in Sect. 4 of [174] and is well understood. Lemma 4.6 of [174] gives the following convergence results:

Lemma 8.44

(i) *Let*

$$V_\epsilon(y) := \frac{\varphi_{\epsilon,x_\epsilon}(x_\epsilon + \epsilon y)}{\varphi_{\epsilon,x_\epsilon}(x_\epsilon)}.$$

Then $V_\epsilon(y) \to V_0(y)$ locally, where $V_0(y)$ solves

$$\Delta u - u = 0, \quad u(0) = 1, u > 0 \text{ in } \mathbb{R}^n. \tag{8.207}$$

For any $\sigma > 0$, we have

$$\sup_{y \in \Omega_\epsilon} e^{-(1+\sigma)|y|}\big(V_\epsilon(y) - V_0(y)\big) \to 0. \tag{8.208}$$

(ii) *As $\epsilon \to 0$, we get*

$$-\epsilon \log\big(\varphi_{\epsilon,x_\epsilon}(x_\epsilon)\big) \to 2d(x_0, \partial\Omega). \tag{8.209}$$

For $P \in \Omega$, we define

$$\Omega_{\epsilon,P} = \{y \,|\, \epsilon y + P \in \Omega\},$$

$$S_\epsilon(u) = \Delta u - u + u^p \quad \text{for } u \in H^1_{rob}(\Omega_{\epsilon,P}), \qquad \partial_j = \frac{\partial}{\partial P_j},$$

$$\mathcal{K}_{\epsilon,P} = \text{span}\{\partial_j w_{\epsilon,P} : j = 1, \ldots, N\} \subset H^1_{rob}(\Omega_{\epsilon,P}),$$

$$\mathcal{K}^\perp_{\epsilon,P} = \left\{ u \in H^1_{rob}(\Omega_{\epsilon,P}) : \int_\Omega u \partial_j w_{\epsilon,P} dx = 0, j = 1, \ldots, N \right\},$$

and

$$\mathcal{C}_{\epsilon,P} = \text{span}\{\partial_j w_{\epsilon,P} : j = 1, \ldots, N\} \subset L^2(\Omega_{\epsilon,P}),$$

$$\mathcal{C}^\perp_{\epsilon,P} = \left\{ u \in L^2(\Omega_{\epsilon,P}) : \int_\Omega u \partial_j w_{\epsilon,P} dx = 0, j = 1, \ldots, N \right\}.$$

We set $Q^0_\epsilon := P_0 + \epsilon\frac{1}{2}d(P_0, \partial\Omega)a$, where P_0 is a nondegenerate peak point (recall that it satisfies (8.199) and (8.200)). For β_0 sufficiently small, let $\Lambda := B_{\beta_0\epsilon}(Q^0_\epsilon)$.

By [14], for each $P \in \Lambda$ there exists a solution $\varphi_{\epsilon,P} \in \mathcal{K}^\perp_{\epsilon,P}$ such that

$$S_\epsilon(w_{\epsilon,P} + \varphi_{\epsilon,P}) \in \mathcal{C}_{\epsilon,P}.$$

Now we state our theorem on small eigenvalues.

Theorem 8.45 *The eigenvalue problem*

$$\epsilon^2 \Delta\phi - \phi + pu^{p-1}_\epsilon\phi = \tau^\epsilon\phi \quad \text{in } \Omega, \qquad \epsilon\frac{\partial\phi}{\partial\nu} + a\phi = 0 \quad \text{on } \partial\Omega \tag{8.210}$$

admits the following set of $o(1)$ eigenvalues:

$$\tau^\epsilon_j = \big(c_0 + o(1)\big)\varphi_{\epsilon,P_0}(P_0)\lambda_j, \quad j = 1, \ldots, N,$$

where λ_j, $j = 1, \ldots, N$, *are the eigenvalues of the matrix* $G(P_0)$ *introduced in* (8.200). *Further, we have*

$$c_0 = 2d^{-2}(P_0, \partial\Omega) \frac{\int_{\mathbb{R}^n} pw^{p-1}w'V_0'(r)dy}{\int_{\mathbb{R}^n}(\partial w/\partial y_1)^2 dy} < 0, \tag{8.211}$$

where $V_0(r)$ *is the unique radial solution of the problem* (8.207). *The eigenfunctions with eigenvalues* τ_j^ϵ, $j = 1, \ldots, N$, *can be written*

$$\phi_j^\epsilon = \sum_{l=1}^{N}(a_{j,l} + o(1))\epsilon \frac{\partial w_{\epsilon,P}}{\partial P_l}\bigg|_{P=P_\epsilon}, \tag{8.212}$$

where $\vec{a}_j = (a_{j,1}, \ldots, a_{j,N})^t$ *is the eigenvector of* $G(P_0)$ *with eigenvalue* λ_j.

Proof We introduce

$$u_\epsilon = w_{\epsilon,Q_\epsilon} + v_{\epsilon,Q_\epsilon}.$$

Let $(\tau^\epsilon, \phi_\epsilon)$ be given by

$$L_\epsilon\phi_\epsilon = \tau^\epsilon\phi_\epsilon \quad \text{in } \Omega, \qquad \epsilon\frac{\partial\phi_\epsilon}{\partial\nu} + a\phi = 0 \quad \text{on } \partial\Omega. \tag{8.213}$$

Assuming that $\tau_\epsilon \to 0$ as $\epsilon \to 0$, we normalise ϕ_ϵ according to $\|\phi_\epsilon\|_\epsilon = 1$.

A scaling and limiting process as in [171, 172, 174] yields $\tilde{\phi}_\epsilon(y) = \phi_\epsilon(Q_\epsilon + \epsilon y) \to \phi_0$, where ϕ_0 solves

$$\Delta v - v + pw^{p-1}v = 0 \quad \text{in } \mathbb{R}^n, v \in H^1(\mathbb{R}^n).$$

By Lemma 4.2 of [172], there are s_j such that $\phi_0 = \sum_{j=1}^{N} s_j \frac{\partial w}{\partial y_j}$.

Thus we decompose $\phi_\epsilon = \sum_{j=1}^{N} s_j^\epsilon \epsilon\partial_j w_{\epsilon,Q_\epsilon} + \bar{\phi}_\epsilon$, where $\bar{\phi}_\epsilon \in \mathcal{K}_{\epsilon,Q_\epsilon}^\perp$ and $|s_j^\epsilon| \leq C$.

Using $\|\phi_\epsilon\|_\epsilon = 1$ gives $\|\bar{\phi}_\epsilon\|_\epsilon \leq C$. Then $\bar{\phi}_\epsilon$ satisfies

$$(L_\epsilon - \tau^\epsilon)\bar{\phi}_\epsilon + \sum_{j=1}^{N} s_j^\epsilon\left[p(u_\epsilon)^{p-1}\epsilon\partial_j w_{\epsilon,Q_\epsilon} - pw^{p-1}\epsilon\partial_j w\right]$$

$$= \tau^\epsilon \sum_{j=1}^{N} s_j^\epsilon\epsilon\partial_j w_{\epsilon,Q_\epsilon}. \tag{8.214}$$

Following the argument in Proposition 6.3 of [246] implies that the linear operator

$$\pi_{\epsilon,Q_\epsilon} \circ (L_\epsilon - \tau^\epsilon) : \mathcal{K}_{\epsilon,Q_\epsilon}^\perp \to \mathcal{C}_{\epsilon,Q_\epsilon}^\perp$$

is invertible. Since $\bar{\phi}_\epsilon \in \mathcal{K}^\perp_{\epsilon,Q_\epsilon}$, we get

$$\|\bar{\phi}_\epsilon\|_{H^1(\Omega_{\epsilon,Q_\epsilon})} = O\left(\left\|\sum_{j=1}^{N} s_j^\epsilon [p(u_\epsilon)^{p-1}\epsilon\partial_j w_{\epsilon,Q_\epsilon} - pw^{p-1}\epsilon\partial_j w]\right\|_{L^2(\Omega_{\epsilon,Q_\epsilon})}\right)$$

$$= O\left((|\varphi_{\epsilon,Q_\epsilon}(Q_\epsilon)|^{(1+\sigma)/2})\sum_{j=1}^{N}|s_j^\epsilon|\right)$$

for some $\sigma > 0$.

Multiplication of (8.214) by $\epsilon\partial_k(w_{\epsilon,Q_\epsilon})$ and integration give

$$\sum_{j=1}^{N} s_j^\epsilon \int_{\Omega_{\epsilon,Q_\epsilon}} [p(u_\epsilon)^{p-1}\epsilon\partial_j w_{\epsilon,Q_\epsilon} - pw^{p-1}\epsilon\partial_j w]\epsilon\partial_k w_{\epsilon,Q_\epsilon}dx$$

$$= \tau^\epsilon \sum_{j=1}^{N} \int_{\Omega_{\epsilon,Q_\epsilon}} s_j^\epsilon\epsilon\partial_j w_{\epsilon,Q_\epsilon}\epsilon\partial_k w_{\epsilon,Q_\epsilon}dx$$

$$+ \int_{\Omega_{\epsilon,Q_\epsilon}} [p(u_\epsilon)^{p-1}\bar{\phi}_\epsilon\epsilon\partial_k(w_{\epsilon,Q_\epsilon}) - pw^{p-1}\bar{\phi}_\epsilon\epsilon\partial_k w]dx$$

$$+ O(|\tau_\epsilon|\|\bar{\phi}_\epsilon\|_{H^1(\Omega_{\epsilon,Q_\epsilon})}). \tag{8.215}$$

To estimate the left-hand side of (8.215), we first calculate

$$-\int_{\Omega_{\epsilon,Q_\epsilon}} \left[pw^{p-1}\epsilon\frac{\partial w}{\partial P_j}\bigg|_{P=Q_\epsilon} - p(w_{\epsilon,Q_\epsilon}+v_{\epsilon,Q_\epsilon})^{p-1}\epsilon\frac{\partial w_{\epsilon,P}}{\partial P_k}\bigg|_{P=Q_\epsilon}\right]dy$$

$$= -\epsilon^2\int_{\Omega_{\epsilon,Q_\epsilon}} \left[pw^{p-1}\frac{\partial w}{\partial P_j}\bigg|_{P=Q_\epsilon} - p(w_{\epsilon,Q_\epsilon})^{p-1}\frac{\partial w_{\epsilon,P}}{\partial P_j}\bigg|_{P=Q_\epsilon}\right]\frac{\partial w_{\epsilon,P}}{\partial P_k}\bigg|_{P=Q_\epsilon}dy$$

$$+ O(|\varphi_{\epsilon,Q_\epsilon}(Q_\epsilon)|^{1+\sigma})$$

$$= -\epsilon^2\int_{\Omega_{\epsilon,Q_\epsilon}} \frac{\partial}{\partial P_j}\bigg|_{P=Q_\epsilon}[w^p - (w_{\epsilon,Q_\epsilon})^p]\frac{\partial w_{\epsilon,P}}{\partial P_k}\bigg|_{P=Q_\epsilon}dy$$

$$+ O(|\varphi_{\epsilon,Q_\epsilon}(Q_\epsilon)|^{1+\sigma})$$

$$= -\epsilon^2\int_{\Omega_{\epsilon,Q_\epsilon}} \frac{\partial}{\partial P_j}\bigg|_{P=Q_\epsilon}[pw^{p-1}\varphi_{\epsilon,Q_\epsilon}(Q_\epsilon+\epsilon y)]\frac{\partial w_{\epsilon,P}}{\partial P_k}\bigg|_{P=Q_\epsilon}$$

$$+ O(|\varphi_{\epsilon,Q_\epsilon}(Q_\epsilon)|^{1+\sigma})$$

$$= 2\varphi_{\epsilon,P_0}(P_0)(1+o(1))$$

$$\times \int_{\mathbb{R}^n} pw^{p-1}\int_{\partial\Omega} e^{\langle(z-P_0)/|z-P_0|,y\rangle}e^{\langle(z-P_0)/|z-P_0|,2(Q_\epsilon-P_0)/\epsilon\rangle}d\mu_{P_0}(z)$$

$$\times \left(\frac{z - P_0}{|z - P_0|}\right)_j \frac{\partial w}{\partial y_k} dy + O\big(|\varphi_{\epsilon, P_0}(Q_\epsilon)|^{1+\sigma}\big)$$

$$= 2\varphi_{\epsilon, P_0}(P_0)\big(1 + o(1)\big)$$

$$\times \int_{\mathbb{R}^n} pw^{p-1} \int_{\partial\Omega} e^{\langle (z - P_0)/|z - P_0|, y\rangle} e^{\langle z - P_0, a\rangle} \left(\frac{z - P_0}{|z - P_0|}\right)_j d\mu_{P_0}(z) \frac{\partial w}{\partial y_k} dy$$

$$+ O\big(|\varphi_{\epsilon, Q_\epsilon}(Q_\epsilon)|^{1+\sigma}\big)$$

$$= \frac{2\gamma}{d^2(P_0, \partial\Omega)} \varphi_{\epsilon, P_0}(P_0)\left(\int_{\partial\Omega} e^{\langle z - P_0, a\rangle}(z - P_0)_i(z - P_0)_k d\mu_{P_0}(z) + o(1)\right),$$

where

$$\gamma := \int_{\mathbb{R}^n} pw^{p-1} w'(y) V_0'(y) dy.$$

Thus for the left-hand side of (8.215), we compute

l.h.s. of (8.215)

$$= \sum_{j=1}^{N} s_j^\epsilon \left(\int_{\Omega_{\epsilon, Q_\epsilon}} \big[p(w_{\epsilon, Q_\epsilon})^{p-1} \epsilon \partial_j w_{\epsilon, Q_\epsilon} - pw^{p-1} \epsilon \partial_j w\big] \epsilon \partial_k w_{\epsilon, Q_\epsilon} dy\right)$$

$$+ O\big(|\varphi_{\epsilon, Q_\epsilon}(Q_\epsilon)|^{1+\sigma}\big)$$

$$= \int_{\partial\Omega} e^{\langle z - P_0, a\rangle} \left\langle \frac{z - P_0}{|z - P_0|}, s^\epsilon \right\rangle \left(\frac{z - P_0}{|z - P_0|}\right)_k \varphi_{\epsilon, P_0}(P_0) d\mu_{P_0}(z)\big(-2\gamma + o(1)\big)$$

where $s^\epsilon = (s_1^\epsilon, \ldots, s_N^\epsilon)$.

For the right-hand side of (8.215), we get

r.h.s. of (8.215)

$$= \tau^\epsilon \sum_{j=1}^{N} s_j^\epsilon \big(B\delta_{jk} + o(1)\big)$$

$$+ O\left(\sum_{j=1}^{N} |s_j^\epsilon| |\varphi_{\epsilon, Q_\epsilon}(Q_\epsilon)|^{(1+\sigma)}\right) + O\left(|\tau^\epsilon|\left(\sum_{j=1}^{N} |s_j^\epsilon| |\varphi_{\epsilon, Q_\epsilon}(Q_\epsilon)|^{(1+\sigma)/2}\right)\right),$$

where $B = \int_{\mathbb{R}^n} (\frac{\partial w}{\partial y_1})^2 dy$.

Combining l.h.s. and r.h.s., we get

$$|\tau^\epsilon| = O\big(\varphi_{\epsilon, Q_\epsilon}(Q_\epsilon)\big) = O\big(\varphi_{\epsilon, P_0}(P_0)\big)$$

and $\tau^\epsilon / \varphi_{\epsilon, P_0}(P_0) \to \tau_0$, $s^\epsilon \to s$, where (τ_0, s) satisfies

$$(-2\gamma)G(P_0)s = Bd^2(P_0, \partial\Omega)\tau_0 s.$$

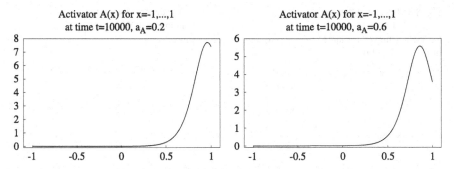

Fig. 8.3 Near-boundary spikes $a_A = 0.2$ and $a_A = 0.6$. The steady states shown have been achieved for large times ($t = 10000$)

Hence $\frac{Bd^2(P_0, \partial\Omega)}{-2\gamma}\tau_0$ is an eigenvalue of $G(P_0)$. This implies that $\tau^\epsilon/\varphi_{\epsilon, P_0}(P_0) \to \tau_j$, $s^\epsilon \to \vec{a}_j$, where

$$\tau_j = \frac{-2\gamma}{Bd^2(P_0, \partial\Omega)}\lambda_j, \qquad G(P_0)\vec{a}_j = \lambda_j\vec{a}_j, \quad j = 1, \ldots, N.$$

By [37], these are the only small eigenvalues of the order $o(1)$ as $\epsilon \to 0$.
The proof of Theorem 8.45 is complete. $\square$

Completion of the Proof of Theorem 8.24   All the small eigenvalues in Theorem 8.45 have negative real part. By a proof similar to that of Theorem 8.37, where we replace w by w_{y_0} and interior spikes by near-boundary spikes, the large eigenvalues all have negative real part. Combining these two results, Theorem 8.24 is proved. $\square$

8.3.3 Numerical Simulations

We present a few numerical simulations which illustrate the instabilities for spikes in the case of Robin boundary conditions. They will confirm our theoretical results.

We choose the Gierer-Meinhardt system (8.150) with Robin boundary conditions (8.151) on $\Omega = (-1, 1)$ for the following selections of parameters: diffusion constants $\epsilon^2 = 0.01$, $D = 10^9$, time relaxation constant $\tau = 10^{-9}$, and Robin boundary condition parameters a_A (which varies) and $a_H = 0$. The reaction constants p and q vary, and we take $r = 2$, $s = 0$.

We begin with the classical Gierer-Meinhardt system for $p = 2$, $q = 1$. We observe stable near-boundary spikes for $a_A = 0.2$ or $a_A = 0.6$ (Fig. 8.3) and an interior spike for $a_A = 0.2$ (Fig. 8.4).

Next we compute near-boundary spikes for $p = 4.85$, $q = 2$, $r = 2$, $s = 0$ and $a_A = 0.8$ (Fig. 8.5). Then, at $p = 4.86$, a rather dramatic change of stability is ob-

Fig. 8.4 Interior spikes for
$a_A = 0.2$. The state shown
has been achieved for large
times ($t = 10000$)

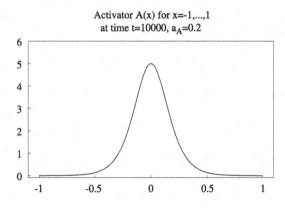

Fig. 8.5 Stable
near-boundary spikes for
$a_A = 0.8, q = 2, r = 2, s = 0$
and $p = 4.85$. The state has
been achieved for large times
($t = 10000$)

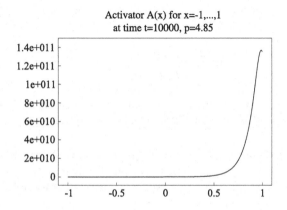

served: The amplitude increases with time and the solution blows up (Fig. 8.6). This
behaviour is similar to phenomena which occur for systems with large reaction rates
[242]. By the Robin boundary condition the instability threshold has been squeezed
to lower reaction rates.

In contrast to Dirichlet or Neumann boundary conditions, for Robin boundary
conditions multiple spikes will not be symmetric and their amplitudes and distances
will differ. Thus the stability thresholds will be shifted due to boundary conditions.

Reaction-diffusion systems with Robin boundary are very important for the modelling of biological situations with partially closed boundaries. To represent the full
geometry of the problems higher-dimensional systems are indispensable. Gaining
a better mathematical understanding of these systems and their dependence on the
choice of boundary conditions will unravel the relevant biological processes. Results on the stability of solutions will help in the selection of relevant patterns from
the whole variety of possible patterns.

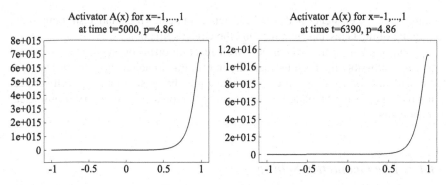

Fig. 8.6 Unstable near-boundary spikes. We choose $a_A = 0.8$, $q = 2$, $r = 2$, $s = 0$ and $p = 4.86$. The near-boundary spike is now numerically unstable. In the time evolution the amplitude increases (shown for $t = 5000$ and $t = 6390$)

8.4 The Gierer-Meinhardt System on Manifolds

8.4.1 Introduction

Reaction-diffusion models on manifolds are very important for biological applications such as the study of cell membranes. This motivates the following study of spiky steady states of the Gierer-Meinhardt system on a compact two-dimensional Riemannian manifold $(\mathcal{S}, g)$ without boundary. The system is stated as follows ([73, 145]):

$$\begin{cases} A_t = d\Delta_g A - A + \frac{A^2}{H} & \text{in } \mathcal{S}, \\ \tau H_t = D\Delta_g H - H + A^2 & \text{in } \mathcal{S}, \end{cases} \tag{8.216}$$

where $A = A(p, t)$ and $H = H(p, t) > 0$ are the concentrations of activator and inhibitor, respectively, at point $p \in \mathcal{S}$ and time $t > 0$; their diffusivities are denoted by d, $D > 0$; $\tau \geq 0$ is the time-relaxation constant of the inhibitor and Δ_g denotes the Laplace-Beltrami operator with respect to the metric tensor g.

We set ϵ and β by $d = \epsilon^2$ and $D = \frac{1}{\beta^2}$.

We will consider the weak coupling case in two dimensions (as in Chap. 6), namely we will choose parameters (ϵ, β) such that $\epsilon, \beta \to 0$ (hence, $d \to 0$ and $D \to \infty$). We assume that

$$\epsilon \text{ is small enough} \tag{8.217}$$

and that

$$\lim_{\epsilon \to 0} \frac{\beta^2}{\epsilon^2} = \kappa > 0. \tag{8.218}$$

Next we introduce a function $F(p)$, $p \in \mathcal{S}$, which is defined as a convex combination of Gaussian curvature and a Green's function and will be crucial in stating our main results on existence and stability. Here κ indicates the relative strength in the coupling of the Gaussian curvature and the Green's function.

There are few results which deal with reaction-diffusion systems on a curved manifold. Possibly the most biologically interesting domain is the two-dimensional Riemannian manifold since it could model a membrane structure, e.g. cell surface, in which the Gierer-Meinhardt system correctly represents the observed biological phenomena.

8.4.2 The Geometric Setting

Before stating the main results in detail, we introduce some notation. Let S be a compact two-dimensional Riemannian manifold without boundary. Let T_pS be the tangent plane to S at p, and given an orthonormal basis $\{e_1(p), e_2(p)\}$ of T_pS, we can obtain, via the exponential map $\exp_p : T_pS \to S$, a natural correspondence $x_1e_1(p) + x_2e_2(p) \mapsto q = \exp_p(x_1e_1(p) + x_2e_2(p))$.

To define an explicit chart, let $E_p : \mathbb{R}^2 \to T_pS$ denote the map $E_p(x_1, x_2) = x_1e_1(p) + x_2e_2(p)$. Then there is a maximal $\delta_p > 0$ such that

$$E_p^{-1} \circ \exp_p^{-1} : B_g(p, \delta_p) \to B(0, \delta_p) \subset \mathbb{R}^2$$

is a diffeomorphism. Moreover, since S is compact, we actually have an injectivity radius $i_g > 0$ so that

$$X_p := E_p^{-1} \circ \exp_p^{-1} : B_g(p, i_g) \to B(0, i_g) \tag{8.219}$$

is a diffeomorphism for every $p \in S$. The values of this natural chart X_p are called (geodesic) normal coordinates about p.

We assume that the exponential map is smooth (C^∞). Moreover, since the tangent bundle TS has a natural differentiable structure, we may choose the basis $\{e_1(p), e_2(p)\}$ of T_pS to be smooth. Thus any smooth function f defined on S by means of the normal coordinates varies smoothly with p as well as the coordinates (x_1, x_2).

Let χ be the cut-off function defined in (2.51). For $p \in S$, we introduce

$$\chi_{\delta_0, p}(q) = \chi\left(2\frac{d_g(p, q)}{\delta_0}\right), \quad q \in S, \tag{8.220}$$

and we choose $\delta_0 = i_g$. We set $\chi_{\delta_0}(x) = \chi(x/\delta_0)$ for $x \in \mathbb{R}^2$.

We denote the geodesic gradient of f by $\nabla_g f$. Written in normal coordinates, the partial derivatives of f with respect to (x_1, x_2) are denoted by ∇f. We will frequently consider rescaled normal coordinates $y = x/\epsilon$.

Next we introduce suitable function spaces. We define

$$L^2(S) = \left\{ u \text{ a measurable function defined on } S \text{ s.t. } \int_S u^2(p) dv_g(p) < \infty \right\},$$

where dv_g denotes the Riemannian measure with respect to the metric g. We further set

$$H^1(\mathcal{S}) = \left\{ u \in L^2(\mathcal{S}) : \nabla_g u \in L^2(\mathcal{S}) \right\}.$$

We will use analogous definitions for other Sobolev spaces.

Let $H_\epsilon^1(\mathcal{S})$ be the Sobolev space $H^1(\mathcal{S})$ equipped with the inner product

$$\langle u, v \rangle_{H_\epsilon^1(\mathcal{S})} = \frac{1}{\epsilon^2} \left(\epsilon^2 \int_{\mathcal{S}} \nabla_g u \cdot \nabla_g v \, dv_g + \int_{\mathcal{S}} u v \, dv_g \right)$$

which induces the norm

$$\|u\|_{H_\epsilon^1(\mathcal{S})}^2 = \frac{1}{\epsilon^2} \left(\epsilon^2 \int_{\mathcal{S}} \nabla_g u \cdot \nabla_g v \, dv_g + \int_{\mathcal{S}} u v \, dv_g \right).$$

In the same way, we define $L_\epsilon^2(\mathcal{S})$, $H_\epsilon^2(\mathcal{S})$ and other Sobolev spaces.

We introduce a Green's function G_0 as follows: $G_0 : \mathcal{S} \times \mathcal{S} \backslash \{(p, q) \in \mathcal{S} \times \mathcal{S} : p = q\} \to \mathbb{R}$ uniquely defined by

$$\begin{cases} \Delta_g G_0(p, q) - \frac{1}{|\mathcal{S}|} + \delta_p(q) = 0 & \text{in } \mathcal{S}, \\ \int_{\mathcal{S}} G_0(p, q) dv_g(q) = 0. \end{cases} \tag{8.221}$$

For basic properties and a constructive proof of its existence we refer to [7]. We denote by

$$\frac{1}{2\pi} \log \frac{1}{d_g(p, q)} \chi_{\delta_0, p}(q) \quad \text{and}$$

$$R_0(p, q) := \frac{1}{2\pi} \log \frac{1}{d_g(p, q)} \chi_{\delta_0, p}(q) - G_0(p, q) \tag{8.222}$$

the singular and regular parts of G_0, respectively, where $d_g(p, q)$ is the geodesic distance between $p \in \mathcal{S}$ and $q \in \mathcal{S}$. Setting

$$R(p) = R_0(p, p), \tag{8.223}$$

we have $R_0 \in C^\infty(\mathcal{S} \times \mathcal{S})$ and $R \in C^\infty(\mathcal{S})$.

Let the function $F : \mathcal{S} \to \mathbb{R}$ be defined by

$$F(p) := c_1 K(p) + c_2 R(p), \tag{8.224}$$

where $K(p)$ denotes the Gauss curvature on $\mathcal{S}$, $R(p)$ the diagonal of the regular part of the Green's function defined in (8.223),

$$c_1 = \frac{\pi}{4} \int_0^\infty (w')^2 r^3 dr, \qquad c_2 = \frac{|\mathcal{S}|\pi}{2} \kappa \int_0^\infty w^2 r \, dr, \qquad w' = \frac{\partial w}{\partial r}.$$

Here w is given by (6.9) and κ has been defined in (8.218). We recall that

$$w(y) \sim |y|^{-1/2} e^{-|y|} \quad \text{as } |y| \to \infty \tag{8.225}$$

and note that $F(p) \in C^\infty(\mathcal{S})$.

Let us write

$$\mathcal{M}(p) = (\nabla^2 F(p)), \tag{8.226}$$

where $\nabla^2 F$ is the Hessian of the function F on $\mathcal{S}$ with respect to normal coordinates, so that $\mathcal{M}(p)$ is a 2×2 matrix with components $\frac{\partial^2 F}{\partial x_j \partial x_k}(p)$, $j, k = 1, 2$. Likewise, the derivatives of the Green's function in normal coordinates are denoted by

$$\nabla_x R_0(p, q) \text{ derivative of the first component,}$$

$$\nabla_z R_0(p, q) \text{ derivative of the second component.}$$

Using (8.223) we have

$$\nabla R(p) = (\nabla_x + \nabla_z) R_0(p, p),$$
$$\nabla^2 R(p) = (\nabla_x^2 + 2\nabla_x \nabla_z + \nabla_z^2) R_0(p, p)$$
$$= 2(\nabla_x^2 + \nabla_x \nabla_z) R_0(p, p)$$

since $R_0(p, q)$ is symmetric in its arguments p and q.

Remark 8.46 Although $\mathcal{M}(p)$ will be evaluated using a normal coordinate system, the eigenvalues of $\mathcal{M}(p)$ (and hence its negative-definiteness, which we will assume) will be independent of the choice of coordinates. Moreover, the entries of $\mathcal{M}(p)$ vary differentiably with p because the basis of the tangent plane $T_p \mathcal{S}$, namely $\{e_1(p), e_2(p)\}$, is chosen to vary differentiably with p.

8.4.3 The Main Results

The stationary system for (8.216) is given by the following system of elliptic equations:

$$\begin{cases} \epsilon^2 \Delta_g A - A + \frac{A^2}{H} = 0, & A > 0 \text{ in } \mathcal{S}, \\ \frac{1}{\beta^2} \Delta H - H + A^2 = 0, & H > 0 \text{ in } \mathcal{S}. \end{cases} \tag{8.227}$$

Our first theorem concerns the existence of single-peaked solutions whose position is determined by an interaction of local curvature and nonlocal Green's function.

Theorem 8.47 *Let $p^0 \in S$ be a nondegenerate critical point of the function $F(p)$ defined in (8.224), i.e.*

$$\nabla F(p^0) = 0, \qquad \det(\nabla^2 F(p^0)) \neq 0. \tag{8.228}$$

Then, under assumptions (8.217) and (8.218), problem (8.216) has a positive spiky steady state (A_ϵ, H_ϵ) with the following properties:

(1) $A_\epsilon(x) = \xi_\epsilon(w(\frac{x-p^\epsilon}{\epsilon}) + O(\epsilon^2))$ *uniformly for $x \in S$, where w is the unique solution of (6.9) and*

$$\xi_\epsilon = \frac{|S|}{\epsilon^2 \int_{\mathbb{R}^2} w^2(y)dy} + O\left(\log \frac{1}{\epsilon}\right). \tag{8.229}$$

Further, $p^\epsilon \to p^0$ as $\epsilon \to 0$.

(2) $H_\epsilon(x) = \xi_\epsilon(1 + O(\epsilon^2))$ *uniformly for $x \in S$.*

To consider the stability of the K-peaked solutions given in Theorem 8.47, we study the eigenvalue problem

$$\mathcal{L}_\epsilon \begin{pmatrix} \phi_\epsilon \\ \psi_\epsilon \end{pmatrix} = \begin{pmatrix} \epsilon^2 \Delta_g \phi_\epsilon - \phi_\epsilon + 2\frac{A_\epsilon}{H_\epsilon}\phi_\epsilon - \frac{A_\epsilon^2}{H_\epsilon^2}\psi_\epsilon \\ \frac{1}{\tau}(\frac{1}{\beta^2}\Delta_g \psi_\epsilon - \psi_\epsilon + 2A_\epsilon\phi_\epsilon) \end{pmatrix} = \lambda_\epsilon \begin{pmatrix} \phi_\epsilon \\ \psi_\epsilon \end{pmatrix}, \tag{8.230}$$

where (A_ϵ, H_ϵ) is given in Theorem 8.47 and $\lambda_\epsilon \in \mathbb{C}$ are some complex numbers.
    Then the stability result is stated as follows:

Theorem 8.48 *Let p^0 is a nondegenerate local maximum point of the function $F(p)$ defined in (8.224), i.e. we assume that*

$$\nabla F(p^0) = 0, \qquad \nabla^2 F(p^0) \text{ is negative definite.} \tag{8.231}$$

Under the assumptions (8.217) and (8.218), let (A_ϵ, H_ϵ) be the single-peaked solution constructed in Theorem 8.47 whose peak approaches p^0.
    Then there exists a unique $\tau_1 > 0$ such that for $\tau < \tau_1$, (A_ϵ, H_ϵ) is linearly stable and for $\tau > \tau_1$, (A_ϵ, H_ϵ) is linearly unstable.

For more details and proofs, we refer to [230].

Remark 8.49 The condition (8.231) on the locations p^0 arises from small eigenvalues of order $o(1)$. For any compact two-dimensional Riemannian manifold without boundary, the functional $F(p)$ defined by (8.224) always admits a global maximum at some $p^0 \in S$ since it is a continuous function defined on a compact set. We expect that for generic manifolds, this global maximum point p^0 is nondegenerate and (8.231) is valid.

8.5 Notes on the Literature

The Gierer-Meinhardt system with finite diffusivity has been studied in [266]. The approach follows [249, 269, 281] but is adapted from the real line to a finite interval.

The main reference for large reaction rates is [118]. Hunding and Engelhardt [95] studied the effect of large reaction rates on Turing instability for a few popular reaction-diffusion systems (Sel'kov model, Brusselator, Schnakenberg model, Gierer-Meinhardt system, Lengyel-Epstein model). Considering an increasing reaction rate, or Hill constant in case of Hill-type kinetics, modelling cooperativity within the system, they showed, using linearised stability analysis, that pattern formation by Turing instability is possible for large reaction rates, even if the ratio of the diffusion constants is close to unity.

In the Hill equation it is assumed that many molecules can interact at the same time, which is not very realistic [283]. Instead, from a biochemical viewpoint it would be more plausible to suppose that one ligand molecule after another is being bound to a receptor. This can happen in essentially two different ways: Firstly, by a sequential binding mechanism, for which the order in which the sites are filled is determined. Secondly, by an independent binding mechanism, for which the sites are occupied independently. Although for these two processes the Hill constant is smaller than the number of sites, the resulting Hill constants can nevertheless be very high.

An interesting case study on the molecular basis of cooperative interactions appears in [233]. The mechanism of binding Calcium ions to Calmodulin, a multi-site and multi-functional protein, has been modelled quantitatively and theoretical predictions have been confirmed by experimental observations.

In [23] evidence has been found for the fact that protein subunits can degrade less rapidly when they are associated in multimeric complexes. This effect is called cooperative stability.

For models of pattern formation induced by gene hierarchy it is reasonable to assume that the reaction rates are large since the degree of cooperativity is high [95]. Even for rather primitive animals and plants like flatworm, ciliates and fungi high cooperativity plays an important role, which has been investigated in *Drosophila*. Key ingredients of the gene hierarchy have been identified, see Sect. 12.3. In *Drosophila* homeobox genes too are crucial in facilitating a high degree of cooperativity [209]. The reason lies in their ability to create proteins which can bind to several other genes, in this process activating or inhibiting them [69].

Experimentally, for several different gene control systems reaction rates exceeding 8 have been found. We remark that in our system even for $p = 8$ the steady state solution is very well approximated by a spike on the real line whose spatial decay is of order $e^{-14|x|}$.

Reaction-diffusion systems with large reaction rates can elucidate the emergence of a variety of complex patterns. Biologically, these systems can recall gradients in the positional information. This is an essential requirement, since previous information must often be used sequentially for example in the head-tail (anterior-posterior) or front-back (dorsal-ventral) gradients in *Drosophila*. The systems also have a switching-type behaviour and can react in an almost on-off manner to very shallow gradients in positional information. This effect is needed to control the cell cycle governing mitosis, where the properties of the system must change qualitatively after it has grown to double size. Other qualitative properties of solutions include time oscillations and multi-stability, which is crucial when modelling cell differentiation.

Reaction-diffusion systems with large reaction rates are important for biological modelling as they widen the range of possible applications for Turing systems, explaining pattern formation in areas where there is no good justification for vastly different reaction rates but where it is known that there are large reaction rates. If there is a high degree of cooperativity between the components, which is often the case for many gene hierarchies, a large reaction rate can often be explained theoretically and measured experimentally, thus opening the door for suitable Turing systems to explain the patterns observed.

Robin boundary conditions for spiky patterns of the shadow Gierer-Meinhardt system have been considered in [144]. The proof of Theorem 8.37 follows Appendix F of [99]. The proof of Theorem 8.42 is similar to Theorem 3.8 of [251]. Robin boundary conditions play a role in understanding osmosis (diffusion through a semi-permeable membrane), whereas Neumann boundary conditions would be the modelling hypothesis for closed systems.

For spikes on Riemannian manifolds we refer to [231].

Chapter 9
The Gierer-Meinhardt System with Saturation

We investigate the shadow Gierer-Meinhardt system with saturation:

$$\begin{cases} A_t = \epsilon^2 \Delta A - A + \frac{A^2}{\xi(1+kA^2)}, & A > 0 \text{ in } \Omega \times (0, \infty), \\ \tau \xi_t = -\xi + \frac{1}{|\Omega|} \int_\Omega A^2(x)dx, & \xi > 0 \text{ in } (0, \infty), \\ \frac{\partial A}{\partial \nu} = 0 & \text{on } \partial\Omega \times (0, \infty). \end{cases} \tag{9.1}$$

Concerning the existence of steady states, we can no longer rescale with respect to the amplitude as we did for the system in case $k = 0$ without saturation. Thus it is impossible to reduce the existence problem for steady states to that of a single partial differential equation alone. Instead, we consider a system of a partial differential equation coupled to an algebraic equation:

$$\begin{cases} \epsilon^2 \Delta A - A + \frac{A^2}{\xi(1+kA^2)} = 0, & A > 0 \text{ in } \Omega, \\ \xi = \frac{1}{|\Omega|} \int_\Omega A^2(x)dx, & \xi > 0, \\ \frac{\partial A}{\partial \nu} = 0 & \text{on } \partial\Omega. \end{cases} \tag{9.2}$$

Firstly, we solve the parametrised ground state equation

$$\begin{cases} \Delta w_\delta - w_\delta + \frac{w_\delta^2}{1+\delta w_\delta^2} = 0, & w_\delta > 0 \text{ in } \mathbb{R}^n, \\ w_\delta(0) = \max_{y \in \mathbb{R}^n} w_\delta(y), & w_\delta(y) \to 0 \text{ as } |y| \to \infty. \end{cases} \tag{9.3}$$

Secondly, we consider the algebraic equation

$$\delta \left(\int_{\mathbb{R}^n} w_\delta^2(y)dy \right)^2 = k_0, \tag{9.4}$$

where

$$k_0 = \lim_{\epsilon \to 0} 4k\epsilon^{-2n}|\Omega|^2. \tag{9.5}$$

We remark that by introducing saturation the type of nonlinearity changes from convex in (5.5) to bistable in (9.3).

J. Wei, M. Winter, *Mathematical Aspects of Pattern Formation in Biological Systems*, 249
Applied Mathematical Sciences 189, DOI 10.1007/978-1-4471-5526-3_9,
© Springer-Verlag London 2014

For the stability part, we study the NLEP

$$\begin{cases} \Delta\phi - \phi + (\frac{2w_\delta}{1+\delta w_\delta^2} - \frac{2\delta w_\delta^3}{(1+\delta w_\delta^2)^2})\phi - 2\frac{\int_{\mathbb{R}^n} w_\delta\phi}{\int_{\mathbb{R}^n} w^2} \frac{w_\delta^2}{1+\delta w_\delta^2} = \lambda\phi & \text{in } \mathbb{R}^n, \\ \phi \in H^1(\mathbb{R}^n), \quad \lambda \in \mathbb{C}. \end{cases} \tag{9.6}$$

In the one-dimensional case, we will conduct a complete study. In higher dimensions, we will derive sufficient conditions on k to ensure the existence and stability of solutions.

We state our result in the one-dimensional case. Setting $\Omega = (0, 1)$, we have

Theorem 9.1 *Assume that*

$$\lim_{\epsilon \to 0} 4k\epsilon^{-2}|\Omega|^2 = k_0 \in [0, +\infty). \tag{9.7}$$

Then for each $k_0 \geq 0$ and for $\epsilon > 0$ small enough, (9.2) has a steady state $(u_\epsilon, \xi_\epsilon)$ such that

(a) $A_\epsilon(x) = (1 + o(1))\xi_\epsilon w_{\delta_\epsilon}(\frac{x}{\epsilon})$, *where $\delta_\epsilon \to \delta$, δ is the unique solution to (9.4) and w_{δ_ϵ} is the unique solution to (9.3),*

(b) $\xi_\epsilon = (2 + o(1))(\epsilon \int_{\mathbb{R}} w_{\delta_\epsilon}^2)^{-1}$.

If τ is small enough, the steady state $(A_\epsilon, \xi_\epsilon)$ is linearly stable for (9.1).

In case of higher dimensions, the statement is more involved. Let $Q \in \partial\Omega$. Denoting the mean curvature function at Q by $H(Q)$, we call Q a nondegenerate critical point of $H(Q)$, if we have

$$\partial_i H(Q) = 0, \quad i = 1, \ldots, n-1, \quad \det(\partial_i\partial_j H(Q)) \neq 0,$$

where ∂_i denotes the i-th tangential derivative. Then we have

Theorem 9.2 *Consider dimensions $n = 2, 3, \ldots$. Assume that*

$$\lim_{\epsilon \to 0} 4k\epsilon^{-2n}|\Omega|^2 = k_0 \in [0, +\infty) \tag{9.8}$$

and that $Q_0 \in \partial\Omega$ is a nondegenerate critical point of $H(Q)$.

Then for each $k_0 \geq 0$ and for ϵ small enough, (9.2) admits a steady-state solution $(A_\epsilon, \xi_\epsilon)$ such that

(a) $A_\epsilon(x) = (1 + o(1))\xi_\epsilon w_{\delta_\epsilon}(\frac{x - Q_\epsilon}{\epsilon})$, *where $\delta_\epsilon \to \delta$, δ is a solution to (9.4) and w_{δ_ϵ} is the unique solution to (9.3),*

(b) $Q_\epsilon \to Q_0$,

(c) $\xi_\epsilon = (2 + o(1))(\epsilon^n \int_{\mathbb{R}^n} w_{\delta_\epsilon}^2)^{-1}$.

If Q_0 is a nondegenerate local maximum point of $H(Q)$, then there is a $\hat{k}_0 > 0$ such that if $n \leq 3$ and τ is small enough, for all $k_0 \in (0, \hat{k}_0)$ the steady state $(A_\epsilon, \xi_\epsilon)$ is linearly stable for (9.1).

9.1   The Parametrised Ground State

In this subsection, we consider (9.3) and (9.4).

First we note that for $\delta = 0$ (9.3) becomes (2.3). For δ, we use the scaling

$$w_\delta(y) = \frac{1}{\sqrt{\delta}} v\left(\frac{y}{\delta^{1/4}}\right) \tag{9.9}$$

and change (9.3) to the equivalent problem

$$\begin{cases} \Delta v + g(v) = 0, & v > 0 \text{ in } \mathbb{R}^n, \\ v(0) = \max_{y \in \mathbb{R}^n} v(y), & v(y) \to 0 \text{ as } |y| \to \infty \end{cases} \tag{9.10}$$

where

$$g(v) = -\sqrt{\delta} v + \frac{v^2}{1 + v^2}. \tag{9.11}$$

Now for each $\delta \in (0, \frac{1}{4})$, the equation $g(v) = 0$ has exactly two roots for $v > 0$ given by

$$t_1(\delta) = \frac{1 - \sqrt{1 - 4\delta}}{2\sqrt{\delta}}, \qquad t_2(\delta) = \frac{1 + \sqrt{1 - 4\delta}}{2\sqrt{\delta}}. \tag{9.12}$$

Next we study

$$c(\delta) = \int_0^{t_2(\delta)} g(s)\,ds. \tag{9.13}$$

We calculate

$$c(\delta) = -\sqrt{\delta}\frac{(t_2(\delta))^2}{2} + t_2(\delta) - \arctan\big(t_2(\delta)\big).$$

To study $c(\delta)$, we consider the function

$$\rho(t) = \frac{t - \arctan(t)}{t^2}$$

which is well-defined for $t \in [0, +\infty)$. Further, $\rho(t)$ has a unique critical point t_* which solves

$$\arctan t = \frac{2t + t^3}{2(1 + t^2)}, \qquad t > 0. \tag{9.14}$$

Numerically we get $t_* = 1.514\ldots < \frac{\pi}{2}$. Setting

$$\delta_* = \big(2\rho(t_*)\big)^2, \tag{9.15}$$

it is easy to see that

$$c(\delta) \begin{cases} > 0 & \text{for } \delta < \delta_*, \\ = 0 & \text{for } \delta = \delta_*, \\ < 0 & \text{for } \delta > \delta_*. \end{cases} \tag{9.16}$$

Next we state a few properties of the function $g(v)$.

Lemma 9.3 *For each $\delta \in (0, \delta_*)$, the function $g(v)$ has the following properties:*

(g1) $g \in C^3(\mathbb{R}, \mathbb{R})$, $g(0) = 0$, $g'(0) < 0$.

(g2) *There exist $b, c > 0$ such that $b < c$, $g(b) = g(c) = 0$, $g(v) > 0$ in $(-\infty, 0) \cup (b, c)$, and $g(v) < 0$ in $(0, b) \cup (c, +\infty)$.*

(g3) $\int_0^c g(v)dv > 0$.

(g4) *Let θ be a number such that $\theta > b$ and $G(\theta) = 0$, where*

$$G(\theta) = \int_0^\theta g(s)ds.$$

Further, let ρ be the smallest number such that $\frac{g(u)}{u-\rho}$ is nonincreasing for $u \in (\rho, c)$. Then either

(i) $\theta \geq \rho$, *or*

(ii) $\theta < \rho$ *and $K_g(u)$ is nonincreasing in (θ, ρ), where*

$$K_g(u) = \frac{ug'(u)}{g(u)}.$$

Further, we have $K_g(u) \geq K_g(\theta)$ for $u \in (b, \theta)$ and $K_g(u) \leq K_g(\rho)$ for $u \in (0, b) \cup (\rho, c)$.

Proof For the proof of Lemma 9.3 we refer to [270]. The proof is elementary and we note that $K_g(u) \to \pm\infty$ as $u \to \pm b$ if $g'(b) > 0$. $\qquad\square$

Next we state some important properties of w_δ.

Lemma 9.4 *For each $\delta \in [0, \delta_*)$, (9.3) possesses a unique solution, denoted by w_δ, such that*

(i) $w_\delta \in C^\infty(\mathbb{R}^n)$.

(ii) $w_\delta > 0$ *is radially symmetric and $w_\delta'(r) < 0$ for $r \neq 0$.*

(iii) w_δ *and its first- and second-order derivatives decay exponentially at infinity, i.e., for every $\tilde{\delta} > 0$ there is a $c_1 > 0$ such that*

$$\left| w_\delta(y) \right| \leq c_1 e^{-(1-\tilde{\delta})|y|},$$

$$\left| \frac{\partial w_\delta}{\partial y_i}(y) \right| \leq c_1 e^{-(1-\tilde{\delta})|y|}, \quad i = 1, \dots, n,$$

$$\left| \frac{\partial^2 w_\delta}{\partial y_i y_j}(y) \right| \leq c_1 e^{-(1-\tilde{\delta})|y|}, \quad i, j, = 1, \dots, n.$$

(iv) *The largest eigenvalue of the operator*

$$L_\delta = \Delta - 1 + \frac{2w_\delta}{1 + \delta w_\delta^2} - \frac{2\delta w_\delta^3}{(1 + \delta w_\delta^2)^2} : H^2(\mathbb{R}^n) \to L^2(\mathbb{R}^n), \qquad (9.17)$$

denoted by $\lambda_1 = \lambda_1(L_\delta)$, is positive and simple. Its eigenfunction ϕ is radially symmetric and it can be chosen to be positive.

(v) *The second largest eigenvalue of L_δ is 0. Its kernel consists of the translation modes and has dimension n. That is, $\lambda_2(L_\delta) = 0$ and*

$$\text{Kernel}\left(\Delta - 1 + \frac{2w_\delta}{1 + \delta w_\delta^2} - \frac{2\delta w_\delta^3}{(1 + \delta w_\delta^2)^2}\right) = \text{span}\left\{\frac{\partial w_\delta}{\partial y_1}, \ldots, \frac{\partial w_\delta}{\partial y_n}\right\}. \quad (9.18)$$

Proof By Lemma 9.3, $g(v) = -\sqrt{\delta}v + \frac{v^2}{1+v^2}$ satisfies conditions (g1)–(g4). By Proposition 1.3 of [9], Lemma 9.4 holds. To prove this lemma, we first establish the statements of Lemma 9.4 for (9.10). Then they follow for the transformed function (9.3). We refer to [36, 194, 195] for related results. □

Now we provide some information concerning the dependence of w_δ on δ, stating some relevant identities.

Lemma 9.5

(1) $w_\delta(y)$ *is C^1 in δ for all $\delta \in (0, \delta_*)$ and $y \in \mathbb{R}^n$.*
(2) $w_\delta(y) \to t_2(\delta_*)/\sqrt{\delta_*}$ *in $C^2_{\text{loc}}(\mathbb{R}^n)$ as $\delta \to \delta_*$.*
(3) *We have*

$$L_\delta w_\delta = \frac{w_\delta^2}{1 + \delta w_\delta^2} - \frac{2\delta w_\delta^4}{(1 + \delta w_\delta^2)^2}, \qquad (9.19)$$

$$L_\delta \frac{dw_\delta}{d\delta} = \frac{w_\delta^4}{(1 + \delta w_\delta^2)^2}, \qquad (9.20)$$

$$L_\delta(y \cdot \nabla w_\delta) = 2\left(w_\delta - \frac{w_\delta^2}{1 + \delta w_\delta^2}\right), \qquad (9.21)$$

$$L_\delta\left(w_\delta + 2\delta\frac{dw_\delta}{d\delta} + \frac{1}{2}y \cdot \nabla w_\delta\right) = w_\delta, \qquad (9.22)$$

$$L_\delta\left(w_\delta + 2\delta\frac{dw_\delta}{d\delta}\right) = \frac{w_\delta^2}{1 + \delta w_\delta^2}. \qquad (9.23)$$

Proof

(1) Lemma 9.4 gives the uniqueness of w_δ and the result follows.
(2) Noting that $w_\delta \le t_2(\delta)/\sqrt{\delta}$ and taking the limit $\delta \to \delta_*$, we have that w_δ converges in $C^2_{\text{loc}}(\mathbb{R}^n)$ to a solution of

$$\Delta u - u + \frac{u^2}{1 + \delta_* u^2} = 0, \quad y \in \mathbb{R}^n, u = u(|y|)$$

which admits only constant solutions. Further, this constant is $t_2(\delta_*)/\sqrt{\delta_*}$ since $w_\delta(0) \to t_2(\delta_*)/\sqrt{\delta_*}$. (2) follows.

(3) The identities (9.19) and (9.20) are computed directly. (9.21) are derived using Pohozaev's identity. Finally, (9.22) and (9.23) follow from (9.19)–(9.21). $\square$

Next we consider an algebraic equation.

Lemma 9.6   *For each fixed $k_0 > 0$, there exists a $\delta \in (0, \delta_*)$ such that*

$$k_0 = \delta \left(\int_{\mathbb{R}^n} w_\delta^2(y) dy \right)^2 \tag{9.24}$$

holds.

Proof Let $\beta(\delta) = \delta (\int_{\mathbb{R}^n} w_\delta^2(y) dy)^2$. Then the function $\beta(\delta)$ is continuous and $\beta(0) = 0$. Next we consider the asymptotic behaviour of w_δ as $\delta \to \delta_*$. By Lemma 9.5(2), we have $w_\delta(|y|) \to t_2(\delta_*)/\sqrt{\delta_*}$ in $C_{\text{loc}}^2(\mathbb{R}^n)$ as $\delta \to \delta_*$. Hence we get

$$\beta(0) = 0, \qquad \beta(\delta) \to \infty \quad \text{as } \delta \to \delta_*. \tag{9.25}$$

Finally, using the mean-value theorem, for each $k_0 \in (0, +\infty)$, there exists a $\delta \in (0, \delta_*)$ such that $\beta(\delta) = k_0$. $\square$

Remark 9.7   To show the uniqueness of the solution $\delta \in (0, \delta^*)$ to (9.24), we compute

$$\frac{d\beta}{d\delta} = \left[\int_{\mathbb{R}^n} w_\delta^2(y) dy + 4\delta \int_{\mathbb{R}^n} w_\delta \frac{dw_\delta}{d\delta} dy \right] \int_{\mathbb{R}^n} w_\delta^2(y) dy. \tag{9.26}$$

Then we claim that

Lemma 9.8

$$\int_{\mathbb{R}^n} w_\delta \frac{dw_\delta}{d\delta} dy \bigg|_{\delta=0} > 0. \tag{9.27}$$

Proof From (9.20) and (9.22), we get

$$\int_{\mathbb{R}^n} w_\delta \frac{dw_\delta}{d\delta} dy \bigg|_{\delta=0} = \int_{\mathbb{R}^n} w_0 L_0^{-1}(w_0^4) dy$$

$$= \int_{\mathbb{R}^n} w_0^4 (L_0^{-1} w_0) dy = \left(1 - \frac{n}{10} \right) \int_{\mathbb{R}^n} w_0^5 dy > 0. \quad \square$$

Thus the solution to (9.24) is unique if k is small enough. We expect that Lemma 9.8 holds for any $\delta \in (0, \delta_*)$ and show that this is true for the one-dimensional case:

Lemma 9.9 *If* $n = 1$, *for any* $\delta \in (0, \delta_*)$ *we have*

$$\frac{d}{d\delta}\left(\int_{\mathbb{R}} w_\delta^2 dy\right) > 0. \tag{9.28}$$

Proof The proof of Lemma 9.9 is technical and we refer to [270].   □

9.2 Stability of Spikes

Let $(A_\epsilon, \xi_\epsilon)$ be the steady state given in Theorems 9.1 and 9.2. Linearising around $(A_\epsilon, \xi_\epsilon)$, we have

$$\epsilon^2 \Delta \phi - \phi + \frac{2A_\epsilon \phi}{\xi_\epsilon (1 + kA_\epsilon^2)} - \frac{2kA_\epsilon^3 \phi}{\xi_\epsilon (1 + kA_\epsilon^2)^2} - \frac{A_\epsilon^2}{\xi_\epsilon^2 (1 + kA_\epsilon^2)} \eta = \lambda \phi, \tag{9.29}$$

$$-\eta + \frac{2}{|\Omega|} \int_\Omega A_\epsilon \phi dx = \tau \lambda \eta, \tag{9.30}$$

where $(\phi, \eta) \in H_N^2(\Omega) \times \mathbb{R}$.

In case $\tau = 0$, we have

$$\eta = \frac{2}{|\Omega|} \int_\Omega A_\epsilon \phi dx. \tag{9.31}$$

Inserting (9.31) into (9.29), rescaling and taking the limit as $\epsilon \to 0$, we obtain the NLEP

$$\Delta \phi - \phi + \frac{2w_\delta \phi}{1 + \delta w_\delta^2} - \frac{2\delta w_\delta^3 \phi}{(1 + \delta w_\delta^2)^2} - \frac{2 \int_{\mathbb{R}^n} w_\delta \phi dy}{\int_{R^2} w_\delta^2 dy} \frac{w_\delta^2}{1 + \delta w_\delta^2} = \lambda \phi. \tag{9.32}$$

To study (9.32), we will derive the following key result:

Theorem 9.10 *Consider the case* $n \leq 3$. *Assume that* $\delta \in [0, \delta_{**})$, *where* $\delta_{**} > 0$ *is defined by*

$$\delta_{**} = \sup\left\{\delta \in (0, \delta_*) : \int_{\mathbb{R}^n} w_s \frac{dw_s}{ds} > 0, \text{ for all } s \in (0, \delta)\right\}. \tag{9.33}$$

Then for all nonzero eigenvalues λ *of (9.32), we have* $\mathrm{Re}(\lambda) \leq -c_0$ *for some* $c_0 > 0$.

Remark 9.11 By Lemma 9.8, we have $\delta_{**} > 0$. By Lemma 9.9, for $n = 1$ we get $\delta_{**} = \delta_*$. Hence we have the following result.

Corollary 9.12 *Let* $n = 1$. *Then for all nonzero eigenvalues* λ *of (9.32) and all* $\delta \in [0, \delta_*)$, *it holds that* $\mathrm{Re}(\lambda) \leq -c_0$ *for some* $c_0 > 0$.

To prove Theorem 9.10, we use a continuation argument. In case $\delta = 0$, Theorem 9.10 has been proved in Chap. 3 and follows from the following key inequality:

Lemma 9.13 (Lemma 5.1 of [249]) *Assume that $n \leq 3$. Then we have*

$$
\int_{\mathbb{R}^n} \left(|\nabla \phi|^2 + |\phi|^2 - 2w_0^2|\phi|^2 \right) dy + \frac{2 \int_{\mathbb{R}^n} w_0 \phi_0 dy \int_{\mathbb{R}^n} w_0^2 \phi dy}{\int_{\mathbb{R}^n} w_0^2 dy}
$$

$$
- \frac{(\int_{\mathbb{R}^n} w_0 \phi dy)^2}{(\int_{\mathbb{R}^n} w_0^2 dy)^2} \int_{\mathbb{R}^n} w_0^3 dy \geq c_1 d_{L^2}(\phi, X_1), \tag{9.34}
$$

where

$$
X_1 = \left\{ w_0, \frac{\partial w_0}{\partial y_j}, j = 1, \dots, n \right\}
$$

and d_{L^2} is the L^2-distance.

Proof of Theorem 9.10 We use the continuation method and begin by restricting ϕ to the Sobolev space of radially symmetric functions given by

$$
\phi \in H_r^2(\mathbb{R}^n) = \{ \phi \in H^2(\mathbb{R}^n) : \phi = \phi(|y|) \}.
$$

This is possible due to the argument in [37] and [281]. Then multiplication of (9.32) by the conjugate function $\overline{\phi}$ of ϕ and integration gives

$$
Q_\delta[\phi_R, \phi_R] + Q_\delta[\phi_I, \phi_I] = -\lambda \int_{\mathbb{R}^n} |\phi|^2 dy, \tag{9.35}
$$

where

$$
Q_\delta[u, u] = \int_{\mathbb{R}^n} \left(|\nabla u|^2 + u^2 - \frac{2w_\delta^2 u^2}{1 + \delta w_\delta^2} + \frac{2\delta w_\delta^3 u^2}{(1 + \delta w_\delta^2)^2} \right) dy
$$

$$
+ 2 \frac{\int_{\mathbb{R}^n} w_\delta u dy}{\int_{\mathbb{R}^n} w_\delta^2 dy} \int_{\mathbb{R}^n} \frac{w_\delta^2 u}{1 + \delta w_\delta^2} dy \tag{9.36}
$$

and $\phi_R = \mathrm{Re}(\phi)$, $\phi_I = \mathrm{Im}(\phi)$ are the real and the imaginary parts of ϕ, respectively.

To prove Theorem 9.10, it remains to show that Q_δ is positive definite for $\delta \in [0, \delta_{**})$. We rewrite Q_δ as follows:

$$
Q_\delta[u, u] = -(\mathcal{L}_\delta u, u),
$$

where

$$
\mathcal{L}_\delta u = \Delta u - u + \frac{2w_\delta}{1 + \delta w_\delta^2} u - \frac{2\delta w_\delta^3}{(1 + \delta w_\delta^2)^2} u - \frac{\int_{\mathbb{R}^n} w_\delta u dy}{\int_{\mathbb{R}^n} w_\delta^2 dy} \frac{w_\delta^2}{1 + \delta w_\delta^2}
$$

$$
- \frac{w_\delta}{\int_{\mathbb{R}^n} w_\delta^2 dy} \int_{\mathbb{R}^n} \frac{w_\delta^2 u}{1 + \delta w_\delta^2} dy. \tag{9.37}
$$

Then we have that

$$Q_\delta \text{ is positive definite} \quad \Longleftrightarrow \quad \mathcal{L}_\delta \text{ has negative spectrum.} \qquad (9.38)$$

By inequality (9.34), the principal eigenvalue of $\mathcal{L}_\delta$ is negative for $\delta = 0$. Considering varying δ, we assume that for some $\delta \in (0, \delta_*)$, the principal eigenvalue of $\mathcal{L}_\delta$ vanishes. Equivalently, for some function $\phi \in H^2_r(\mathbb{R}^n)$ we have

$$\mathcal{L}_\delta \phi = 0. \qquad (9.39)$$

Next we rewrite (9.39) as

$$L_\delta \phi = \frac{\int_{\mathbb{R}^n} w_\delta \phi \, dy}{\int_{\mathbb{R}^n} w_\delta^2 \, dy} \frac{w_\delta^2}{1 + \delta w_\delta^2} + \int_{\mathbb{R}^n} \frac{w_\delta^2 \phi}{1 + \delta w_\delta^2} \, dy \frac{w_\delta}{\int_{\mathbb{R}^n} w_\delta^2 \, dy}.$$

Now by Lemma 9.4 the inverse operator L_δ^{-1} exists and we have

$$\phi = \frac{\int_{\mathbb{R}^n} w_\delta \phi \, dy}{\int_{\mathbb{R}^n} w_\delta^2 \, dy} \left(L_\delta^{-1} \frac{w_\delta^2}{1 + \delta w_\delta^2} \right) + \int_{\mathbb{R}^n} \frac{w_\delta^2 \phi}{1 + \delta w_\delta^2} \, dy \frac{L_\delta^{-1} w_\delta}{\int_{\mathbb{R}^n} w_\delta^2 \, dy}. \qquad (9.40)$$

To solve (9.40), we set $A = \int_{\mathbb{R}^n} w_\delta \phi \, dy$ and $B = \int_{\mathbb{R}^n} \frac{w_\delta^2 \phi}{1 + \delta w_\delta^2} \, dy$. Then we get

$$\begin{cases} A = \dfrac{\int_{\mathbb{R}^n} w_\delta L_\delta^{-1}(w_\delta^2/(1 + \delta w_\delta^2)) \, dy}{\int_{\mathbb{R}^n} w_\delta^2 \, dy} A + \dfrac{\int_{\mathbb{R}^n} w_\delta L_\delta^{-1} w_\delta \, dy}{\int_{\mathbb{R}^n} w_\delta^2 \, dy} B \\[4mm] B = \dfrac{\int_{\mathbb{R}^n} (w_\delta^2/(1 + \delta w_\delta^2)) L_\delta^{-1}(w_\delta^2/(1 + \delta w_\delta^2)) \, dy}{\int_{\mathbb{R}^n} w_\delta^2 \, dy} A \\[4mm] \qquad + \dfrac{\int_{\mathbb{R}^n} (w_\delta^2/(1 + \delta w_\delta^2)) L_\delta^{-1} w_\delta \, dy}{\int_{\mathbb{R}^n} w_\delta^2 \, dy} B. \end{cases} \qquad (9.41)$$

Using Lemma 9.4 and noting that $\phi \in H^2_r(\mathbb{R}^n)$, we cannot have $L_\delta \phi = 0$ and $\phi \neq 0$. This implies $A^2 + B^2 \neq 0$.

Then (9.41) has nontrivial solutions if and only if

$$\begin{vmatrix} 1 - \dfrac{\int_{\mathbb{R}^n} w_\delta L_\delta^{-1}(w_\delta^2/(1 + \delta w_\delta^2)) \, dy}{\int_{\mathbb{R}^n} w_\delta^2 \, dy} & -\dfrac{\int_{\mathbb{R}^n} w_\delta L_\delta^{-1} w_\delta \, dy}{\int_{\mathbb{R}^n} w_\delta^2 \, dy} \\[4mm] -\dfrac{\int_{\mathbb{R}^n} (w_\delta^2/(1 + \delta w_\delta^2)) L_\delta^{-1}(w_\delta^2/(1 + \delta w_\delta^2)) \, dy}{\int_{\mathbb{R}^n} w_\delta^2 \, dy} & 1 - \dfrac{\int_{\mathbb{R}^n} (w_\delta^2/(1 + \delta w_\delta^2)) L_\delta^{-1} w_\delta \, dy}{\int_{\mathbb{R}^n} w_\delta^2 \, dy} \end{vmatrix} = 0, \qquad (9.42)$$

which is equivalent to

$$\left(1 - \frac{\int_{\mathbb{R}^n} (w_\delta^2/(1 + \delta w_\delta^2)) L_\delta^{-1} w_\delta \, dy}{\int_{\mathbb{R}^n} w_\delta^2 \, dy} \right)^2$$

$$- \frac{1}{(\int_{\mathbb{R}^n} w_\delta^2 \, dy)^2} \left(\int_{\mathbb{R}^n} w_\delta L_\delta^{-1} w_\delta \, dy \right) \left(\int_{\mathbb{R}^n} \frac{w_\delta^2}{1 + \delta w_\delta^2} L_\delta^{-1} \frac{w_\delta^2}{1 + \delta w_\delta^2} \, dy \right) = 0. \quad (9.43)$$

Using the identities (9.19)–(9.23), we compute

$$\int_{\mathbb{R}^n} \frac{w_\delta^2}{1+\delta w_\delta^2} L_\delta^{-1} w_\delta \, dy = \int_{\mathbb{R}^n} w_\delta L_\delta^{-1} \frac{w_\delta^2}{1+\delta w_\delta^2} \, dy$$

$$= \int_{\mathbb{R}^n} w_\delta \left(w_\delta + 2\delta \frac{dw_\delta}{d\delta} \right) dy$$

$$= \int_{\mathbb{R}^n} w_\delta^2 \, dy + 2\delta \int_{\mathbb{R}^n} w_\delta \frac{dw_\delta}{d\delta} \, dy, \qquad (9.44)$$

$$\int_{\mathbb{R}^n} w_\delta L_\delta^{-1} w_\delta \, dy = \int_{\mathbb{R}^n} w_\delta \left(w_\delta + 2\delta \frac{dw_\delta}{d\delta} + \frac{1}{2} y \cdot \nabla w_\delta \right) dy$$

$$= \left(1 - \frac{n}{4} \right) \int_{\mathbb{R}^n} w_\delta^2 \, dy + 2\delta \int_{\mathbb{R}^n} w_\delta \frac{dw_\delta}{d\delta} \, dy, \quad (9.45)$$

$$\int_{\mathbb{R}^n} \frac{w_\delta^2}{1+\delta w_\delta^2} L_\delta^{-1} \frac{w_\delta^2}{1+\delta w_\delta^2} \, dy = \int_{\mathbb{R}^n} \frac{w_\delta^2}{1+\delta w_\delta^2} \left(w_\delta + 2\delta \frac{dw_\delta}{d\delta} \right) dy$$

$$= \int_{\mathbb{R}^n} \frac{w_\delta^3}{1+\delta w_\delta^2} \, dy + 2\delta \int_{\mathbb{R}^n} \frac{w_\delta^2}{1+\delta w_\delta^2} \frac{dw_\delta}{d\delta} \, dy. \quad (9.46)$$

Multiplication of (9.21) by $\frac{1}{2}\frac{dw_\delta}{d\delta}$, use of (9.20) and integration gives

$$\int_{\mathbb{R}^n} \frac{w_\delta^2}{1+\delta w_\delta^2} \frac{dw_\delta}{d\delta} \, dy - \int_{\mathbb{R}^n} w_\delta \frac{dw_\delta}{d\delta} \, dy = \int_{\mathbb{R}^n} \frac{w_\delta^4}{(1+\delta w_\delta^2)^2} \left(-\frac{1}{2} y \cdot \nabla w_\delta \right) dy$$

and so

$$\int_{\mathbb{R}^n} \frac{w_\delta^2}{1+\delta w_\delta^2} \frac{dw_\delta}{d\delta} \, dy = \int_{\mathbb{R}^n} w_\delta \frac{dw_\delta}{d\delta} \, dy + \frac{n}{2} \int_{\mathbb{R}^n} \gamma_\delta(w_\delta) \, dy, \qquad (9.47)$$

where

$$\gamma_\delta(w_\delta) = \int_0^{w_\delta} \frac{s^4}{(1+\delta s^2)^2} \, ds.$$

Finally, using

$$h(\delta) := \left(2\delta \int_{\mathbb{R}^n} w_\delta \frac{dw_\delta}{d\delta} \, dy \right)^2 - \left(\left(1 - \frac{n}{4} \right) \int_{\mathbb{R}^n} w_\delta^2 \, dy + 2\delta \int_{\mathbb{R}^n} w_\delta \frac{dw_\delta}{d\delta} \, dy \right)$$

$$\times \left(\int_{\mathbb{R}^n} \frac{w_\delta^3}{1+\delta w_\delta^2} \, dy + n\delta \int_{\mathbb{R}^n} \gamma_\delta(w_\delta) \, dy + 2\delta \int_{\mathbb{R}^n} w_\delta \frac{dw_\delta}{d\delta} \, dy \right)$$

$$= -2\delta \int_{\mathbb{R}^n} w_\delta \frac{dw_\delta}{d\delta} \, dy$$

$$\times \left(\left(1 - \frac{n}{4}\right) \int_{\mathbb{R}^n} w_\delta^2 dy + \int_{\mathbb{R}^n} \frac{w_\delta^3}{1 + \delta w_\delta^2} dy + n\delta \int_{\mathbb{R}^n} \gamma_\delta(w_\delta) dy \right)$$

$$- \left(1 - \frac{n}{4}\right) \int_{\mathbb{R}^n} w_\delta^2 dy \left(\int_{\mathbb{R}^n} \frac{w_\delta^3}{1 + \delta w_\delta^2} dy + n\delta \int_{\mathbb{R}^n} \gamma_\delta(w_\delta) dy \right), \quad (9.48)$$

(9.43) can be written as

$$h(\delta) = 0. \tag{9.49}$$

We remark that

$$1 - \frac{n}{4} > 0 \text{ since we are considering the case } n \leq 3.$$

Now, for $0 \leq \delta \leq \delta_{**}$, we have $h(\delta) < 0$ and so we must have $\delta > \delta_{**}$. Since we have assumed that $\delta \in [0, \delta_{**})$ we arrive at a contradiction.

Theorem 9.10 follows. $\qquad \square$

Remark 9.14

(1) By the proof of Theorem 9.10, the number δ_{**} can be replaced by

$$\delta_{***} = \sup\{\delta \in (0, \delta_0) : h(s) < 0, s \in (0, \delta)\}. \tag{9.50}$$

(2) Another sufficient condition for stability can be stated as follows. Note that

$$\int_{\mathbb{R}^n} \frac{w_\delta^3}{1 + \delta w_\delta^2} dy = \int_{\mathbb{R}^n} w_\delta^2 dy + \int_{\mathbb{R}^n} |\nabla w_\delta|^2 dy > \int_{\mathbb{R}^n} w_\delta^2 dy. \tag{9.51}$$

Thus we have

$$\frac{(1 - n/4) \int_{\mathbb{R}^n} w_\delta^2 dy (\int_{\mathbb{R}^n} (w_\delta^3/(1 + \delta w_\delta^2)) dy + \int_{\mathbb{R}^n} n\delta\gamma_\delta(w_\delta) dy)}{(1 - n/4) \int_{\mathbb{R}^n} w_\delta^2 dy + \int_{\mathbb{R}^n} (w_\delta^3/(1 + \delta w_\delta^2)) dy + \int_{\mathbb{R}^n} n\delta\gamma_\delta(w_\delta) dy}$$

$$> \frac{(1 - n/4) \int_{\mathbb{R}^n} w_\delta^2 dy}{(2 - n/4)} = \frac{4 - n}{8 - n} \int_{\mathbb{R}^n} w_\delta^2 dy.$$

Now $h(\delta) < 0$ is guaranteed if

$$\frac{4 - n}{8 - n} \int_{\mathbb{R}^n} w_\delta^2 dy + 2\delta \int_{\mathbb{R}^n} w_\delta \frac{dw_\delta}{d\delta} dy > 0. \tag{9.52}$$

Therefore, setting

$$\delta_{****} = \sup \left\{ \delta \in (0, \delta_*) : \frac{4 - n}{8 - n} \int_{\mathbb{R}^n} w_s^2 dy + 2s \int_{\mathbb{R}^n} w_s \frac{dw_s}{ds} dy > 0, \right.$$

$$\left. \text{for all } s \in (0, \delta) \right\}, \tag{9.53}$$

Theorem 9.10 is valid for $\delta \in (0, \delta_{****})$.

Proof of Theorem 9.1 and Theorem 9.2 Now we finish the proofs of our main theorems. Concerning the existence of solutions to (9.2), we use the scaling

$$A = \xi u, \qquad \xi^{-1} = \frac{1}{|\Omega|} \int_\Omega u^2 dx. \tag{9.54}$$

Then (9.2) is equivalent to

$$\begin{cases} \epsilon^2 \Delta u - u + \frac{u^2}{1+\delta u^2} = 0, & u > 0 \text{ in } \Omega, \\ \frac{\partial u}{\partial \nu} = 0 & \text{on } \partial\Omega \end{cases} \tag{9.55}$$

coupled with the algebraic equation

$$\delta \left(2\epsilon^{-n} \int_\Omega u^2 dx \right)^2 = k_\epsilon := 4k\epsilon^{-2n} |\Omega|^2. \tag{9.56}$$

By assumption (9.7), $\lim_{\epsilon \to 0} k_\epsilon = k_0 \in [0, +\infty)$. Lemma 9.6 implies that there exists a $\delta_1 \in (0, \delta_*)$ such that

$$\delta_1 \left(\int_{\mathbb{R}^n} w_{\delta_1}^2 dy \right)^2 = k_0. \tag{9.57}$$

Next we observe that w_δ is uniformly bounded in $H^1(\mathbb{R}^n)$ for $\delta \in (0, \delta_1)$, where the bound may depend on δ_1.

By Lemma 9.4, for each fixed $\delta \in (0, \delta_1)$ we have that w_δ is nondegenerate. Then Theorem 1.1 of [256] and Theorem 1.1 of [253] (see also Theorem 4.5 of [8]) imply that for ϵ small enough problem (9.55) has a single boundary spike steady state $u_{\epsilon,\delta}$ which is unique, nondegenerate and possesses a unique local maximum point $Q_{\epsilon,\delta}$ which converges to Q_0 as $\epsilon \to 0$. Note that in the one-dimensional case, this follows from the implicit function theorem, whereas in higher dimensions we use Liapunov-Schmidt reduction.

Finally, we solve the algebraic equation

$$\beta_\epsilon(\delta) := \delta \left(2\epsilon^{-n} \int_\Omega u_{\epsilon,\delta}^2 dx \right)^2 = k_\epsilon. \tag{9.58}$$

Using $\beta_\epsilon(0) = 0$ and

$$\lim_{\epsilon \to 0} \beta_\epsilon(\delta) \to \beta(\delta) = \delta \left(\int_{\mathbb{R}^n} w_\delta^2 dy \right)^2,$$

where the convergence is uniform in $\delta \in (0, \delta_1)$, we derive that $\lim_{\epsilon \to 0} \beta_\epsilon(\delta_1) \to \delta_1 (\int_{\mathbb{R}^n} w_{\delta_1}^2)^2 = k_0$. Since $u_{\epsilon,\delta}$ is unique and nondegenerate, β_ϵ is a continuous function of δ. Using the mean-value theorem and considering ϵ small enough, for $k_\epsilon \in (0, k_0)$ there is $\delta_\epsilon \in (0, \delta_1)$ such that $\beta_\epsilon(\delta_\epsilon) = k_\epsilon$. Note that δ_ϵ may not be unique. Since $k_0 \in [0, \infty)$ may be chosen arbitrarily, we get a solution for any $k_\epsilon \in [0, \infty)$.

Then $A_\epsilon = \xi_\epsilon u_{\epsilon,\delta_\epsilon}$, $\xi_\epsilon = (\frac{1}{|\Omega|} \int_\Omega u_{\epsilon,\delta_\epsilon}^2 dx)^{-1}$ is a solution required in Theorems 9.1 and 9.2, respectively.

The existence part of the proof follows.

To investigate the stability of the solution $(A_\epsilon, \xi_\epsilon)$, we consider the eigenvalue problem

$$
\begin{cases}
\epsilon^2 \Delta\phi_\epsilon - \phi_\epsilon + (\frac{2A_\epsilon}{\xi_\epsilon(1+kA_\epsilon^2)} - \frac{2kA_\epsilon^3}{\xi_\epsilon(1+kA_\epsilon^2)^2})\phi_\epsilon \\
\quad - \frac{A_\epsilon^2}{\xi_\epsilon^2(1+kA_\epsilon^2)}\eta_\epsilon = \lambda_\epsilon \phi_\epsilon \quad \text{in } \Omega, \\
-\eta_\epsilon + \frac{2}{|\Omega|}\int_\Omega A_\epsilon \phi_\epsilon dx = \tau\lambda_\epsilon \eta_\epsilon.
\end{cases}
\tag{9.59}
$$

Following the method in [249], we consider two cases separately. In Case 1, let $\lambda_\epsilon \to \lambda_0 \in \mathbb{C}$ with $\lambda_0 \neq 0$, the so-called large eigenvalues. Then, similarly to Chap. 4, we show that λ_0 satisfies

$$
\Delta\phi_0 - \phi_0 + \left(\frac{2w_\delta}{1+\delta w_\delta^2} - \frac{2\delta w_\delta^3}{(1+\delta w_\delta^2)^2}\right)\phi_0
$$

$$
- \frac{2}{1+\tau\lambda_0}\frac{w_\delta^2}{1+\delta w_\delta^2}\frac{\int_{\mathbb{R}^n} w_\delta \phi_0 dy}{\int_{\mathbb{R}^n} w_\delta^2 dy} = \lambda_0 \phi_0.
\tag{9.60}
$$

By Theorem 9.10, for $n \leq 3$ and $\delta \in (0, \delta_{**})$, (9.60) is stable for τ small enough, i.e., for all eigenvalues of (9.60) with $\lambda_0 \neq 0$ we have $\text{Re}(\lambda_0) \leq -c_0$ for some $c_0 > 0$. For $n = 1$, by Corollary 9.12, we may take $\delta_{**} = \delta_*$. This shows that the large eigenvalues are have negative real part.

Finally, we consider Case 2, for which $\lambda_\epsilon \to 0$, the small eigenvalues. In the one-dimensional case, λ_ϵ is bounded away from zero. Thus we only have to consider the higher-dimensional case. Then the proof follows closely Theorem 1.3 of [249].

The stability part of the proof is complete. $\qquad\square$

9.3 Notes on the Literature

The main reference for spikes for the shadow Gierer-Meinhardt system with saturation is [270]. Spikes for the full Gierer-Meinhardt system with saturation have been studied in [123, 162].

Saturation for the full Gierer-Meinhardt system is related to the question of stripes versus spots: For low saturation, spikes are the preferred pattern. However, as saturation increases, stripes will prevail. This is related to the issue of spot splitting which has been studied in a reaction-diffusion system with Schnakenberg kinetics [115].

In the next chapter, we will present results on spikes for reaction-diffusion systems with reaction kinetics which are not of activator-inhibitor type.

Chapter 10
Spikes for Other Two-Component Reaction-Diffusion Systems

In this chapter, we will study spikes for other reaction-diffusion systems. We begin with two examples of so-called activator-substrate models: The Schnakenberg and Gray-Scott models. We will consider the Schnakenberg model in one space dimension and the Gray-Scott model in two space dimensions. We will conclude by considering flow-distributed spikes. We approach this issue by considering the influence of convection on existence and stability of spikes in the case of the Schnakenberg model.

In this chapter we will focus on the main results and their biological relevance but skip most of the proofs to avoid technical complications. When we do give proofs, they will be less detailed than in previous chapters and we will present only the main steps.

10.1 The Schnakenberg Model

The Schnakenberg can be stated as follows:

$$\begin{cases} u_t = \epsilon^2 u'' - u + vu^2 & \text{in } (-1, 1), \\ \epsilon v_t = Dv'' + \frac{1}{2} - \frac{c}{\epsilon}vu^2 & \text{in } (-1, 1), \\ u'(-1) = u'(1) = v'(-1) = v'(1) = 0, \end{cases} \tag{10.1}$$

where $\epsilon > 0$ is a positive constant which is small enough. Further, $D > 0$ and $c > 0$ are constants which are independent of ϵ.

Here u is the activator which consumes the substrate v at a quadratic rate. On the other hand, the substrate is fed into the system at a constant rate. Further, the activator decays at a linear rate. Note that the diffusivity of the activator is much smaller than that of the substrate.

This activator-substrate model can be derived from chemical reaction kinetics using the mass balance laws and has been named after Schnakenberg [214]. It has been also been suggested by Gierer and Meinhardt [73].

The main existence result can be stated as follows:

J. Wei, M. Winter, *Mathematical Aspects of Pattern Formation in Biological Systems*, Applied Mathematical Sciences 189, DOI 10.1007/978-1-4471-5526-3_10, © Springer-Verlag London 2014

Theorem 10.1 *Assume that $\epsilon > 0$ is small enough, $D > 0$ and $c > 0$ are constants which are independent of ϵ and N is a positive integer. Then (10.1) admits a steady state $(u_{\epsilon,N}, v_{\epsilon,N})$ with the following properties:*

$$u_{\epsilon,N}(x) = \xi_\epsilon^{-1}\left(\sum_{j=1}^N w\left(\frac{x-x_j}{\epsilon}\right) + o(1)\right) \tag{10.2}$$

in $H_N^2(-\frac{1}{\epsilon}, \frac{1}{\epsilon})$, where

$$x_j = -1 + \frac{2j-1}{N}, \quad j = 1, \ldots, N, \tag{10.3}$$

and ξ_ϵ satisfies

$$\xi_0 = \lim_{\epsilon \to 0} \xi_\epsilon = cN \int_{-\infty}^\infty w^2 dy. \tag{10.4}$$

Further,

$$v_{\epsilon,N}(x_j) = \xi_\epsilon, \quad j = 1, \ldots, N \tag{10.5}$$

and $v_{\epsilon,N} \to v_0$ in $H_N^2(-1, 1)$, where v_0 satisfies

$$\begin{cases} D\Delta v_0 + \frac{1}{2} - \frac{1}{N}\sum_{j=1}^N \delta_{x_j} = 0, \\ v_0(x_j) = cN \int_{-\infty}^\infty w^2 dy = \xi_0, \quad v_0'(-1) = v_0'(1) = 0. \end{cases} \tag{10.6}$$

The main result on stability is the following:

Theorem 10.2 *Assume that D is finite. For $N \geq 2$, let*

$$D_N := \frac{1}{2c \int w^2 dy} \frac{1}{N^3}. \tag{10.7}$$

Then for $\epsilon > 0$ small enough,

(a) *the symmetric one-spike steady state $(u_{\epsilon,1}, v_{\epsilon,1})$ is stable for any constants $D > 0$ and $c > 0$ which are independent of ϵ;*

(b) *for any constant $c > 0$ (independent of ϵ) there is a threshold $D_N(\epsilon)$ which tends to a limit as $\epsilon \to 0$ such that, for $D < D_N(\epsilon)$ the symmetric N-spike steady state $(u_{\epsilon,N}, v_{\epsilon,N})$ is stable while for $D > D_N$ the N-spike steady state $(u_{\epsilon,N}, v_{\epsilon,N})$ is unstable, where*

$$\lim_{\epsilon \to 0} D_N(\epsilon) = \frac{1}{2c \int w^2 dy} \frac{1}{N^3}.$$

We omit the proofs of these results.

We encourage the reader to compare the results on existence and stability for the Schnakenberg model with those for the one-dimensional Gierer-Meinhardt system studied in Chaps. 2 and 4, respectively.

10.2 The Gray-Scott Model

Next we consider another very popular activator-substrate model, the so-called Gray-Scott model, which can be stated as follows:

$$\begin{cases} v_t = \epsilon^2 \Delta v - v + Auv^2 & \text{in } \Omega, \\ \tau u_t = D \Delta u + 1 - u - uv^2 & \text{in } \Omega, \\ \frac{\partial u}{\partial \nu} = \frac{\partial v}{\partial \nu} = 0 & \text{on } \partial \Omega. \end{cases} \tag{10.8}$$

We assume that

$\epsilon > 0$ is small enough and independent of x,

$\tau > 0$ does not depend on x or ϵ,

$D, A > 0$ both do not depend on x (but may depend on ϵ),

$D \ll e^{C/\epsilon}$ for some $C < 1$.

This is a particular example of an activator-substrate model and has been named after Gray and Scott [75, 76]. It has also been suggested by Gierer and Meinhardt [73].

We state results on the existence and stability of steady states with symmetric multiple spikes in two dimensions, i.e. we assume that $\Omega \subset \mathbb{R}^2$.

We will use the following two parameters:

$$\eta_\epsilon = \frac{|\Omega|}{2\pi D_\epsilon} \log \frac{1}{\epsilon}, \qquad \alpha_\epsilon = \frac{\epsilon^2 \int_{R^2} w^2}{A_\epsilon^2 |\Omega|}. \tag{10.9}$$

Note that the index ϵ denotes the dependence of D_ϵ and A_ϵ on ϵ. For their limits we make the following assumptions:

$$\eta_0 = \lim_{\epsilon \to 0} \eta_\epsilon \in [0, +\infty], \qquad \alpha_0 = \lim_{\epsilon \to 0} \alpha_\epsilon \in [0, +\infty]. \tag{10.10}$$

We assume that

$$\lim_{\epsilon \to 0} 4(\eta_\epsilon + K)\alpha_\epsilon < 1 \tag{10.11}$$

and

$$\lim_{\epsilon \to 0} \frac{(2\eta_\epsilon + K)^2}{\eta_\epsilon} \alpha_\epsilon \neq 1. \tag{10.12}$$

Let $\beta_\epsilon^2 = \frac{1}{D_\epsilon}$ and let G_0 be defined in (6.1).

Then we have the following main existence result:

Theorem 10.3 *Suppose that $\lim_{\epsilon \to 0} \beta_\epsilon = 0$. Assume that (10.11) and (10.12) hold. Further, let $\mathbf{P}_0 = (P_1^0, \ldots, P_K^0) \in \overline{\Lambda}$ be a nondegenerate critical point of $F(\mathbf{P})$ which is defined in (6.4). Then, for $\epsilon > 0$ small enough, (10.8) has two steady states $(v_\epsilon^\pm, u_\epsilon^\pm)$ with the following properties:*

(1) $v_\epsilon^\pm(x) = \sum_{j=1}^{K} \frac{1}{A\xi_\epsilon^\pm}(w(\frac{x-P_j^\epsilon}{\epsilon}) + O(h(\epsilon,\beta)))$ *uniformly for* $x \in \bar{\Omega}$ *as* $\epsilon \to 0$,
   where

$$\xi_\epsilon^\pm = \begin{cases} \frac{1\pm\sqrt{1-4K\alpha_0}}{2} + O(k(\epsilon,\beta)) & \text{if } \eta_\epsilon \to 0, \\ \frac{1\pm\sqrt{1-4\lim_{\epsilon\to 0}\eta_\epsilon\alpha_\epsilon}}{2} + O(k(\epsilon,\beta)) & \text{if } \eta_\epsilon \to \infty, \\ \frac{1\pm\sqrt{1-4(K+\eta_0)\alpha_0}}{2} + O(k(\epsilon,\beta)) & \text{if } \eta_\epsilon \to \eta_0 \in (0,\infty), \end{cases} \qquad (10.13)$$

$$k(\epsilon,\beta) = \begin{cases} \eta_\epsilon\alpha_\epsilon & \text{if } \eta_\epsilon \to 0, \\ \alpha_\epsilon & \text{if } \eta_\epsilon \to \infty, \\ \beta^2\alpha_\epsilon & \text{if } \eta_\epsilon \to \eta_0 \in (0,+\infty) \end{cases} \qquad (10.14)$$

and

$$h(\epsilon,\beta) = \min\left\{\frac{1}{\log(1/\epsilon)}, \beta^2\right\}. \qquad (10.15)$$

(2) $u_\epsilon^\pm(x) = \xi_\epsilon^\pm(1 + O(h(\epsilon,\beta)))$ *uniformly for* $x \in \bar{\Omega}$ *as* $\epsilon \to 0$.
(3) $P_j^\epsilon \to P_j^0$ *as* $\epsilon \to 0$ *for* $j = 1,\ldots,K$.

In Theorem 10.3, we have obtained two solutions. We call $(v_\epsilon^-, u_\epsilon^-)$ the small solution and $(v_\epsilon^+, u_\epsilon^+)$ the large solution.

Next we state the main stability result:

Theorem 10.4 (Stability of K-spot solutions) *Assume that* (10.11) *and* (10.12) *hold. Let* $\epsilon > 0$ *be small enough. Further, assume that*

$$\mathbf{P}^0 \text{ is a nondegenerate local maximum point of } F(\mathbf{P}). \qquad (*)$$

Then for the K-*spot solutions* (u_ϵ, v_ϵ) *in Theorem* 10.3 *we have the following stability result.*

The large solutions are linearly unstable for all $\tau \geq 0$. *For the small solutions we have*

Case 1. $\eta_\epsilon \to 0$.
   If $K = 1$, *there is a unique* $\tau_1 > 0$ *such that for* $\tau < \tau_1$, (u_ϵ, v_ϵ) *is linearly stable and for* $\tau > \tau_1$ *it is linearly unstable.*
   If $K > 1$, (u_ϵ, v_ϵ) *is linearly unstable for all* $\tau \geq 0$.

Case 2. $\eta_\epsilon \to \infty$.
   In this case (u_ϵ, v_ϵ) *is linearly stable for all* $\tau \geq 0$.

Case 3. $\eta_\epsilon \to \eta_0 \in (0,\infty)$.
   If $\alpha_0 < \frac{\eta_0}{(2\eta_0+K)^2}$, (u_ϵ, v_ϵ) *is linearly stable for* τ *small enough or* τ *large enough.*
   If $K = 1$ *and* $\alpha_0 > \frac{\eta_0}{(2\eta_0+1)^2}$, *there exist* $\tau_2 > 0$ *and* $\tau_3 > 0$ *such that* (u_ϵ, v_ϵ) *is linearly stable for* $\tau < \tau_2$ *and linearly unstable for* $\tau > \tau_3$.
   If $K > 1$ *and* $\alpha_0 > \frac{\eta_0}{(2\eta_0+K)^2}$, *then* (u_ϵ, v_ϵ) *is linearly unstable for all* $\tau \geq 0$.

We omit the proof of these results. It is interesting to compare these results on existence and stability with those for the two-dimensional Gierer-Meinhardt system studied in Chap. 6: While the general picture remains the same, the fine details in terms of the analytical form of the thresholds has changed.

10.3 Flow-Distributed Spikes

We study the effect of convection on a Turing system, in particular on spiky patterns. This issue has been studied from the Turing instability point of view in [213].

We consider the Schnakenberg model with convection

$$\begin{cases} \epsilon a_t = D a_{xx} - D\alpha a_x + \frac{1}{2} - \frac{c}{\epsilon} ab^2, & x \in (-1, 1), \\ b_t = \epsilon^2 b_{xx} - \epsilon^2 \alpha b_x - b + ab^2, & x \in (-1, 1). \end{cases} \tag{10.16}$$

Two types of boundary conditions are studied:

(i) In case of homogeneous Neumann boundary conditions which model zero diffusive flux

$$a_x = b_x = 0 \quad \text{for } x = -1 \text{ or } x = 1 \tag{10.17}$$

we have the following result:

Theorem 10.5 *For $\epsilon > 0$ small enough, there is a spiky steady state (a_ϵ, b_ϵ) of system (10.16) with Neumann boundary conditions (10.17). Its shape is given by*

$$b_\epsilon(x) = \frac{1}{\xi_\epsilon} w\left(\frac{x - x_1^\epsilon}{\epsilon}\right) + O(\epsilon) \quad \text{in } H_\epsilon^2(-1, 1), \tag{10.18}$$

$$a_\epsilon\left(x_1^\epsilon\right) = \xi_\epsilon. \tag{10.19}$$

The amplitude satisfies

$$\xi_\epsilon = \xi_0 + O(\epsilon) \quad \text{with } \xi_0 = \frac{6c\alpha}{e^{\alpha x_1^0} \sinh \alpha}. \tag{10.20}$$

For the position, we have

$$x_1^\epsilon = x_1^0 + O(\epsilon) \quad \text{with } x_1^0 = \frac{1}{\alpha} \ln\left(1 + \sqrt{1 + 24 D c \alpha^3 \coth \alpha}\right) - \frac{1}{\alpha} \ln(2 \cosh \alpha). \tag{10.21}$$

(ii) In case of Robin boundary conditions which model zero flux:

$$a_x - \alpha a = b_x - \alpha b = 0, \quad x = -1, x = 1, \tag{10.22}$$

we have

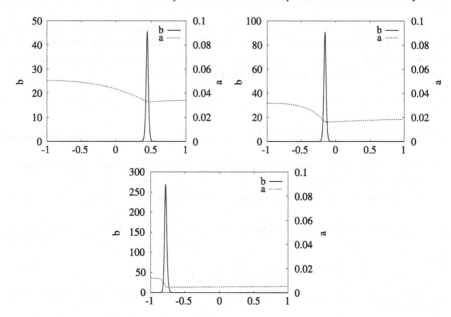

Fig. 10.1 A spiky steady state for Neumann boundary conditions with $\alpha = 1.0, 4.0, 30$. The other constants are chosen as $\epsilon = 0.01$, $D = 50$, $c = 0.01$. The spike begins by moving to the right, then it turns around and is shifted to the left until it finally approaches the left boundary

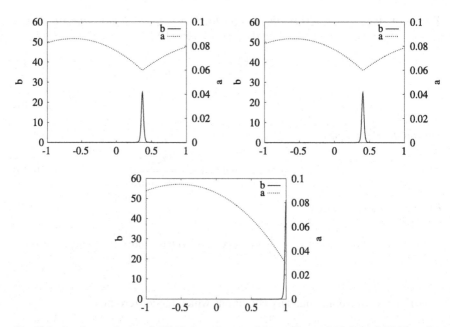

Fig. 10.2 A spiky steady state for Robin boundary conditions with $\alpha = 0.22, 0.24, 0.25$. The other constants are chosen as $\epsilon = 0.01$, $D = 10$, $c = 0.01$. The spike moves to the right quickly after α exceeds 0.20

Theorem 10.6 *For $\epsilon > 0$ small enough, there is a spiky steady state (a_ϵ, b_ϵ) of the system (10.16) with Robin boundary conditions (10.22). The shape of this solution is given by*

$$b_\epsilon(x) = \frac{1}{\xi_\epsilon} w\left(\frac{x - x_1^\epsilon}{\epsilon}\right) + O(\epsilon) \quad in \ H_\epsilon^2(-1, 1), \tag{10.23}$$

$$a_\epsilon\left(x_1^\epsilon\right) = \xi_\epsilon. \tag{10.24}$$

The amplitude satisfies

$$\xi_\epsilon = \xi_0 + O(\epsilon) \quad with \ \xi_0 = 6c. \tag{10.25}$$

For the position we have

$$x_1^\epsilon \to x_1^0 \quad with \ x_1^0 = 18Dc\alpha. \tag{10.26}$$

Remark 10.7 It has been shown that the solutions in Theorems 10.5 and 10.6 are linearly stable [279].

We illustrate these solutions by the following numerical computations:

We will present single-spike patterns and will observe that due to the boundary conditions its position has been moved. For Neumann boundary conditions the spike can be shifted in the same or opposite direction of the convective flow depending on the size of the convective flow (Fig. 10.1), whereas for Robin boundary conditions it is always shifted in the same direction as the flow (Fig. 10.2).

We first consider Neumann boundary conditions. We observe that with increasing size of the convective flow, the spike begins by moving to the right, then it turns around and is shifted to the left until it finally approaches the left boundary.

Next we present some computations with Robin boundary conditions. Now with increasing size of the flow, the spike always moves to the right and does not change direction. After the size of the flow has crossed a certain threshold, the spike moves very quickly and approaches the right boundary.

10.4   Notes on the Literature

Spikes for the Schnakenberg model have been considered in [100, 241] for one dimension and [274] for two dimensions. We state results on the existence and stability of steady states with symmetric multiple spikes in one dimension and follow [100]. Asymmetric spikes for one dimension have been studied in [241]. The two-dimensional case has been investigated in [274].

The presentation of spikes for the Gray-Scott model follows [267]. This model has been studied using various approaches. We refer to [203] for numerical computation of pulse splitting, [46, 47, 51] for rigorous results on the real line, [163, 164]

for a formal approach based on asymptotic expansions, [85, 86] for equal diffusivities, [112, 113] for spikes in one dimension, [250, 254] for a rigorous study
of the shadow system, [184] for a skeleton structure of self-replicating dynamics,
[49, 112, 113, 181, 182, 184, 186, 223] for a rigorous study of instability mechanisms of multi-spots and [185] for spatio-temporal chaos. Asymmetric spikes for
the Gray-Scott model in two dimensions have been studied rigorously in [268]. The
one-dimensional Gray-Scott model has been investigated using formal asymptotics
in [112, 113].

Our main reference for flow-distributed spikes is [279]. Both the existence results
and numerical computations are taken from there.

Turing bifurcations for the Schnakenberg model with spatially varying diffusion
coefficients are considered for the two-dimensional square in [12]. The authors show
how spatial variations can be used to partially reduce the degeneracy of Turing bifurcation. Interesting phenomena are investigated such as stable subcritical striped
patterns and stripes losing stability super-critically to give way to stable spotted
patterns.

In [77, 286] normal forms for the Brusselator model are studied and results on
dynamics are derived including chaotic behaviour. In [106, 107] secondary bifurcations are considered for the Brusselator model and an activator-inhibitor system.

The motion of spots has been analysed in [56, 182]. For absolute instability we
refer to [212]. For chaotic behaviour see [185]. Singular limit eigenvalue problems
for reaction-diffusion equations have been considered in [183].

The splitting of two-dimensional spots has been studied in [115] for both the
cases of one spot and multiple spots. A formal asymptotic analysis is used to derive
an ODE differential algebraic system (DAE) for the source strengths of positions
of each spot which involves the Neumann Green's function of the Laplacian. By
numerically examining the stability thresholds for a single spot solution, a specific
criterion in terms of the source strengths is then formulated to theoretically predict
the initiation of a spot-splitting event. Instabilities and dynamics of spots have also
been considered in [25, 26, 52, 57–59, 98].

Chapter 11
Reaction-Diffusion Systems with Many Components

In this chapter, we will consider some large reaction-diffusion systems which consist of more than two components. We begin with a system which has an arbitrary number of components, the so-called hypercycle of Eigen and Schuster. For this system we determine the maximum number of components for which a stable cluster is possible. Next we study a five-component system for which we will prove the existence and stability of mutually exclusive spikes, i.e. spikes which for different components are located at different positions. Then we consider systems with multiple activators and substrates and derive conditions for stable spiky patterns. Finally, we investigate a consumer chain model, which is a three-component system with two small diffusion constants and prove the existence and stability of a new type of clustered spiky pattern.

11.1 The Hypercycle of Eigen and Schuster

The hypercycle of Eigen and Schuster is a model for pre-biotic evolution which has been introduced as a system of time-dependent ordinary differential equations [60–63].

We consider a reaction-diffusion system which can be stated as follows:

$$\begin{cases} \partial_t X_i = \epsilon^2 \Delta X_i - X_i + AM \sum_{j=1}^{N} k_{ij} X_i X_j, & i = 1, \ldots, N, \\ \tau \partial_t M = \Delta M + 1 - M - M \sum_{i,j=1}^{N} k_{ij} X_i X_j, \end{cases} \tag{11.1}$$

where $t > 0$, $x \in \mathbb{R}$, $\epsilon > 0$ is a constant which is small enough, and $A > 0$ and $\tau \geq 0$ are constants which are independent of ϵ.

We assume that the connection matrix k_{ij} satisfies the following three hypotheses:

J. Wei, M. Winter, *Mathematical Aspects of Pattern Formation in Biological Systems*,
Applied Mathematical Sciences 189, DOI 10.1007/978-1-4471-5526-3_11,
© Springer-Verlag London 2014

(H1d) The equation

$$\sum_{j=1}^{N} k_{ij}\xi_j = 1, \quad i = 1, \dots, N \tag{11.2}$$

has a unique solution $(\xi_1, \dots, \xi_N)$ with $\xi_i > 0$.

(H2d) We have

$$\sum_{i=1}^{N} k_{ij}\xi_i = 1, \quad j = 1, \dots, N, \tag{11.3}$$

where ξ_j is given by (11.2).

The third assumption concerns the eigenvalue problem

$$\begin{cases} \Delta\phi - \phi + \mu w\phi = 0, \\ \phi \in H^1(\mathbb{R}). \end{cases} \tag{11.4}$$

By Lemma 4.1 of [245], (11.4) possesses the eigenvalues

$$\mu_1 = 1, \quad \mu_2 = 2, \quad 2 < \mu_3 \leq \mu_4 \leq \cdots. \tag{11.5}$$

In fact, we have the following explicit values for μ_n (see Appendix A of [263]):

$$\mu_n = \frac{(1+n)(2+n)}{6}, \quad n = 1, 2, \dots. \tag{11.6}$$

Setting

$$\mathcal{B} = (b_{ij}), \quad \text{where } b_{ij} = (\xi_i k_{ij}), \tag{11.7}$$

by (H1d) the matrix $\mathcal{B}$ has eigenvalue 1 with associated eigenvector $\boldsymbol{\xi} = (\xi_1, \dots, \xi_N)^\tau$. Thus we have $\mathcal{B}\boldsymbol{\xi} = \boldsymbol{\xi}$. Using the Jordan decomposition of $\mathcal{B}$, which is given by

$$\mathcal{B} = \mathcal{P}\mathcal{D}\mathcal{P}^{-1}, \tag{11.8}$$

where $\mathcal{P}$ is an orthogonal matrix and $\mathcal{D}$ has Jordan normal form, we have

$$b_{ij} = \sum_{k,l=1}^{N} p_{ik} d_{kl} p_{lj}^{-1}.$$

We recall that $\mathcal{D}$ is composed of Jordan blocks of the type

$$\begin{pmatrix} \sigma_k & 1 & 0 & \cdots & 0 \\ 0 & \sigma_k & 1 & \cdots & 0 \\ 0 & 0 & \sigma_k & \cdots & \vdots \\ \vdots & \vdots & \vdots & \vdots & 1 \\ 0 & 0 & 0 & \cdots & \sigma_k \end{pmatrix}$$

with eigenvalues $\sigma_k \in \mathbb{C}$.

Then we assume that

(H3d)

$$\begin{cases} [1 + \mathrm{spec}(\mathcal{B})] \cap \mathrm{spec}(\mathrm{EVP}) = \{2\}, \\ 1 \text{ is a simple eigenvalue of } \mathcal{B}. \end{cases} \tag{11.9}$$

Let us denote the eigenvalues of $\mathcal{B}$ by

$$\sigma_1 = 1, \qquad \sigma_2, \ldots, \sigma_N, \tag{11.10}$$

where σ_j may be complex. Then assumption (H3d) is equivalent to

$$\sigma_j \neq \frac{(1+n)(2+n)}{6} - 1 \quad \text{for } j = 2, 3, \ldots, n = 1, 2, \ldots . \tag{11.11}$$

Since $\boldsymbol{\xi} = (\xi_1, \ldots, \xi_N)^\tau$ is an eigenvector of $\mathcal{B}$ with eigenvalue 1, by assumption (H3d) we may assume that

$$\mathcal{P} = (\mathbf{p}_1, \ldots, \mathbf{p}_N) \quad \text{with } \mathbf{p}_1 = \frac{1}{\|\boldsymbol{\xi}\|} \boldsymbol{\xi}, \ \|\boldsymbol{\xi}\| = \sqrt{\sum_{i=1}^{N} \xi_i^2}. \tag{11.12}$$

A special case of connection matrix which satisfies these three hypotheses is a hypercyclical $N \times N$ matrix given by

$$\left(k_{ij}^{hyper} \right) = \begin{pmatrix} 0 & 0 & 0 & \cdots & k_0 \\ k_0 & 0 & 0 & \cdots & 0 \\ 0 & k_0 & 0 & \cdots & 0 \\ \cdots & \cdots & \cdots & \cdots & 0 \\ 0 & 0 & \cdots & k_0 & 0 \end{pmatrix}_{N \times N}, \quad k_0 > 0. \tag{11.13}$$

The system (11.1) with the matrix (k_{ij}^{hyper}) and without diffusion is termed an "elementary hypercycle" by Eigen and Schuster [60–63] since each component interacts with its successive neighbour in a cyclical way.

This system arises in the modelling of catalytic networks under the assumption that RNA-like polymers catalyse the replication of each other in a cyclical manner. Typical examples are the Krebs cycle for biosynthesis in the living cell and the Bethe-Weizsäcker cycle for high-rate energy production in massive stars.

We define

$$L = L(A, \epsilon) := \frac{1}{2A^2 \sum_{i=1}^{N} \xi_i} \epsilon \int_{\mathbb{R}} (w(y))^2 dy \tag{11.14}$$

and note that, for all $0 < L < \frac{1}{4}$, the equation

$$\eta(1 - \eta) = L \tag{11.15}$$

has two solutions which are given by $0 < \eta^s < \frac{1}{2} < \eta^l < 1$.

Now the main result on existence can be stated as follows:

Theorem 11.1 *Suppose that* (H1d) *holds.*
   Assume that

$$\epsilon > 0 \text{ is small enough} \tag{11.16}$$

and

$$\epsilon \ll L < \frac{1}{4} - \delta_0 \tag{11.17}$$

for some $\delta_0 > 0$ independent of ϵ.
   Then problem (11.1) *admits two single-cluster steady states* $(X_\epsilon^s, M_\epsilon^s) = (X_{\epsilon,1}^s, \ldots, X_{\epsilon,N}^s, M_\epsilon^s)$ *and* $(X_\epsilon^l, M_\epsilon^l) = (X_{\epsilon,1}^l, \ldots, X_{\epsilon,N}^l, M_\epsilon^l)$ *with the following properties:*

(1) *all components are even functions.*
(2) $X_{\epsilon,i}^s = \frac{\xi_i}{A M_\epsilon^s(0)}(1 + o(1))w(\frac{|x|}{\epsilon})$, $i = 1, \ldots, N$, $X_{\epsilon,i}^l = \frac{\xi_i}{A M_\epsilon^l(0)}(1 + o(1))w(\frac{|x|}{\epsilon})$,
   $i = 1, \ldots, N$ *uniformly on* $\mathbb{R}$, *where w is the unique ground state solution of*
   (2.3).
(3) $M_\epsilon^s(x) \to 1$, $M_\epsilon^l(x) \to 1$ *for all $x \neq 0$ and $M_\epsilon^s(0)$, $M_\epsilon^l(0)$ satisfy*

$$M_\epsilon^s(0) \to \eta^s, \qquad M_\epsilon^l(0) \to \eta^l, \quad \text{where } 0 < \eta_s < \frac{1}{2} < \eta_l < 1. \tag{11.18}$$

(4) *There exist $0 < a < 1, 0 < b < 1, C > 0$ such that*

$$0 < 1 - M_\epsilon^s(x) \le C e^{-a|x|},$$
$$0 < 1 - M_\epsilon^l(x) \le C e^{-a|x|},$$
$$0 < X_{\epsilon,i}^s(x) \le C\left(A M_\epsilon^s(0)\right)^{-1} e^{-b|x|/\epsilon}, \tag{11.19}$$
$$0 < X_{\epsilon,i}^l(x) \le C\left(A M_\epsilon^l(0)\right)^{-1} e^{-b|x|/\epsilon}.$$

   Finally, if ϵ is small enough and $L > \frac{1}{4} + \delta_0$ (in the same sense as in (11.17)) *then there are no single-cluster solutions.*

   Next we give our main result on stability.

Theorem 11.2 *Suppose that the matrix (k_{ij}) satisfies the three hypotheses* (H1d), (H2d) *and* (H3d). *Assume that*

$$\epsilon > 0 \text{ is small enough} \quad \text{and} \quad \epsilon \ll L < \frac{1}{4} - \delta_0, \tag{11.20}$$

in the same sense as in (11.17).
   Let $(X_\epsilon^s, M_\epsilon^s)$ *and* $(X_\epsilon^l, X_\epsilon^l)$ *be the solutions constructed in Theorem 11.1.*
   Let $\sigma = \sigma_R + i\sigma_I$ be an eigenvalue of $\mathcal{B}$. Define

$$f(\sigma) := (12\sigma_R + 5)^2 (3\sigma_R^2 + 2\sigma_R) - 3\sigma_I^2. \tag{11.21}$$

Then we have the following:

(1) *(Stability) Assume that for all eigenvalues σ of $\mathcal{B}$ with $\sigma \neq 1$ and $\sigma_R > 0$, we have $f(\sigma) < 0$. Then there exists a $\tau_0 > 0$ which is independent of ϵ such that for all τ with $0 \leq \tau < \tau_0$ the steady state $(X_\epsilon^s, M_\epsilon^s)$ is linearly stable.*
(2) *(Instability) Assume that there exists an eigenvalue σ of $\mathcal{B}$ with $\sigma \neq 1$ and $\sigma_R > 0$ such that $f(\sigma) > 0$. Then $(X_\epsilon^s, M_\epsilon^s)$ is linearly unstable for all $\tau \geq 0$.*
(3) *(Instability) $(X_\epsilon^l, M_\epsilon^l)$ is linearly unstable for all $\tau \geq 0$.*

The proof of Theorem 11.2 is based on Nonlocal Eigenvalue Problems (see Chap. 3), in particular on hypergeometric functions (see Sect. 3.4).

Theorem 11.2 applies to many particular classes of matrices. In particular, they include the hypercycle case, $(k_{ij}) = (k_{ij}^{hyper})$, where (k_{ij}^{hyper}) is given by (11.13). Then we have

Theorem 11.3 *Consider the hypercycle case, where (k_{ij}) is given in (11.13). Assume that (11.20) holds. Let $(X_\epsilon^s, M_\epsilon^s)$ and $(X_\epsilon^l, X_\epsilon^l)$ be the solutions constructed in Theorem 11.1.*

Then we have the following:

(1) *(Stability) Assume that $N \leq 4$. Then there exists a $\tau_0 > 0$ which is independent of ϵ such for all τ with $0 \leq \tau < \tau_0$ the steady state $(X_\epsilon^s, M_\epsilon^s)$ is linearly stable.*
(2) *(Instability) Assume that $N > 4$. Then $(X_\epsilon^s, M_\epsilon^s)$ is linearly unstable for all $\tau \geq 0$.*
(3) *(Instability) $(X_\epsilon^l, M_\epsilon^l)$ is linearly unstable for all $\tau \geq 0$.*

This theorem implies that for the hypercycle system a cluster is stable up to system size four and unstable for systems of larger size.

Next we consider another interaction of spikes, the so-called mutual exclusion phenomenon, which means that spikes for different components appear at different locations. We will show that this is possible and the resulting configurations are stable steady states of a suitable reaction-diffusion system.

11.2 Mutual Exclusion of Spikes

Mutual exclusion can be studied in a five-component model introduced by Meinhardt and Gierer to model phenomena from sociology [145, 154] or segmentation of *cockroach* legs to explain the famous experiment of Bohn on intercalary regeneration [20].

Segmentation is a pattern widely observed in biology. The mutual exclusion effect described in this section can be considered as segmentation in the simplest case of only two segments. Typical examples of biological segmentation are the body or leg segments of insects. Segments usually resemble each other strongly, but they may also have slight differences. They sometimes have internal polarity which

may be visible in the form of external bristles or hairs. Internal patterns can depend
on the segment position within a sequence. In some biological cases a good understanding of the connection between segment position and internal structure has been
achieved, such as for surgical experiments on insects. In the case of *cockroach* legs
these experiments have been particularly successful [20]. Creating a discontinuity
in the normal neighbourhood of structures by cutting a leg and pasting two partial
legs together creates a discontinuity in the segment structure. In particular, some
segments are now missing their natural neighbours. This results in the emergence
of new stable patterns in the *cockroach* leg which allow all segments to have their
natural neighbours. However, the resulting pattern can be strikingly different from
any naturally occurring pattern.

For example, in *cockroach* legs, assume that the normal sequence of structures
within a segment is $123\ldots9$. Then the addition of two partial legs with segments
12345678 and 456789 results in the structure 12345678456789, whose sequence
has a jump discontinuity between the numbers 8 and 4. In the next step, by segment
regulation the piece 765 will be added thus removing the discontinuity and leading
to the final structure 12345678**765**456789. Note that this sequence is different from
the original natural structure, yet each segment has the same neighbours as in the
natural structure.

In this example, which was experimentally verified by Bohn [20], it is not the
natural sequence but the normal neighbourhood which is regulated. Using an extension of the model from [154] considered in this section, it will be possible to study
such neighbourhood structures and investigate the possible patterns they support.

We conclude with a sociological application of mutual exclusion (see [154]):
Consider two families who may be relatives or friends. Living in the same house
or flat would lead to overcrowding and is therefore undesirable. However, living in
the same street or area, they will be able to support, nurture and benefit each other.
Such a collaborative behaviour in a close neighbourhood but not at exactly the same
location can result in a rather stable situation. Mathematically, stable coexisting
states with concentration peaks remaining close but keeping a certain characteristic
distance from each other is a typical phenomenon which is observed in quantitative
models of systems for mutual exclusion behaviour.

The system can be stated as follows:

$$\begin{cases} g_{1,t} = \epsilon^2 g_{1,xx} - g_1 + \frac{cs_2 g_1^2}{r}, \qquad g_{2,t} = \epsilon^2 g_{2,xx} - g_2 + \frac{cs_1 g_2^2}{r}, \\ \tau r_t = D_r r_{xx} - r + cs_2 g_1^2 + cs_1 g_2^2, \\ \tau s_{1,t} = D_s s_{1,xx} - s_1 + g_1, \qquad \tau s_{2,t} = D_s s_{2,xx} - s_2 + g_2. \end{cases} \tag{11.22}$$

Here $0 < \epsilon$ is a constant which is assumed to be small enough, $D_r > 0$ and $D_s > 0$
are diffusion constants, and $c > 0$ is a reaction constant. Finally, $\tau \geq 0$ is a time-
relaxation constant (in [154] the choice $\tau = 1$ was made).

The interactions in the system can be summarised as follows: The first two components, the activators g_1 and g_2, activate themselves locally due to the terms g_1^2
and g_2^2, respectively, in the first two equations. Note that the negative terms in the
first two equations lead to decay as is the case for the remaining equations.

The fourth and fifth components, the lateral activators s_1 and s_2, can be considered as nonlocal versions of g_1 and g_2, respectively due to last two equations. Now s_1 acts as an activator to g_2 and s_2 acts as in activator to g_1 due to the factors s_2 in the first and s_1 in the second equation within positive terms.

Finally, lateral activation is coupled with overall inhibition. The third component r acts as an inhibitor to both activators g_1 and g_2 due to the factor $\frac{1}{r}$ in the first and second equations within positive terms. Further, both the local and lateral activators give a positive feedback on r due to the terms $s_2 g_1^2$ and $s_1 g_2^2$ in the third equation.

We begin with an existence result for a spiky pattern, where both g_1 has a spike and g_2 has spike, but the two spikes have different locations. We assume:

(i) the diffusivities of the two lateral activators are large compared to the inhibitor diffusivity; and
(ii) the inhibitor diffusivity is large compared to the diffusivities of the two local activators.

Conditions (i) and (ii) are stated as follows:

We assume that $\epsilon^2 \ll C_1(L)D_r < D_s$ for some constant $2 < C_1(L) < 4$. (11.23)

Theorem 11.4 *Assume that* (11.23) *holds. Then there exist mutually exclusive, spiky steady states of* (11.22) *in* $(-L, L)$ *with Neumann boundary conditions such that*

$$g_1^\epsilon(x) = t_1^\epsilon w\left(\frac{x - x_1^\epsilon}{\epsilon}\right)(1 + O(\epsilon)),$$

$$g_2^\epsilon(x) = t_1^\epsilon w\left(\frac{x + x_1^\epsilon}{\epsilon}\right)(1 + O(\epsilon)) \tag{11.24}$$

with

$$t_1^\epsilon = \frac{1}{\epsilon \int_R w\,dy(G_{D_r}(x_1, x_1) + G_{D_r}(x_1, -x_1))} + O(1) \tag{11.25}$$

and $x_1^\epsilon \to x_1$ *as* $\epsilon \to 0$, *where*

$$\frac{G_{D_s,x_1}(x_1, -x_1)}{G_{D_s}(x_1, -x_1)} - \frac{G_{D_r,x_1}(x_1, -x_1) - H_{D_r,x_1}(x_1, x_1)}{G_{D_r}(x_1, x_1) + G_{D_r}(x_1, -x_1)} = 0. \tag{11.26}$$

If $D_s/D_r > 4$, (11.26) *has a unique solution* $x_1 \in (0, L/2]$ *and no solution in* $(L/2, L]$. *Further,* $x_1 \to L/2$ *as* $\theta_s \to 0$.

Finally, we compute the equation for x_1 in the limit $L \to \infty$. Then we have

$$\frac{\theta_s}{\theta_r} = \frac{e^{-2\theta_r x_1}}{1 + e^{-2\theta_r x_1}} + O\left(e^{-CL}\right)$$

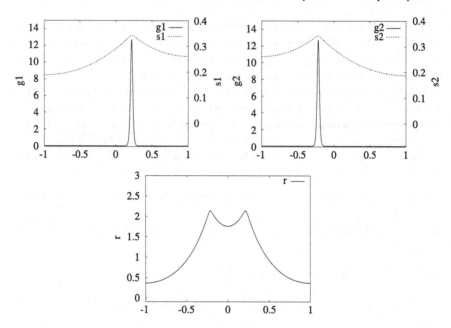

Fig. 11.1 The stable, mutually exclusive, two-spike steady state. *The figures* show the five components g_1, g_2, r, s_1, s_2. The parameter values are $\epsilon^2 = 0.0001$, $D_r = 0.1$, $D_s = 1$, $c = 1$, $\tau = 0.1$

for some $C > 0$ independent of x_1. This is equivalent to

$$e^{2|x_1|/\sqrt{D_r}} = \sqrt{\frac{D_s}{D_r} - 1} + O\left(e^{-CL}\right). \tag{11.27}$$

Further, this solution is stable if (i)

$$\frac{\cosh\theta_r(L+x_1) - \cosh\theta_r(L-x_1)}{\cosh\theta_r(L+x_1) + \cosh_r\theta_r(L-x_1)} > \frac{1}{7}(1 + \sqrt{15}) \tag{11.28}$$

(condition for $O(1)$ eigenvalues) and if (ii) $D_s/D_r > 4$ (condition for $o(1)$ eigenvalues). The details are as in Chap. 4.

Remark 11.5 In the limit $L \to \infty$, the condition (11.28) reduces to $D_s/D_r > 43.33$.

The stability result is stated as follows:

Theorem 11.6 *Suppose that the following two conditions are valid*: (i) *given by* (11.28), *and* (ii) *given by*

$$\frac{D_s}{D_r} > 4.$$

Then, for $\epsilon > 0$ small enough there exists a $\tau_0 > 0$ such that for $0 \leq \tau \leq \tau_0$, the mutually exclusive, spiky steady state given in Theorem 11.4 is linearly stable.

Results of a finite-element simulation for a mutually exclusive spiky patterns are shown in Fig. 11.1.

We have considered a five-component system with two activators both of which have local and nonlocal versions coupled with an inhibitor. This system is an extension of the Gierer-Meinhardt system. In the next section, we will study spiky patterns for an activator-substrate system with multiple activators and substrates which are coupled by a suitable set of connection matrices. That system is a generalisation of the Gray-Scott model (see Sect. 10.2).

11.3 Multiple Activators and Substrates

Biological systems are often very complex and they may consist of multiple activators and substrates. Following [227], we consider an extension of the Gray-Scott model to multiple activators and substrates which at the same time is an extension of the hypercycle model (see Sect. 11.1) to multiple substrates. The system can be stated as follows:

$$\begin{cases} \tau \frac{\partial u_i}{\partial t} = D\Delta u_i + 1 - u_i - \frac{A}{\epsilon} u_i \sum_{j,k} W_i^{(j,k)} v_j v_k, & i = 1, \ldots, M, x \in \mathbb{R}, \\ \frac{\partial v_i}{\partial t} = \epsilon^2 \Delta v_i - v_i + v_i \sum_{j,k} W_j^{(i,k)} u_j v_k, & i = 1, \ldots, N, x \in \mathbb{R}, \end{cases} \tag{11.29}$$

where u_i and v_i denote the concentrations of the substrates and the activators, respectively. Here $\epsilon > 0$ (sufficiently small) and $D > 0$ (independent of ϵ) are two diffusion constants. The constants A (positive) and τ (nonnegative) will be treated as parameters (independent of ϵ).

We first introduce some assumptions and notations.

For the connection matrices $W_j^{(i,k)} \geq 0$, we assume

Assumption 11.1

$$\sum_{j,k} W_j^{(i,k)} = T, \quad i = 1, \ldots, N \text{ for some } T > 0. \tag{11.30}$$

In particular, (11.30) implies that $\sum_{j,k} W_j^{(i,k)}$ is independent of i.

Assumption 11.2

$$\sum_{j,k} W_i^{(j,k)} = S, \quad i = 1, \ldots, M \text{ for some } S > 0. \tag{11.31}$$

In particular, (11.31) implies that $\sum_{j,k} W_i^{(j,k)}$ is independent of i.

The third assumption is the "transpose" of Assumption 11.1.

Assumption 11.3

$$\sum_{i,j} W_j^{(i,k)} = T, \quad k = 1, \dots, N, \tag{11.32}$$

where T has been defined in (11.30). In particular, (11.32) implies that $\sum_{i,j} W_j^{(i,k)}$ is independent of k.

Next we consider the simplest case: two activators and two substrates. For $M = N = 2$, we will construct steady states of (11.29) which are even functions:

$$1 - u_i = 1 - u_i(|x|) \in H^1(\mathbb{R}), \quad i = 1, 2,$$

$$v_i = v_i(|x|) \in H^1(\mathbb{R}), \quad i = 1, 2.$$

We now state the existence result.

Theorem 11.7 *Suppose that Assumptions 11.1–11.3 hold. Let $M = N = 2$. Assume that*

$$\epsilon > 0 \text{ is small enough} \tag{11.33}$$

and

$$\epsilon \ll \frac{12AS}{\sqrt{D}T^2} < 1 - \delta_0. \tag{11.34}$$

Then problem (11.29) admits two "single-spike" steady states $(u_\epsilon^s, v_\epsilon^s) = (u_{\epsilon,1}^s, u_{\epsilon,2}^s, v_{\epsilon,1}^s, v_{\epsilon,2}^s)$ and $(u_\epsilon^l, v_\epsilon^l) = (u_{\epsilon,1}^l, u_{\epsilon,2}^l, v_{\epsilon,1}^l, v_{\epsilon,2}^l)$ with the following properties:

(i) *all components are even functions.*
(ii) *$u_{\epsilon,i}^s(x) \to 1, u_{\epsilon,i}^l(x) \to 1$ as $\epsilon \to 0$ for all $x \neq 0$ and $u_{\epsilon,i}^s(0), u_{\epsilon,i}^l(0)$ satisfy*

$$u_{\epsilon,i}^s(0) \to u^s, \qquad u_{\epsilon,i}^l(0) \to u^l, \quad i = 1, 2,$$

$$0 < u^s < \frac{1}{2} < u^l < 1. \tag{11.35}$$

(iii) *$v_{\epsilon,i}^s = \xi^s(1 + o(1))w(\frac{|x|}{\epsilon}), v_{\epsilon,i}^l = \xi^s(1 + o(1))w(\frac{|x|}{\epsilon}), i = 1, 2,$ as $\epsilon \to 0$, where w is the unique solution of (2.3) and*

$$\xi^s = \frac{1}{Su^s}, \qquad \xi^l = \frac{1}{Su^l} \tag{11.36}$$

with S as defined in (11.31) and where u^s, u^l solve

$$u = \frac{1}{2} \pm \sqrt{\frac{1}{4} - \frac{3AS}{\sqrt{D}T^2}} \quad \text{with } 0 < u^s < \frac{1}{2} < u^l < 1.$$

(iv) *There exist $c_1 > 0, 0 < c_2 < 1, 0 < c_3 < 1$ such that*

$$0 < 1 - u^s_{\epsilon,i}(x) \le c_1 e^{-c_2|x|/\sqrt{D}},$$

$$0 < 1 - u^l_{\epsilon,i}(x) \le c_1 e^{-c_2|x|/\sqrt{D}},$$

$$0 < v^s_{\epsilon,i}(x) \le c_1 e^{-c_3|x|/\epsilon}, \tag{11.37}$$

$$0 < v^l_{\epsilon,i}(x) \le c_1 e^{-c_3|x|/\epsilon}.$$

Finally, if ϵ is small enough and $12 A S D^{-1/2} T^{-2} > 1 + \delta_0$, there are no single-spike solutions which satisfy (i)–(iv).

We now state our main result on stability.

Theorem 11.8 *Suppose that Assumptions 11.1–11.3 hold. Let $M = N = 2$. Assume that*

$$\epsilon > 0 \text{ is small enough}$$

and

$$\epsilon \ll \frac{12 A S}{\sqrt{D} S T^{-2}} < 1 - \delta_0.$$

Let $(u^s_\epsilon, v^s_\epsilon) = (u^s_{\epsilon,1}, u^s_{\epsilon,2}, v^s_{\epsilon,1}, v^s_{\epsilon,2})$ and $(u^l_\epsilon, v^l_\epsilon) = (u^l_{\epsilon,1}, u^l_{\epsilon,2}, v^l_{\epsilon,1}, v^l_{\epsilon,2})$ be the solutions constructed in Theorem 11.7.
Let

$$a = \frac{1}{T} \sum_{j=1}^{2} \left(W_j^{(1,1)} - W_j^{(1,2)} \right)$$

and

$$0 \le b = \frac{2(1 - u)}{T S} \sum_{j,k} W_j^{(1,k)} \left[W_j^{(1,1)} - W_j^{(2,2)} \right]$$

where $u = u^s$ or $u = u^l$, respectively.
Then we have the following:

(1) *(Stability) Assume that $b > a$. Then there exists a $\tau_0 > 0$ which is independent of ϵ such for all τ with $0 \le \tau < \tau_0$ we have $(u^s_\epsilon, v^s_\epsilon) = (u^s_{\epsilon,1}, u^s_{\epsilon,2}, v^s_{\epsilon,1}, v^s_{\epsilon,2})$ is linearly stable.*
(2) *(Instability) Assume that $b < a$. Then $(u^s_\epsilon, v^s_\epsilon) = (u^s_{\epsilon,1}, u^s_{\epsilon,2}, v^s_{\epsilon,1}, v^s_{\epsilon,2})$ is linearly unstable for all $\tau \ge 0$.*
(3) *(Instability) The solution $(u^l_\epsilon, v^l_\epsilon) = (u^l_{\epsilon,1}, u^l_{\epsilon,2}, v^l_{\epsilon,1}, v^l_{\epsilon,2})$ is linearly unstable for all $\tau \ge 0$.*

In the following two cases the conditions of Theorem 11.8 can be made explicit and interpreted in biological terms.

Case 1: Assume that

$$W_j^{(i,k)} \text{ is independent of } j \text{ for } i, k = 1, 2, \tag{11.38}$$

which implies

$$W_j^{(i,k)} = \frac{c+d}{2} + (-1)^{i+k}\frac{c-d}{2} \quad \text{for some } c > 0, d > 0.$$

Then we compute $S = T = 2(c + d)$. The condition $b > a$ in Theorem 11.8 takes the form $d > c$.

This means that the single spike is stable if off-diagonal interaction of the activators is dominated by self-interaction.

Case 2: We assume that

$$W_j^{(i,k)} = 0 \text{ if } i \neq k \text{ for } i, j, k = 1, 2, \tag{11.39}$$

which implies

$$W_j^{(i,k)} = \left(\frac{c+d}{2} + (-1)^{i+j}\frac{c-d}{2}\right)\delta_{ik} \quad \text{for some } c > 0, d > 0.$$

Then we compute $S = T = c+d$. Setting $c = \alpha S, d = (1-\alpha)D$ for some $0 \leq \alpha \leq 1$, the condition $b > a$ in Theorem 11.8 takes the form

$$8\alpha(1 - \alpha) < -c_4 + \sqrt{c_4^2 + 4c_4}, \quad \text{where } c_4 = 1 - \frac{12A}{\sqrt{DS}} > 0.$$

The condition is satisfied if α is sufficiently close to 0 or 1. This means that the single spike is stable if each activator has its own preferred substrate with which it interacts more strongly than the other by a certain amount.

In the next section we consider an extension of the Schnakenberg model. In all previous cases the substrate directly supplied all the activators, but now the supply is through a chain-like structure where the substrate supplies an activator, and this activator plays the role of substrate to a second activator.

11.4 Exotic Spiky Patterns for a Consumer Chain Model

Models involving a chain of components play an important role in biology, chemistry, the social sciences and many other fields. Well-known examples include food chains, consumer chains, genetic signaling pathways and autocatalytic chemical reactions or nuclear chain reactions. For food chains it is commonly assumed that there is only a limited supply of resources resulting in a saturation effect. On the other hand, for autocatalytic chemical or nuclear chain reactions the systems shows self-enhancing behaviour and after an initial cue the concentrations of the components are able to grow by themselves. Consumer chains in a special case could be food chains but in general consumption of various commodities such as water, energy, raw materials, technology and information will be studied. A complex, realistic consumer chain model will have to take into account both the limited amount of resources and the cooperation of consumers. If the consumption rate is small and resources are plentiful their limited amount can be ignored. Utilising mutual cooperation, consumers will be able to consume other constituents of the chain with increased efficiency for increasing concentration and some of the nonlinear terms in the system may be superlinear. Bettencourt and West [16] support this hypothesis of superlinear growth with a huge amount of empirical data on various activities in large cities in different continents, such as scientific publications or patents, GDP or the number of educational institutions as positives, and crime, traffic congestion or the occurrence of certain diseases as negatives. They established a universal growth rate which applies to most of the activities in major cities independent of geographic location or ethnicity of the population and cultural background. In our model we use quadratic terms and so we have assumed that the limited amount of resources is not felt and consumers cooperate to utilise resources efficiently.

The system is also a realistic model for a sequence of irreversible autocatalytic reactions taking place in a container which is constantly supplied by a well-stirred reservoir. For models similar to ours, we refer to Chap. 8 of Volpert et al. [236] and the references therein.

We consider a reaction-diffusion system which serves as a cooperative consumer chain model. It models the interaction of three components, one producer and two consumers, which supply each other in a linear sequence. This is an extension of the Schnakenberg model (see Sect. 10.1) for which there is only one producer and one consumer. In this consumer chain model there is a middle component which plays a hybrid role: it acts both as consumer and producer. It is assumed that the producer diffuses much faster than the first consumer and the first consumer much faster than the second consumer.

The system can be written as follows:

$$
\begin{cases}
\tau \frac{\partial S}{\partial t} = D\Delta S + 1 - \frac{a_1}{\epsilon_1} S u_1^2, & x \in \Omega, t > 0, \\
\tau_1 \frac{\partial u_1}{\partial t} = \epsilon_1^2 \Delta u_1 - u_1 + S u_1^2 - a_2 \frac{\epsilon_1}{\epsilon_2} u_1 u_2^2, & x \in \Omega, t > 0, \\
\frac{\partial u_2}{\partial t} = \epsilon_2^2 \Delta u_2 - u_2 + u_1 u_2^2, & x \in \Omega, t > 0,
\end{cases}
\tag{11.40}
$$

where S and u_i denote the concentrations of source and two consumers, respectively. Here $0 < \epsilon_2^2 \ll \epsilon_1^2 \ll 1$ and $0 < D$ are three positive diffusion constants. The constants a_1, a_2 (positive) for the feed rates and τ, τ_1 (nonnegative) for the time relaxation will be assumed to be independent of ϵ_1, ϵ_2. Note that the overall feed rates $\frac{a_1}{\epsilon_1}$ and $a_2 \frac{\epsilon_1}{\epsilon_2}$ are large, and they increase as the diffusion constant ϵ_1^2 and the ratio of the two small diffusion constants $\frac{\epsilon_2^2}{\epsilon_1^2}$, respectively, tend to zero.

As domain we choose the interval $\Omega = (-1, 1)$ and consider Neumann boundary conditions

$$\frac{dS}{dx}(-1) = \frac{dS}{dx}(1) = 0,$$

$$\frac{du_1}{dx}(-1) = \frac{du_1}{dx}(1) = 0, \qquad \frac{du_2}{dx}(-1) = \frac{du_2}{dx}(1) = 0. \tag{11.41}$$

We first state a result on the existence of single cluster solutions in an interval. Their profile is as follows: The second consumer is a spike, the middle component consists of two partial spikes glued together and connected by a thin transition layer, and the source has a shallow profile with a dip in the neighbourhood of the cluster.

These steady states may be stable or unstable and we will derive conditions to give more detailed information.

Using the notation $\epsilon = (\epsilon_1, \frac{\epsilon_2}{\epsilon_1})$, by $\epsilon \to 0$ we will mean that both $\epsilon_1 \to 0$ and $\frac{\epsilon_2}{\epsilon_1} \to 0$.

We first state the main existence result.

Theorem 11.9 *Assume that*

$$\epsilon_1 \ll 1, \qquad \frac{\epsilon_2}{\epsilon_1} \ll 1, \qquad D = const. \tag{11.42}$$

and

$$a_1^2 a_2 < c_0 - \delta_0, \tag{11.43}$$

where

$$c_0 = \max_{0 < z < 1} \frac{z(1 - z^2)^2}{(6 + 9z - 3z^3)^2} \approx 0.0025. \tag{11.44}$$

Then (11.40) *admits two "single-cluster" steady states* $(S_\epsilon^s, u_{1,\epsilon}^s, u_{2,\epsilon}^s)$ *and* $(S_\epsilon^l, u_{1,\epsilon}^l, u_{2,\epsilon}^l)$ *with the following properties:*

(1) *all components are even functions.*
(2) $S_\epsilon^s(0), S_\epsilon^l(0)$ *satisfy*

$$S_\epsilon^s(0) \to S^s, \qquad S_\epsilon^l(0) \to S^l, \quad \text{where } 0 < S^s < S^l. \tag{11.45}$$

(3)

$$u_{1,\epsilon}(x) = \xi_1^\epsilon \left[w\left(\frac{|x|}{\epsilon_1} - L_0 - r_\epsilon\right)\left(1 - \chi\left(\frac{|x|}{\epsilon_1 r_\epsilon}\right)\right) \right.$$
$$\left. + \left(u_{1,\epsilon}(0) + \frac{\epsilon_2}{\epsilon_1} u_{1a,\epsilon}\left(\frac{|x|}{\epsilon_2}\right)\right)\chi\left(\frac{|x|}{\epsilon_1 r_\epsilon}\right) \right], \tag{11.46}$$

$$u_{2,\epsilon}(x) = \xi_2^\epsilon w\left(\frac{|x|}{\epsilon_2}\right), \tag{11.47}$$

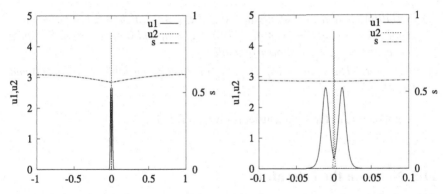

Fig. 11.2 *Left*: The steady state $(S^s_\epsilon, u^s_{1,\epsilon}, u^s_{2,\epsilon})$ for parameters $D = 10$, $\epsilon_1^2 = 10^{-4}$, $\epsilon_2^2 = 10^{-8}$, $\frac{a_1}{\epsilon} = 10$, $a_2\frac{\epsilon_1}{\epsilon_2} = 1$. *Right*: Same as *left panel*, but with spatial scale zoomed in

where w is the unique solution of (2.3) and

$$
\begin{aligned}
\xi_1^\epsilon &= \frac{1}{S_\epsilon(\epsilon_1 r_\epsilon)} = \frac{1}{S_\epsilon(0)} + O(\epsilon_1 r_\epsilon), \\
\xi_2^\epsilon &= \frac{1}{u_{1,\epsilon}(0)} = \frac{1}{\xi_1^\epsilon w(-L_0)} + O(r_\epsilon) = \frac{S_\epsilon(0)}{w(-L_0)} + O(r_\epsilon).
\end{aligned}
\tag{11.48}
$$

Further, χ is a suitably chosen smooth cutoff function,

$$
r_\epsilon = 10\frac{\epsilon_2}{\epsilon_1} \log \frac{\epsilon_1}{\epsilon_2}
\tag{11.49}
$$

and $u_{1a,\epsilon}$ tends to a limit as $\epsilon \to 0$.

Finally, if ϵ_1 and $\frac{\epsilon_2}{\epsilon_1}$ are small enough, and $a_1^2 a_2 > c_0 + \delta_0$ for some $\delta_0 > 0$ independent of ϵ_1 and $\frac{\epsilon_2}{\epsilon_1}$, then there are no single-cluster solutions which satisfy (1)–(4).

Remark 11.10 The shift L_0 and amplitude $S = S^s$ and $S = S^l$, respectively, are determined by solving the system

$$
S = a_1\left(6 + 9z - 3z^3\right),
\tag{11.50}
$$

$$
S = \frac{\sqrt{3}}{2\sqrt{a_2}}\sqrt{z}\left(1 - z^2\right).
\tag{11.51}
$$

Next we state our main result on stability.

Theorem 11.11 *Assume that conditions (11.42) and (11.43) are satisfied. Let $(S^s_\epsilon, u^s_{1,\epsilon}, u^s_{2,\epsilon})$ and $(S^l_\epsilon, u^l_{1,\epsilon}, u^l_{2,\epsilon})$ be the solutions given in Theorem 11.9. Then we have*

(1) (*Stability*) *There are constants* $\tau_0 > 0$ *and* $\tau_{1,0} > 0$ *which may be chosen independently of* ϵ_1 *and* $\frac{\epsilon_2}{\epsilon_1}$ *such that for* $0 \leq \tau < \tau_0$ *and* $0 \leq \tau_1 < \tau_{1,0}$, *the spike cluster* $(S_\epsilon^s, u_{1,\epsilon}^s, u_{2,\epsilon}^s)$ *is linearly stable.*

(2) (*Instability*) *The solution* $(S_\epsilon^l, u_{1,\epsilon}^l, u_{2,\epsilon}^l)$ *is linearly unstable for all* $\tau \geq 0$ *and* $\tau_1 \geq 0$.

A computation of the spike cluster is given in Fig. 11.2.

11.5   Notes on the Literature

Numerical simulations of patterns for the hypercycle of Eigen and Schuster are given in [19, 32, 33]. Results on the stability of spiky patterns in full space have been derived for two dimensions [260] and one dimension [263]. Using hypergeometric functions (see Sect. 3.4), the results in one dimension are more explicit than in two dimensions. The hypercycle system with nonlinear rates has been studied in [264].

The main reference for rigorous results on mutually exclusive spikes is [275].

Existence and stability of reaction-diffusion systems with multiple activators and substrates has been considered in [278].

The main reference for clusters in a consumer model is [280]. Although there we do not consider genetic signalling pathways it is generally understood that their typical behaviour includes activator and inhibitor feedback loops between different components. Some work has been done on modelling their dynamics including stochastic approaches. On the other hand, in the vast majority of studies their spatial components are ignored. However, the spatial dependence is important for many settings, e.g. for the Wnt signalling pathway which describes the interaction of cells and the passing of signals from the surface of a cell to its nucleus via a complex signaling pathway. It plays a role in embryonic development, cell differentiation and cell polarity generation.

Further aspects which should be included in the model when applying it to genetic signalling include typical length scales, mutual switch-on-off mechanisms of genetic components, strength of interaction classified into ranges of growth and plateau levels, interaction of neighbouring cells resulting in spatio-temporal patterns at cellular or intracellular level, and complex combinations of activator and inhibitor loops. Here we just consider a chain of two activators. This could lead to progress in the modelling of complex signaling pathways and mathematical analysis of spatiotemporal structures for genetic signaling pathways which are generally represented by complex networks of multiple activators and inhibitors or substrates.

Chapter 12
Biological Applications

12.1 Biological, Chemical and Ecological Applications of Reaction-Diffusion Systems

Turing systems [232] play an important role as fundamental models in many areas of biology. In this chapter, we present a few selected but typical examples where formation of pattern or shape could be very well explained by reaction-diffusion systems and agreement with experiments has been extremely convincing.

Biological applications of reaction-diffusion systems such as the Gierer-Meinhardt system and its generalisations to such diverse biological processes as the formation of head in *hydra*, body axis in *newt* and *Drosophila melanogaster*, segmentation in *Drosophila*, pigment patterns on sea shells, fish and mammals as well as the development of organs or nerve cells and brain activity have been described in [145, 152, 165, 166]. For reference we also mention the excellent textbooks [22, 54] and the lecture notes [169]. Biological and chemical applications have been described in [83, 180] and self-organisation (synergetics) has been described and analysed using order parameters (the slaving principle). Reviews on pattern formation for reaction-diffusion systems and its applications in biology chemistry and ecology have been given in [142, 239].

In chemistry, open systems in which chemicals are supplied to the system from outside are often realistic. The Gray-Scott and Schnakenberg models belong to this class [234, 235]. The Gray-Scott system as a model for the ferrocyanide-iodate-sulfate (FIS) reaction has been investigated both numerically and experimentally [128, 129]. Self-replication of spots has been investigated numerically and by formal analysis [207, 208]. For chemistry the assumption that the diffusion constants of the chemicals differ substantially is often questionable. Using the Turing instability approach, it has been shown that chemical patterns with equal diffusion coefficients are possible [234]. These results have then been applied to the Belousov-Zhabotinskii reaction [204]. On the other hand, using mathematical analysis of multi-spots in one space-dimension for the Gray-Scott system it has been proved that they are unstable for equal diffusion constants [65, 66, 85, 86].

J. Wei, M. Winter, *Mathematical Aspects of Pattern Formation in Biological Systems*,
Applied Mathematical Sciences 189, DOI 10.1007/978-1-4471-5526-3_12,
© Springer-Verlag London 2014

For the CIMA reaction some patterns found experimentally [24, 38, 194, 195] could be explained successfully by reaction-diffusion modelling and mathematical investigations [130].

Even though the Schnakenberg model is a strong simplification of the processes occurring in actual chemical reactors, many of the patterns observed experimentally can be computed and explained mathematically with the Schnakenberg model, such as multi-spots forming hexagonal arrays, stripes and wiggled stripes [53].

We would also like to comment on ecology. In [131] it is argued that pattern and scale form the basis of central processes in ecology, spanning from population biology to ecosystems science, and from pure to applied ecology. The main task therefore is to understand mechanisms leading to pattern formation which are able to act on multiple scales. Such multiple-scale patterns can be explained by Turing instabilities [215] and Turing patterns have the advantages that no genetic information is required in the model and a wide variety of patterns are possible. This wealth of patterns is of fundamental importance in ecology as it serves to explain diversity. Thus the study of Turing patterns in an ecological context is very important for a basic understanding of real-life applications.

12.2 *Hydra*: Transplantation of Head

The Gierer-Meinhardt system has originally been proposed to model head transplantation experiments in *hydra*, a fresh-water polyp [73], see also Chap. 6 of [145]. It has a tubular body which is about 1–2 mm long and consists of circa 100000 cells. However, it is segmented into only about seven different types of cells, namely the head H including mouth and tentacles for food consumption, gastric column consisting of four regions (named 1, 2, 3, 4 starting from the head), one or more buds and the basal disc which has adhesive properties. *Hydra* has two main cell layers: epidermis, the outer layer, and gastrodermis, the inner layer. The nerves are arranged in a net-like structure which is predominantly unpolarised. *Hydra*'s ability to regenerate missing parts has been studied for a long time, dating back to the famous experiments of Abraham Trembley in 1744 [229]. Due to its simple body architecture, *hydra* allows for the investigation of many important features of biological development. It is able to form multiple heads, tentacles can be regenerated and the development of nerve cells is well understood. The Gierer-Meinhardt system was first applied to experiments performed by Wolpert [288]. It has been introduced based on simple and molecularly plausible interactions of two morphogens such as synthesis supported by catalysis, supply from sources, spreading and degradation. Possible interactions have been identified which, under the assumption that their diffusion rates are different, lead to the establishment of peaks in the slowly diffusing morphogen which, due to its role in the interactions has been termed an activator. The second morphogen has been termed an inhibitor or substrate, depending on the class of interactions chosen. Examples have been computed to explain various experimental biological outcomes. For simplicity, *hydra* is first modelled by the concentration of morphogens in an interval, and it is assumed that the head will form at the position of a maximum peak of activator concentration. Pre-existing nerve cells are supposed to be the source for activator and inhibitor production. Firstly, the model is able to explain the formation of a head at one end of the interval but not the other. An activator peak is able to form at that end, starting from a shallow gradient. Hence this mechanism provides a rational basis for the origin of polarity. Secondly, if the diffusivity of the inhibitor is reduced, it is observed that there will be a periodic pattern whose length scale will be determined by this diffusivity once its extent is smaller than the total interval.

Now we describe various regeneration phenomena for which experimental results and theoretical predictions are in good agreement. Firstly, separating the neighbouring gastric regions 1/2 or 2/3 a new head will form at the side of 1 and 2, respectively. The explanation for this effect is that the sources will ensure the regeneration of the original polarity, even in the absence of activator. Secondly, if region 1 is attached to the head of another *hydra*, resulting in the order 1/1/2/3/4, only one head will form since the two regions 1 of the gastric column with high activator concentration are too close together. Thirdly, if we consider 1/2/1/2/3/4 there will now be two new heads since the two regions 1 of the gastric column with high activator concentration are separated by one region 2 and so they are sufficiently far apart to allow for this to happen. Fourthly, if we consider H/1/2/1/2/3/4 there will be only one head since the presence of the original head now suppresses the formation of a second one. Fifthly, for H/1/2/3/1/2/3/4 the distance between the two areas of high activator concentration located at H/1 and the second region 1 is now large enough to allow for the formation of a second head. Next some polarity reversal experiments due to [284, 285] are considered. The crucial phenomena are the timescale of inhibitor diffusion and stability of source distribution. Firstly, if the head is removed and attached to the opposite end of the gastric column, a new head will form at the previous head region, resulting in a double-headed *hydra*. The explanation for this behaviour is that it takes more time for the inhibitor to diffuse than for a signal initiating a new head to form. Secondly, if a head is attached to the other end of the gastric column and the old head is only removed after four hours or later, no new head will form at the previous head region, resulting in a single-headed *hydra*. This is explained by allowing enough time for the inhibitor to diffuse and so suppressing the formation of a new head. Thirdly, if the new head of the *hydra* in the previous experiment is removed, it will grow at its original location and not at the new position. This memory effect can be explained by assuming that it is the slowly varying source distribution and not the rapidly changing inhibitor concentration which is the deciding factor here.

A recent update and survey on the development of *hydra* is given in [72] and [153] including a discussion of the molecular basis and relations to other organisms. In [147, 153] the formation of secondary structures such as buds and tentacles is considered.

These biological applications in *hydra* development are related to our mathematical results in Chaps. 2–6. To understand these experiments it will be enlightening to think about the length scales which are implied by our theoretical results.

A head can only form if there is enough space for an activator peak to form. Roughly speaking, the *hydra* body must be large enough compared to the diffusion rate of the activator so that a graded profile can fit in. The results in Sect. 8.1 state this connection with mathematical rigour: For Gierer-Meinhardt kinetics we have derived the inequality $L > \pi$ if the activator diffusivity is scaled to unity, see (8.6). Note that this result even holds if the inhibitor does not have a profile at all, but it has "infinite" diffusion rate (shadow system) implying that it is forced to be constant throughout the body.

Further, the results in Chaps. 2–6 show that in a one- or two-dimensional setting a peak of the activator signal can form in a stable manner for very small activator diffusivity, even if the inhibitor is spread throughout the body and does not have any profile. However, here we do not strive to make the activator diffusivity as large as possible, in contrast to the result of Sect. 8.1.

Next we consider the question of when a second peak of the activator signal can form in a stable way. Our results imply that this is only possible if the inhibitor is localised in the following sense: The inhibitor maximum needed to stabilise the first activator peak decays and is small enough at the position of the second peak; conversely, the inhibitor maximum needed to support the second activator peak almost vanishes at the position of the first activator peak. This means that the diffusivity of the inhibitor must be small enough. On the other hand, if the inhibitor is forced to be constant throughout the body, the formation of a second activator peak will mean that the amount of inhibitor concentration would have to be doubled everywhere, thus in turn destabilising the double-peak configuration and leading to the escape of the system to a different morphogenetic pattern. Possibilities for the further dynamics include forcing one activator peak to vanish and a return to the single-peak situation, or extinguishing the spatial profile altogether, leading to spatially homogeneous morphogen concentrations.

The same principle as just discussed for the formation of a second activator peak applies to an arbitrary number of peaks. If the inhibitor concentration required to stabilise one of the peaks decays at the positions of all the other peaks the whole configuration will be stable. Therefore a relation between the number of peaks and the inhibitor diffusivity is required (see Remark 4.4 for the one-dimensional case and Remark 6.10 for the two-dimensional case). Each activator peak needs a certain amount of minimal space (meaning length in one dimension and area in two dimensions) to be sustainable. If this condition is not satisfied, some of the peaks will have to vanish so that finally a stable pattern is formed. Which of the possible outcomes happens will depend on the fine details of the morphogen concentrations such as the distances of the peaks from each other and from the boundary, their amplitudes and exact profiles. Interpreting these conditions in terms of length scale, in the one-dimensional case the distance between neighbouring peaks in a periodic arrangement must at least be proportional to the square root of the inhibitor diffusivity, whereas in two dimensions they must at least be proportional to the square root of the inhibitor diffusivity divided by the square root of a factor which increases for small activator diffusivity at logarithmic order. These relations are similar to the length scales expected from Turing instability for which it is assumed that activator and inhibitor diffusivity are of the same order, although in general their ratio will be given by a certain constant to be determined by the initial destabilisation of the homogenous state. There it is found that the length scales are proportional to the square root of the two diffusivities.

Thus our results will help to predict single, double and general periodic patterns as seen in many biological species. Mathematically well-studied examples of exactly or approximately periodic patterns include pigment patterns on sea shells, fish, mammalian coats or butterfly wings [145, 152, 165, 166]. Some particular examples will be discussed in Sect. 12.4.

12.3 Embryology: Formation of Body Axes for Newt and *Drosophila*, Segmentation for *Drosophila*

It is now clear that the development of an organism occurs in stages which are in many respects independent of each other, as can be seen in the following examples. In the early development of an embryo, organising centres play an important role in directing the onset of the growth of specialised tissues or organs. Spemann discovered these effects experimentally by grafting centres consisting of a group of cells from one *newt* embryo to another and observing that secondary centres were formed, see [219, 220]. Using reaction-diffusion modelling, Meinhardt was able to explain the behaviour of the Spemann centre by assuming that it secretes morphogens which diffuse in the embryo and influence genes in other cells by their local concentration there [149, 150]. The formation of the organiser is based on the mutual inhibition of the molecular components *BMP* and *Chordin*, as reviewed in [88], thus producing a system which allows for positive autoregulation. By complementing this system with a third component which acts as an antagonist (a possible candidate could be Anti-Dorsalising Morphogenetic Protein (*ADMP*) [161]) the behaviour seen in nature can be recaptured: *BMP* and *Chordin* appear at opposite ends and are stabilised by ADMP. Simulations with pre-existing maternal asymmetry indicate avoidance of multiple organisers and speeding up of organiser formation. Implantation of activated tissue can lead to the formation of a second organiser if it is located far enough from the existing one and the total field is large enough [149, 151].

Pioneering work into the genetic and molecular basis of embryological development in *Drosophila melanogaster* (fruit fly) has been achieved by Nüsslein-Volhard and Wieschaus. Performing extensive screening experiments, they studied how mutations would affect the segmentation pattern of the larvae. Many important changes to this pattern were discovered and related to their genetic basis, such as missing parts of segments or whole segments [190]. A review of these results for non-specialists and much more background on development and genetics can be found in [189].

The establishment of the body axis in the early embryo of *Drosophila melanogaster* is influenced by a maternal effect. It has been found that there are about 30 genes which affect this process. They determine four systems, out of which three are related to the anterior-posterior (AP, head-tail) axis and one is connected with the dorsal-ventral (DV, front-back) axis. The first two AP systems regulate the segmented parts of head and abdominal regions, respectively, while the third determines the unsegmented tips at the head and tail ends. The AP systems describe the morphogen concentration in different zones of the egg, whereas the DV system acts by controlling morphogen concentration in the nucleus. The formation of body axes is closely related to segmentation. These systems together determine at least seven distinct segments along the AP axis and at least three separate regions along the DV axis [97, 188, 202].

A morphogenetic model for the basis of segment formation was suggested by Meinhardt [146] before many molecular details were known, and its main properties were essentially confirmed by later findings. Four stages in the process of segment formation which are based on possible molecular interactions have been specified. Together they facilitate the reliable subdivision of a large area into a given number of segments along the AP axis. During each stage a new structure is added which acts on a smaller scale than the previous one. Firstly, an overall gradient in the AP direction is formed by some maternal asymmetry such as the maternal genes *bicoid* and *caudal*. Secondly, this gradient signal leads to division into primary regions by concentration-dependent local activation of the gap genes such as *giant, huckebein, hunchback, knirps, krüppel* and *tailless*. Thirdly, they trigger the emergence of two binary sequences of cell states 121212 and 343434 which are related to pair rule genes such as *even-skipped, hairy, odd-skipped, paired, fushitarazu* and *runt*. These sequences have the same period but are out of phase. They are able to transmit polarity, and polarity can be retained even if one of the four cell states is missing. These binary sequences define segments. Fourthly, each piece of four pair rule genes activates two pieces of segment polarity genes each consisting of three genes, sometimes on their own and sometimes by interaction with their neighbours. This is a rather intricate process, and if one of the pair rule genes is missing, then all three segment polarity genes are still present in the piece, but some regions are lost. This hierarchy explains the phenotypes observed for mutations, where certain parts of every second segment are missing. To conclude, we remark that a missing gap gene will result in a gap consisting of seven segments, whereas a missing pair rule gene will lead to the lack of an element in every other segment.

To understand segmentation from a mathematical point of view the results in Sect. 11.2 are useful since the mutual exclusion phenomenon studied there can be considered as segmentation in the special case of only two segments. In particular, by the mutual exclusion principle the existence of a segment in a certain position prevents formation of another segment there. In Bohn's example of segmented *cockroach* legs [20] the segments sharing a common border must also be neighbouring in a certain given order. This is reflected in the model of Meinhardt and Gierer [154] since activation occurs in a pairwise manner between neighbouring morphogens. This effect is combined with mutual exclusion which is achieved by assuming that all morphogens are supported by a common substrate and are therefore in competition with each other. This is a feature shared with the *Drosophila* model.

12.4 Pigmentation Patterns on Sea Shells, Fish and Mammals

Stripes on sea mollusks could correspond to one-dimensional periodic spiky patterns, the second direction representing time since they grow along one of their borders [152]. Stripes could also be implied by spatially homogenous temporal oscillations if the stripes are orthogonal to the direction of growth. Examples of periodically arranged two-dimensional spots are often seen on fish skins [166].

Our results will help to understand the occurrence of periodic pigmentation patterns. In particular, they will predict the minimal length scales observed by linking them to the diffusivities. (See also the comments at the end of Sect. 12.2.) From Turing's theory [232] we know that for patterns to appear reliably from an initially homogeneous situation the ratio between the inhibitor and activator diffusivities must exceed unity by a certain amount. Making the extra assumption that activator diffusivity is very small (compared to domain size) activator concentration between peaks will decay very sharply. Thus different peaks communicate mainly through their more slowly decreasing inhibitor concentration. The inhibitor concentration must decay between peaks to a certain amount, otherwise it will be too high at neighbouring peaks which will force some of them to vanish. On the other hand, the inhibitor cannot decay too sharply, since then a new activator peak might form between existing ones. Although our formulas do not predict this to happen, it could be favourable for slightly different models (for example with saturation included in the nonlinearity of the activator as in Chap. 9), in the case of finite activator diffusivity, or in the presence of stochastic perturbations which appear to be ubiquitous in biological systems. This second effect will result in the occurrence of maximal length scales so that altogether typical length scales are expected.

To understand if spots or stripes appear on an animal skin or coat it is important to investigate the Gierer-Meinhardt system with saturation as with increasing saturation spots tend to become unstable and stripes are preferred instead. The destabilisation of spikes by saturation has been shown in Chap. 9.

Often mixed forms of stripes and spots are encountered such as spots arranged along lines or wiggly lines which are in the process of breaking into spots. Further, spots can be destabilised by self-replication whereby a spot splits into several spots. Thus varying the saturation parameter, intermediate patterns appear which share properties of both stripes and spots. Conversely, the saturation parameter for an observed pattern can be estimated by computing the pattern and then choosing the parameter for which the computed and observed pattern are most similar.

It has been noted, and explained by Turing instabilities, that on a mammalian tail, in particular at its narrow end, stripes are preferred over spots [166].

Recently in [216] hair follicle arrangements in mice have been modelled by a reaction-diffusion system, where the *wnt* and *dkk* proteins serve as activator and inhibitor, respectively, and experiments are compared with numerical computations. See also the perspective in [143].

12.5 Patterns on Growing Domains: Stripes on Angelfish and Tooth Formation in Alligators

It has been suggested to use a Turing model to explain the development of pigmentation patterns on certain species of growing angelfish such as *Pomacanthus semicirculantus*, where coloured stripes are observed which change their number, size and orientation [119]. One interesting point in question is the addition of new stripes to a multi-stripe pattern in a growing two-dimensional domain. The two main competing mechanisms for this to happen is the splitting of existing stripes and the insertion of new ones. It has been observed in nature and computed numerically that, assuming growth is slow enough, thus allowing stripes to adjust such that they are always (almost) periodically arranged, at a certain stage of development between each pair of neighbouring stripes new ones are inserted (almost) simultaneously and their number is doubled. This event is triggered by their mutual distances becoming too large which is remedied by the system through intercalation of new stripes thus resulting in closer proximity of distances to the typical length scale. There has been some discussion as to whether the reaction-diffusion model for this doubling effect should be studied in a one-dimensional or two-dimensional setting [140]. It is expected that each have their own advantages in elucidating different aspects of the problem. After this model was refined by including chemotactic cell movement, further effects could be explained such as aggregation of pigment cells of different types [198]. Further, stripes of various thickness as suggested in [148] could be handled. Background on chemotaxis can be found in the following papers: In [108] the most popular model, called the Keller-Segel model, was introduced. In [102] it is shown that blow-up in the Keller-Segel model can occur, see also [94]. In [221] the chemotactic system is derived as a limit of a many-particle approximation. In [6, 82, 93] travelling fronts are considered. In [55] exponential attractors are constructed for the chemotactic-growth model introduced in [160]. In [40, 218] the existence and stability of spikes is considered. In [237] a generalised chemotaxis model with volume filling effect is considered. Finally, [91] provides a recent overview.

Growing domains also play an important role in understanding the order of tooth formation in *alligators* whose initiation has successfully been predicted by a reaction-diffusion system on a growing domain [120, 121]. For other reaction-diffusion systems on growing domains we refer to [28, 29, 138, 139].

Further research will be required to elucidate more details of the behaviour of reaction-diffusion systems on growing domains. Obviously this is of fundamental importance since all organisms grow at certain stages of their life. In particular, during growth spurts, when the speed of growth is high, they play a more prominent role than during periods of relative quiescence, and it is very often at these times that changes in the route to development take place, which decide future paths of activity. For spiky and spotty patterns it is important to gain a better understanding of the issue of insertion versus splitting as a means of preserving a typical length scale during growth. More generally, gaining a better understanding of the adaptation of the length scale of a certain type of pattern to the increasing size of the domain, which could include such complex behaviours as topological change of pattern, will require further rigorous mathematical studies in the future. These investigations should always be closely connected with corresponding biological observations in order to constantly improve the models under investigation and finally serve the ultimate goal of unravelling the natural processes leading to biological development.

Chapter 13
Appendix

13.1 Sobolev Spaces and Linear Operators

Sobolev spaces and linear operators are important tools which are used throughout this monograph. Therefore we state their definition and most important properties here. For a more detailed discussion we refer to the excellent books [74, 87].

Let Ω be a bounded, open, smooth domain in $\mathbb{R}^n$, where $n \geq 1$. For $p \geq 1$, let $L^p(\Omega)$ denote the Lebesgue space consisting of measurable functions defined on Ω such that

$$\|u\|_p := \|u\|_{L^p(\Omega)} = \left(\int_\Omega |u|^p dx \right)^{1/p} < \infty.$$

Then $L^p(\Omega)$ is a Banach space with the norm $\|u\|_p$. Further, the space $L^2(\Omega)$ is a Hilbert space with the scalar product

$$(u, v) := \int_\Omega uv dx.$$

For $k = 1, 2, \ldots$, we define

$$W^{k,p}(\Omega) := \left\{ u \in L^p(\Omega) : D^\alpha u \in L^p(\Omega) \text{ for all } |\alpha| \leq k \right\},$$

where

$$\alpha = (\alpha_1, \ldots, \alpha_n), \quad \alpha_i \in \{0, 1, \ldots\}, |\alpha| = \sum_{i=1}^n \alpha_i,$$

$$D^\alpha u := \frac{\partial^{|\alpha|} u}{\partial_{x_1}^{\alpha_1} \cdots \partial_{x_n}^{\alpha_n}}.$$

J. Wei, M. Winter, *Mathematical Aspects of Pattern Formation in Biological Systems*,
Applied Mathematical Sciences 189, DOI 10.1007/978-1-4471-5526-3_13,
© Springer-Verlag London 2014

We also denote $H^k(\Omega) := W^{k,2}(\Omega)$, and $H^k(\Omega)$ is a Hilbert space with the scalar product

$$(u, v)_k := \int_\Omega \sum_{|\alpha| \le k} D^\alpha u D^\alpha v \, dx.$$

A Banach space $\mathcal{B}_1$ is said to be continuously embedded in a Banach space $\mathcal{B}_2$ if there exists a bounded, linear, one-to-one mapping of $\mathcal{B}_1$ into $\mathcal{B}_2$. Using the notation $\mathcal{B}_1 \to \mathcal{B}_2$, we have the following continuous Sobolev embeddings:

$$W_0^{k,p}(\Omega) \to L^{1/(1/p-k/n)}(\Omega) \quad \text{for } kp < n,$$

$$W_0^{k,p}(\Omega) \to C^m(\bar{\Omega}) \quad \text{for } 0 \le m < k - \frac{n}{p},$$

where $W_0^{k,p}(\Omega)$ is the Banach space which is given by the closure of $C_0^k(\Omega)$ in $W^{k,p}(\Omega)$. Here $C_0^k(\Omega)$ is the set of continuous functions u defined on Ω with compact support in Ω for which the partial derivatives $D^\alpha u$, $|\alpha| \le k$ are also continuous. Further, $C^k(\bar{\Omega})$ is the set of all functions in $C^k(\Omega)$ for which all derivatives $D^\alpha u$, $|\alpha| \le k$ have continuous extensions to the closure $\bar{\Omega}$ of Ω.

Next we present an elliptic regularity theorem.

Theorem 13.1 (Elliptic regularity) *Let $u \in W^{2,p}(\Omega)$ solve the equation*

$$-\Delta u = f - \bar{f} \quad \text{in } \Omega,$$

$$\frac{\partial u}{\partial v} = 0 \quad \text{on } \partial\Omega,$$

where $\bar{f} = \frac{1}{|\Omega|} \int_\Omega f(x) dx$.
   Assume that $f \in L^p(\Omega)$. Then there exists some $c > 0$ such that

$$\|u - \bar{u}\|_{2,p} \le c \|f - \bar{f}\|_p. \tag{13.1}$$

The Fredholm Alternative holds for compact linear operators from a linear space into itself.

Theorem 13.2 (Fredholm Alternative. Theorem 5.3 in [74]) *A linear mapping T of a normed linear space into itself is called compact if L maps bounded sequences into sequences which contain convergent subsequences. Let T be a compact linear mapping of a normed linear space L into itself. Then either*

(i) *the homogeneous equation*

$$x - Tx = 0$$

   has a nontrivial solution $x \in L$ or

(ii) *for each $y \in L$ the equation*

$$x - Tx = y$$

has a uniquely determined solution $x \in L$. Further, in case (ii) *the "solution operator"* $(I - T)^{-1}$ *is bounded.*

Finally, we recall the mapping degree (see [35]). If $\Omega \subset \mathbb{R}^n$ is a bounded region, $f : \bar{\Omega} \to \mathbb{R}^n$ smooth, p a regular value of f and $p \notin f(\partial \Omega)$, then the degree $\deg(f, \Omega, p)$ is defined as follows:

$$\deg(f, \Omega, p) := \sum_{y \in f^{-1}(p)} \text{sign} \det Df(y),$$

where $Df(y)$ is the Jacobi matrix of f in y. This definition of degree may be naturally extended to non-regular values p such that $\deg(f, \Omega, p) = \deg(f, \Omega, p')$, where p' is a point close to p.

The degree satisfies the following five properties and is uniquely characterised by them.

(i) If $\deg(f, \bar{\Omega}, p) \neq 0$, then there exists an $x \in \Omega$ such that $f(x) = p$.
(ii) $\deg(Id, \Omega, y) = 1$ for all $y \in \Omega$.
(iii) *Decomposition property*:

$$\deg(f, \Omega, y) = \deg(f, \Omega_1, y) + \deg(f, \Omega_2, y),$$

where $\Omega_1 \cap \Omega_2 = \emptyset$, $\Omega = \Omega_1 \cup \Omega_2$ and $y \notin f(\bar{\Omega} \setminus (\Omega_1 \cup \Omega_2))$.
(iv) *Homotopy invariance*: If f and g are homotopy equivalent via a continuous homotopy $F(t)$ such that $F(0) = f$, $F(1) = g$ and $p \notin F(t)(\partial \Omega)$ for all $0 < t < 1$, then $\deg(f, \Omega, p) = \deg(g, \Omega, p)$.
(v) The function $p \mapsto \deg(f, \Omega, p)$ is locally constant on $\mathbb{R}^n \setminus f(\partial \Omega)$.

13.2 Uniqueness, Nondegeneracy and Spectrum of the Ground State

In this section, we give a self-contained proof of the existence, nondegeneracy and spectrum of the ground state. These are basic properties which we shall use throughout the book.

First, we have the following existence result.

Lemma 13.3 *Let* $1 < p < \frac{n+2}{n-2}$. *Consider the following minimisation problem*

$$c_p := \inf_{u \in H^1(\mathbb{R}^n) \setminus \{0\}} \frac{\int_{\mathbb{R}^n} (|\nabla u|^2 + u^2) dx}{(\int_{\mathbb{R}^n} u^{p+1} dx)^{2/(p+1)}}. \tag{13.2}$$

Then c_p can be attained by a radially symmetric function $w = w(r)$ which satisfies

$$\Delta w - w + w^p = 0, \quad w > 0, w \in H^1(\mathbb{R}^N). \tag{13.3}$$

Further, we have that $w > 0$ and $w'(r) < 0$ for $r > 0$.

Proof By Sobolev's embedding theorem, $0 < c_p < \infty$. Let $\{u_k\}$ be a minimising sequence. By scaling invariance, we may assume that $\int_{\mathbb{R}^n} u_k^{p+1} dx = 1$ and hence $\int_{\mathbb{R}^n} (|\nabla u_k|^2 + u_k^2) dx \to c_p$. By the rearrangement inequality (see [105, 205]), we have

$$\int_{\mathbb{R}^n} (u_k^*)^{p+1} dx = \int_{\mathbb{R}^n} (u_k)^{p+1} dx = 1$$

and

$$\int_{\mathbb{R}^n} \left(|\nabla u_k^*|^2 + (u_k^*)^2\right) dx \le \int_{\mathbb{R}^n} (|\nabla u_k|^2 + u_k^2) dx,$$

where u_k^* is the Schwarz rearrangement of u_k. Moreover, it holds that

$$\int_{\mathbb{R}^n} \left(|\nabla |u_k||^2 + |u_k|^2\right) dx = \int_{\mathbb{R}^n} (|\nabla u_k|^2 + u_k^2) dx.$$

Because of these two facts, we may assume that the minimising sequence $\{u_k\}$ is radially symmetric and strictly decreasing. By Strauss's lemma (see [222]), for $u = u(r)$, $u'(r) < 0$, we have

$$|u(r)| \le c r^{-(n-1)/2} \|u\|_{H^1(\mathbb{R}^n)}. \tag{13.4}$$

From (13.4), we deduce that the space of radially symmetric function in $H^1(\mathbb{R}^n)$, denoted by $H_r^1(\mathbb{R}^n)$, is continuously embedded into $L^{p+1}(\mathbb{R}^n)$. Thus $\{u_k(r)\}$ contains a convergent subsequence $\{u_k(r)\}$ in $L^{p+1}(\mathbb{R}^n)$. Assume that its limit is w, then $\int_{\mathbb{R}^n} w^{p+1} dx = 1$. By Fatou's Lemma, we have

$$\int_{\mathbb{R}^n} (|\nabla w|^2 + w^2) dx \le \liminf_{k \to \infty} \int_{\mathbb{R}^n} \left(|u_k|^2 + u_k^2\right) dx = c_p.$$

On the other hand, $w \in H^1(\mathbb{R}^n)$ and hence

$$\int_{\mathbb{R}^n} (|\nabla w|^2 + w^2) dx \ge c_p \left(\int_{\mathbb{R}^n} w^{p+1} dx\right)^{1/(p+1)} = c^p.$$

This shows that w is in fact a minimiser of the problem (13.2).

Let

$$E[u] = \frac{\int_{\mathbb{R}^n} (|\nabla u|^2 + u^2) dx}{(\int_{\mathbb{R}^n} u^{p+1} dx)^{2/(p+1)}}.$$

Then

$$E[w + t\phi] \geq 0 \quad \text{for all } t \in \mathbb{R} \text{ and for all } \phi \in C_0^\infty(\mathbb{R}^n).$$

The Euler-Lagrange equation of w implies that w is a weak solution of (13.3). Since $w^{p-1} \in L_{\text{loc}}^{n/2+\epsilon_0}(\mathbb{R}^n)$ for some $\epsilon_0 > 0$, the elliptic regularity theorem (Theorem 8.17 in [74]), yields that w is bounded. By L^p and Schauder estimates, we get $w \in C_{\text{loc}}^{2,\alpha}(\mathbb{R}^n)$ for some $\alpha > 0$. Thus w is a classical solution of (13.3). By the strong Maximum Principle, we finally have $w > 0$ in $\mathbb{R}^n$. $\qquad\square$

Next, we prove the nondegeneracy of w.

Lemma 13.4 *Let ϕ be a bounded solution of*

$$\Delta\phi - \phi + pw^{p-1}\phi = 0, \quad |\phi| \leq 1. \tag{13.5}$$

Then

$$\phi = \sum_{j=1}^n c_j \frac{\partial w}{\partial x_j} \quad \text{for some real constants } c_j, j = 1, \ldots, n.$$

Proof We divide the proof into four steps.

Step 1. ϕ decays exponentially to 0 at infinity, more precisely

$$|\phi(x)| \leq Ce^{(1-\delta)|x|} \quad \text{for some } C, \delta > 0. \tag{13.6}$$

In fact, let ψ be the unique solution of

$$\Delta\psi - \psi + pw^{p-1}\psi = 0.$$

Since $w^{p-1}\psi$ decays exponentially, so does ψ, both in the sense of (13.6). Then the difference $u = \phi - \psi$ satisfies

$$\Delta u - u = 0, \quad u \text{ is bounded.} \tag{13.7}$$

Thus $u \equiv 0$ and so $\phi \equiv \psi$.

Step 2. Assume that $\phi = \phi(r)$. Then $\phi \equiv 0$.

This is the key step. We follow the proof given in [15].

First, we show that $\lambda_2 \geq 0$, where λ_2 denotes the second eigenvalue of the operator

$$\Delta\phi - \phi + pw^{p-1}\phi + \lambda\phi = 0, \quad |\phi| \leq 1.$$

In fact, expanding the minimality condition

$$\frac{d^2}{dt^2}\bigg|_{t=0} \left(E(w + t\phi) - E(w)\right) \geq 0,$$

we obtain that

$$\int_{\mathbb{R}^n} \left(|\nabla\phi|^2 + \phi^2 - pw^{p-1}\phi^2\right)dx + (p-1)\frac{(\int_{\mathbb{R}^n} w^p\phi dx)^2}{\int_{\mathbb{R}^n} w^{p+1}dx} \geq 0. \qquad (13.8)$$

By the Courant-Fischer theorem (see [67]), we deduce that $\lambda_2 \geq 0$. In particular, $\lambda_{2,r} \geq 0$, where $\lambda_{2,r}$ is the second eigenvalue in the radial class.

Since

$$\Delta w - w + pw^{p-1}w = (p-1)w^p,$$

we see that $\int_{\mathbb{R}^n} w^p\phi dx = 0$. Thus ϕ can change sign only once. Without loss of generality, we may assume that for some $r_0 > 0$ we have $\phi \leq 0$ for $r \leq r_0$ and $\phi \geq 0$ for $r \geq r_0$.

Now as in [126] (see also [124, 125]), we consider the function

$$\eta(r) = rw' - \beta w.$$

Then η satisfies

$$\Delta\eta - \eta + pw^{p-1}\eta = 2w - \left(2 + \beta(p-1)\right)w^p. \qquad (13.9)$$

We may choose β such that $1 = (1 + \frac{\beta(p-1)}{2})w^{p-1}(r_0)$ so that

$$2w - \left(2 + \beta(p-1)\right)w^p \leq 0 \quad \text{for } r \leq r_0$$

and

$$2w - \left(2 + \beta(p-1)\right)w^p \geq 0 \quad \text{for } r \geq r_0.$$

Multiplying (13.9) by ϕ and (13.5) by η, we arrive at

$$\int_{\mathbb{R}^n} \phi\left(2w - \left(2 + \beta(p-1)\right)w^p\right)dx = 0$$

which is impossible by the properties of ϕ unless $\phi \equiv 0$. This proves Step 2.

Step 3. Finally, we decompose ϕ into Fourier modes

$$\phi(x) = \sum_{j=1}^{\infty} \phi_j(r)\psi_j(\theta),$$

where $\psi_0 = 1$ and $\psi_j(\theta)$ are the normalised eigenfunctions on S^{n-1} with eigenvalues λ_j. Thus $\lambda_1 = 1, \lambda_2 = \cdots = \lambda_{n+1} = n-1, \lambda_{n+2} > n-1, \ldots$. Then ϕ_j satisfies

$$\Delta\phi_j - \phi_j + pw^{p-1}\phi_j = \frac{\lambda_j}{r^2}\phi_j. \qquad (13.10)$$

We claim that

$$\phi_j \equiv 0 \quad \text{for } j \geq n+2. \tag{13.11}$$

Proof of (13.11) For $j \geq n+2$, we have $\phi_j(0) = 0$. Let r_0 be the first positive zero of ϕ_j (which may be $+\infty$) and we may assume, without loss of generality, that $\phi_j > 0$ for $r \in (0, r_0)$. Multiplying (13.10) by w' and integrating by parts, we obtain

$$\int_{B_{r_0}} \frac{\lambda_j - (n-1)}{r^2} \phi_j w' dx = \int_{\partial B_{r_0}} \left(w' \frac{\partial \phi_j}{\partial r} - \phi_j w'' \right) ds = \int_{\partial B_{r_0}} w' \frac{\partial \phi_j}{\partial r} ds.$$

Now note that the l.h.s. is strictly negative while the r.h.s. is strictly positive unless $\phi_j \equiv 0$. Step 3 follows. □

Finally, we prove

Step 4. For $j = 2, \ldots, n+1$, $\phi_j = c_j w'$ for some $c_j \neq 0$.
   Note that ϕ_j satisfies

$$\Delta \phi_j - \phi_j + p w^{p-1} \phi_j = \frac{n-1}{r^2} \phi_j, \quad \phi_j(0) = 0. \tag{13.12}$$

Then by the uniqueness of the solutions of ordinary differential equations, we have

$$\phi_j = \frac{\phi_j'(0)}{w''(0)} w'$$

and Step 4 is completed.
   Combining Steps 1–4, we have proved the lemma. □

A corollary of Lemma 13.4 is the following result on the spectrum of w.

Lemma 13.5 *Let w be the least energy solution given in Lemma 13.4. Then we have*

(1) *The inequality*

$$\int_{\mathbb{R}^n} \left(|\nabla \phi|^2 + \phi^2 - p w^{p-1} \phi^2 \right) dx + (p-1) \frac{(\int_{\mathbb{R}^n} w^p \phi dx)^2}{\int_{\mathbb{R}^n} w^{p+1} dx} \geq 0$$

   holds for all $\phi \in H^1(\mathbb{R}^n)$.
(2) *The eigenvalue problem*

$$\Delta \phi - \phi + p w^{p-1} \phi + \lambda \phi = 0, \quad \phi \in H^1(\mathbb{R}^n)$$

   satisfies

$$\lambda_1 < 0, \quad \lambda_2 = \cdots = \lambda_{n+1} = 0, \quad \lambda_{n+2} > 0,$$

where the eigenfunction corresponding to λ_1 is simple, radially symmetric and it can be made positive.

Proof We just need to prove the statement on λ_1 and its eigenfunction. In fact, we consider

$$\lambda_1 = \inf_{\phi \in H^1(\mathbb{R}^n) \setminus \{0\}} \frac{\int_{\mathbb{R}^n} (|\nabla \phi|^2 + \phi^2 - pw^{p-1}\phi^2)dx}{\int_{\mathbb{R}^n} \phi^2 dx}.$$

Then we have

$$\lambda_1 < \frac{\int_{\mathbb{R}^n} (|\nabla w|^2 + w^2 - pw^{p+1})dx}{\int_{\mathbb{R}^n} w^2 dx} < 0$$

and the corresponding eigenfunction is simple, radially symmetric and it can be made positive. The rest follows from Lemma 13.4 and its proof, using that by (13.8) we have $\lambda_2 \geq 0$. $\square$

References

1. Abramowitz, M., Stegun, I.A.: Handbook of Mathematical Functions with Formulas, Graphs, and Mathematical Tables. Applied Mathematics. National Bureau of Standards, Washington (1964)
2. Adimurthi, Pacella, F., Yadava, S.L.: Interaction between the geometry of the boundary and positive solutions of a semilinear Neumann problem with critical nonlinearity. J. Funct. Anal. **113**, 318–350 (1993)
3. Adimurthi, Mancinni, G., Yadava, S.L.: The role of mean curvature in a semilinear Neumann problem involving the critical Sobolev exponent. Commun. Partial Differ. Equ. **20**, 591–631 (1995)
4. Adimurthi, Pacella, F., Yadava, S.L.: Characterization of concentration points and L^∞-estimates for solutions involving the critical Sobolev exponent. Differ. Integral Equ. **8**, 41–68 (1995)
5. Ao, W., Wei, J., Zeng, J.: An optimal bound on the number of interior spike solutions for the Lin-Ni-Takagi problem. J. Funct. Anal. **265**, 1324–1356 (2013)
6. Aronson, D.G., Weinberger, H.F.: Multidimensional nonlinear diffusion arising in population genetics. Adv. Math. **30**, 33–76 (1978)
7. Aubin, T.: Some Nonlinear Problems in Riemannian Geometry. Springer Monographs in Mathematics. Springer, Berlin (1998)
8. Bates, P., Shi, J.: Existence and instability of spike layer solutions to singular perturbation problems. J. Funct. Anal. **196**, 211–264 (2002)
9. Bates, P., Dancer, E.N., Shi, J.: Multi-spike stationary solutions of the Cahn-Hilliard equation in higher-dimension and instability. Adv. Differ. Equ. **4**, 1–69 (1999)
10. Benson, D.L., Maini, P.K., Sherratt, J.A.: Analysis of pattern formation in reaction-diffusion models with spatially inhomogeneous diffusion coefficients. Math. Comput. Model. **17**, 29–34 (1993)
11. Benson, D.L., Sherratt, J.A., Maini, P.K.: Diffusion driven instability in an inhomogeneous domain. Bull. Math. Biol. **55**, 365–384 (1993)
12. Benson, D.L., Maini, P.K., Sherratt, J.A.: Unravelling the Turing bifurcation using spatially varying diffusion coefficients. J. Math. Biol. **37**, 381–417 (1998)
13. Berding, C., Haken, H.: Pattern formation in morphogenesis. Analytical treatment of the Gierer-Meinhardt system on a sphere. J. Math. Biol. **14**, 133–151 (1982)
14. Berestycki, H., Wei, J.: On singular perturbation problems with Robin boundary condition. Ann. Sc. Norm. Super. Pisa, Cl. Sci. **5**, 199–230 (2003)
15. Berestycki, H., Wei, J.: On least energy solutions to a semilinear elliptic equation in a strip. Discrete Contin. Dyn. Syst. **28**, 1083–1099 (2010)
16. Bettencourt, L., West, G.: A unified theory of urban living. Nature **467**, 912–913 (2010)
17. Blair, S.S.: Limb development: marginal fringe benefits. Curr. Biol. **7**, R686–R690 (1997)

18. Blair, S.S.: Developmental biology: notching the hindbrain. Curr. Biol. **14**, R570–R572 (2004)
19. Boerlijst, M.C., Hogeweg, P.: Spiral wave structure in pre-biotic evolution: hypercycles stable against parasites. Physica D **48**, 17–28 (1991)
20. Bohn, H.: Interkalare Regeneration und segmentale Gradienten bei den Extremitäten von Leucophaea-Larven. Wilhelm Roux' Arch. **165**, 303–341 (1970)
21. Boozer, A.H.: Equations for studies of feedback stabilization. Phys. Plasmas **5**, 3350 (1998)
22. Britton, N.F.: Essential Mathematical Biology, 2nd edn. Springer Undergraduate Mathematics Series. Springer, London (2003)
23. Buchler, N.E., Gerland, U., Hwa, T.: Nonlinear protein degradation on the functions of genetic circuits. Proc. Natl. Acad. Sci. USA **102**, 9559–9564 (2005)
24. Castets, V., Dulos, E., Boissonade, J., De Kepper, P.: Experimental evidence of a sustained standing Turing-type nonequilibrium chemical pattern. Phys. Rev. Lett. **64**, 2953–2956 (1990)
25. Chen, X., Kowalczyk, M.: Slow dynamics of interior spikes in the shadow Gierer-Meinhardt system. Adv. Differ. Equ. **6**, 847–872 (2001)
26. Chen, X., Kowalczyk, M.: Dynamics of an interior spike in the Gierer-Meinhardt system. SIAM J. Math. Anal. **33**, 172–193 (2001)
27. Chen, X., del Pino, M., Kowalczyk, M.: The Gierer and Meinhardt system: the breaking of homoclinics and multi-bump ground states. Commun. Contemp. Math. **3**(3), 419–439 (2001)
28. Crampin, E.J., Gaffney, E.A., Maini, P.K.: Reaction and diffusion on growing domains: scenarios for robust pattern formation. Bull. Math. Biol. **61**, 1093–1120 (1999)
29. Crampin, E.J., Gaffney, E.A., Maini, P.K.: Mode doubling and tripling in reaction-diffusion patterns on growing domains: a piece-wise linear model. J. Math. Biol. **44**, 107–128 (2002)
30. Crandall, M.G., Rabinowitz, P.H.: Bifurcation from simple eigenvalues. J. Funct. Anal. **8**, 321–340 (1971)
31. Crandall, M.G., Rabinowitz, P.H.: Bifurcation, perturbation of simple eigenvalues and linearized stability. Arch. Ration. Mech. Anal. **52**, 161–180 (1973)
32. Cronhjort, M.B., Blomberg, C.: Hypercycles versus parasites in a two-dimensional partial differential equations model. J. Theor. Biol. **169**, 31–49 (1994)
33. Cronhjort, M.B., Blomberg, C.: Cluster compartmentalization may provide resistance to parasites for catalytic networks. Physica D **101**, 289–298 (1997)
34. Cross, M., Hohenberg, P.: Pattern formation outside of equilibrium. Rev. Mod. Phys. **65**, 851–1112 (1993)
35. Dancer, E.N.: Degree theory on convex sets and applications to bifurcation. In: Buttazzo, G., Marino, A., Murphy, M.K.V. (eds.) Calculus of Variations and Partial Differential Equations, Topics on Geometrical Evolution Problems and Degree Theory. Springer, Berlin (1999)
36. Dancer, E.N.: A note on asymptotic uniqueness for some nonlinearities which change sign. Bull. Aust. Math. Soc. **61**, 305–312 (2000)
37. Dancer, E.N.: On stability and Hopf bifurcations for chemotaxis systems. Methods Appl. Anal. **8**, 245–256 (2001)
38. de Kepper, P., Castets, V., Dulos, E., Boissonade, J.: Turing-type chemical pattern in the chlorite-iodide-malonic acid reaction. Physica D **49**, 161–169 (1991)
39. del Pino, M.: A priori estimates and applications to existence—nonexistence for a semilinear elliptic system. Indiana Univ. Math. J. **43**, 77–129 (1994)
40. del Pino, M., Wei, J.: Collapsing steady states of the Keller-Segel system. Nonlinearity **19**, 661–684 (2006)
41. del Pino, M., Felmer, P., Wei, J.: On the role of mean curvature in some singularly perturbed Neumann problems. SIAM J. Math. Anal. **31**, 63–79 (1999)
42. del Pino, M., Felmer, P., Wei, J.: On the role of the distance function in some singularly perturbed Neumann problems. Commun. Partial Differ. Equ. **25**, 155–177 (2000)
43. del Pino, M., Kowalczyk, M., Chen, X.: The Gierer-Meinhardt system: the breaking of homoclinics and multi-bump ground states. Commun. Contemp. Math. **3**, 419–439 (2001)

44. del Pino, M., Kowalczyk, M., Wei, J.: Multi-bump ground states of the Gierer-Meinhardt system in $\mathbb{R}^2$. Ann. Inst. Henri Poincaré, Anal. Non Linéaire **20**, 53–85 (2003)
45. Doelman, A., van der Ploeg, H.: Homoclinic stripe patterns. SIAM J. Appl. Dyn. Syst. **1**, 65–104 (2002)
46. Doelman, A., Kaper, T.J., Zegeling, P.A.: Pattern formation in the one-dimensional Gray-Scott model. Nonlinearity **10**, 523–563 (1997)
47. Doelman, A., Gardner, R.A., Kaper, T.J.: Stability analysis of singular patterns in the 1-D Gray-Scott model: a matched asymptotic approach. Physica D **122**, 1–36 (1998)
48. Doelman, A., Eckhaus, W., Kaper, T.J.: Slowly modulated two-pulse solutions in the Gray-Scott model I: asymptotic construction and stability. SIAM J. Appl. Math. **61**, 1080–1102 (2000)
49. Doelman, A., Gardner, R.A., Kaper, T.J.: Large stable pulse solutions in reaction-diffusion equations. Indiana Univ. Math. J. **50**, 443–507 (2001)
50. Doelman, A., Kaper, T.J., van der Ploeg, H.: Spatially periodic and aperiodic multi-pulse patterns in the one-dimensional Gierer-Meinhardt equation. Methods Appl. Anal. **8**, 387–414 (2001)
51. Doelman, A., Gardner, R.A., Kaper, T.J.: A stability index analysis of 1-D patterns of the Gray-Scott model. Mem. Am. Math. Soc. **155**(737), xii + 64 pp. (2002)
52. Doelman, A., Kaper, T.J., Peletier, L.A.: Homoclinic bifurcations at the onset of pulse self-replication. J. Differ. Equ. **231**, 359–423 (2006)
53. Dufiet, V., Boissonade, J.: Conventional and unconventional Turing patterns. J. Chem. Phys. **96**, 664–673 (1992)
54. Edelstein-Keshet, L.: Mathematical Models in Biology. SIAM Classics in Applied Mathematics, vol. 46. Society for Industrial and Applied Mathematics, Philadelphia (2005)
55. Efendiev, M., Yagi, A.: Continuous dependence on a parameter of exponential attractors for chemotaxis-growth system. J. Math. Soc. Jpn. **57**, 167–181 (2005)
56. Ei, S.: The motion of weakly interacting pulses in reaction-diffusion systems. J. Dyn. Differ. Equ. **14**, 85–87 (2002)
57. Ei, S., Nishiura, Y., Ueda, K.: 2^n splitting or edge splitting: a manner of splitting in dissipative systems. Jpn. J. Ind. Appl. Math. **18**, 181–205 (2001)
58. Ei, S., Mimura, M., Nagayama, M.: Pulse-pulse interaction in reaction-diffusion systems. Physica D **165**, 176–198 (2002)
59. Ei, S., Mimura, M., Nagayama, M.: Interacting spots in reaction-diffusion systems. Discrete Contin. Dyn. Syst. **14**, 31–62 (2006)
60. Eigen, M., Schuster, P.: The hypercycle. A principle of natural self organization. Part A. Emergence of the hypercycle. Naturwissenschaften **64**, 541–565 (1977)
61. Eigen, M., Schuster, P.: The hypercycle. A principle of natural self organization. Part B. The abstract hypercycle. Naturwissenschaften **65**, 7–41 (1978)
62. Eigen, M., Schuster, P.: The hypercycle. A principle of natural self organization. Part C. The realistic hypercycle. Naturwissenschaften **65**, 341–369 (1978)
63. Eigen, M., Schuster, P.: The Hypercycle: A Principle of Natural Selforganization. Springer, Berlin (1979)
64. Ermentrout, B.: Stripes or spots? Non-linear effects in bifurcation of reaction-diffusion equations on a square. Proc. R. Soc. Lond. A **434**, 413–417 (1991)
65. Fife, P.C.: Stationary patterns for reaction-diffusion systems. In: Nonlinear Diffusion. Research Notes in Mathematics, vol. 14, pp. 81–121. Pitman, London (1977)
66. Fife, P.C.: Large time behaviour of solutions of bistable nonlinear diffusion equations. Arch. Ration. Mech. Anal. **70**, 31–46 (1979)
67. Fischer, E.: Über quadratische Formen mit reellen Koeffizienten. Monatshefte Math. Phys. **16**, 234–249 (1905)
68. Floer, A., Weinstein, A.: Nonspreading wave packets for the cubic Schrödinger equation with a bounded potential. J. Funct. Anal. **69**, 397–408 (1986)
69. Gehring, W.J.: The homeobox in perspective. Trends Biochem. Sci. **17**, 277–280 (1992)

70. Ghergu, M., Radulescu, V.: On a class of singular Gierer-Meinhardt systems arising in morphogenesis. C. R. Math. **344**, 163–168 (2007)

71. Gidas, B., Ni, W.-M., Nirenberg, L.: Symmetry of positive solutions of nonlinear elliptic equations in $\mathbb{R}^n$. Adv. Math. Suppl. Stud. **7A**, 369–402 (1981)

72. Gierer, A.: The Hydra model—a model for what? Int. J. Dev. Biol. **56**, 437–445 (2012)

73. Gierer, A., Meinhardt, H.: A theory of biological pattern formation. Kybernetik (Berlin) **12**, 30–39 (1972)

74. Gilbarg, D., Trudinger, S.: Elliptic Partial Differential Equations of Second Order, 2nd edn. Die Grundlehren der mathematischen Wissenschaften in Einzeldarstellungen, vol. 224. Springer, Berlin (1983)

75. Gray, P., Scott, S.K.: Autocatalytic reactions in the isothermal, continuous stirred tank reactor: isolas and other forms of multistability. Chem. Eng. Sci. **38**, 29–43 (1983)

76. Gray, P., Scott, S.K.: Autocatalytic reactions in the isothermal, continuous stirred tank reactor: oscillations and instabilites to the system $A + 2B \to 3B$, $B \to C$. Chem. Eng. Sci. **39**, 1087–1097 (1984)

77. Guckenheimer, J.: On a codimension two bifurcation. In: Dynamical Systems and Turbulence, Warwick 1980, Coventry, 1979/1980. Lecture Notes in Mathematics, vol. 898, pp. 99–142. Springer, Berlin (1981)

78. Gui, C.: Multipeak solutions for a semilinear Neumann problem. Duke Math. J. **84**, 739–769 (1996)

79. Gui, C., Ghoussoub, N.: Multi-peak solutions for a semilinear Neumann problem involving the critical Sobolev exponent. Math. Z. **229**, 443–474 (1998)

80. Gui, C., Wei, J.: Multiple interior peak solutions for some singularly perturbed Neumann problems. J. Differ. Equ. **158**, 1–27 (1999)

81. Gui, C., Wei, J., Winter, M.: Multiple boundary peak solutions for some singularly perturbed Neumann problems. Ann. Inst. Henri Poincaré, Anal. Non Linéaire **17**, 47–82 (2000)

82. Hadeler, K.P., Rothe, F.: Travelling fronts in nonlinear diffusion equations. J. Math. Biol. **2**, 251–263 (1975)

83. Haken, H.: Synergetics, an Introduction: Nonequilibrium Phase Transitions and Self-organization in Physics, Chemistry, and Biology, 3rd rev. enl. edn. Springer, New York (1983)

84. Haken, H., Olbrich, H.: Analytical treatment of pattern formation in the Gierer-Meinhardt model of morphogenesis. J. Math. Biol. **6**, 317–331 (1978)

85. Hale, J.K., Peletier, L.A., Troy, W.C.: Stability and instability of the Gray-Scott model: the case of equal diffusion constants. Appl. Math. Lett. **12**, 59–65 (1999)

86. Hale, J.K., Peletier, L.A., Troy, W.C.: Exact homoclinic and heteroclinic solutions of the Gray-Scott model for autocatalysis. SIAM J. Appl. Math. **61**, 102–130 (2000)

87. Han, Q., Lin, F.: Elliptic Partial Differential Equations. Courant Lecture Notes. Am. Math. Soc., Providence (2000)

88. Harland, R., Gerhard, J.: Formation and function of Spemann's organizer. Annu. Rev. Cell Dev. Biol. **13**, 661–667 (1997)

89. Harrison, L.G.: Kinetic Theory of Living Pattern. Cambridge University Press, New York (1993)

90. Herschowitz-Kaufman, M.: Bifurcation analysis of nonlinear reaction-diffusion equations II, steady-state solutions and comparison with numerical simulations. Bull. Math. Biol. **37**, 589–636 (1975)

91. Hillen, T., Painter, K.J.: A user's guide to PDE models for chemotaxis. J. Math. Biol. **58**, 183–217 (2009)

92. Holloway, D.M.: Reaction-diffusion theory of localized structures with application to vertebrate organogenesis. Ph.D. thesis, University of British Columbia (1995)

93. Horstmann, D., Stevens, A.: A constructive approach to traveling waves in chemotaxis. J. Nonlinear Sci. **14**, 1–25 (2004)

94. Horstmann, D., Winkler, M.: Boundedness vs. blow-up in a chemotaxis system. J. Differ. Equ. **215**, 52–107 (2005)

95. Hunding, A., Engelhardt, R.: Early biological morphogenesis and nonlinear dynamics. J. Theor. Biol. **173**, 401–413 (1995)
96. Ikeda, K.: Stability analysis for a stripe solution in the Gierer-Meinhardt system. In: Asymptotic Analysis and Singularities—Elliptic and Parabolic PDEs and Related Problems. Advanced Studies in Pure Mathematics, vol. 47-2, pp. 573–587. Math. Soc. Japan, Tokyo (2007)
97. Ingham, P.W.: The molecular genetics of embryonic pattern formation in *Drosophila*. Nature **335**, 25–34 (1988)
98. Iron, D., Ward, M.J.: The dynamics of multi-spike solutions to the one-dimensional Gierer-Meinhardt system. SIAM J. Appl. Math. **62**, 1924–1951 (2002)
99. Iron, D., Ward, M.J., Wei, J.: The stability of spike solutions to the one-dimensional Gierer-Meinhardt model. Physica D **150**, 25–62 (2001)
100. Iron, D., Wei, J., Winter, M.: Stability analysis of Turing patterns generated by the Schnakenberg model. J. Math. Biol. **49**, 358–390 (2004)
101. Irvine, K.D., Rauskolb, C.: Boundaries in development: formation and function. Annu. Rev. Cell Dev. Biol. **17**, 189–214 (2001)
102. Jäger, W., Luckhaus, S.: On explosions of solutions to a system of partial differential equations modelling chemotaxis. Trans. Am. Math. Soc. **329**, 819–824 (1992)
103. Jiang, H.: Global existence of solutions of an activator-inhibitor system. Discrete Contin. Dyn. Syst. **14**, 737–751 (2006)
104. Jiang, H., Ni, W.-M.: A priori estimates of stationary solutions of an activator-inhibitor system. Indiana Univ. Math. J. **56**, 681–732 (2007)
105. Kawohl, B.: Rearrangements and Convexity of Level Sets in PDE. Lecture Notes in Mathematics, vol. 1150. Springer, New York (1985)
106. Keener, J.P.: Secondary bifurcation in nonlinear diffusion reaction equations. Stud. Appl. Math. **55**, 187–211 (1976)
107. Keener, J.P.: Activators and inhibitors in pattern formation. Stud. Appl. Math. **59**, 1–23 (1978)
108. Keller, K.F., Segel, L.A.: Initiation of slime mold aggregation viewed as an instability. J. Theor. Biol. **26**, 399–415 (1970)
109. Kolokolnikov, T., Ren, X.: Smoke-ring solutions of Gierer-Meinhardt system in R^3. SIAM J. Appl. Dyn. Syst. **10**, 251–277 (2011)
110. Kolokolnikov, T., Ward, M.J.: Reduced wave Green's functions and their effect on the dynamics of a spike for the Gierer-Meinhardt model. Eur. J. Appl. Math. **14**, 513–545 (2003)
111. Kolokolnikov, T., Ward, M.J.: Bifurcation of spike equilibria in the near-shadow Gierer-Meinhardt model. Discrete Contin. Dyn. Syst., Ser. B **4**, 1033–1064 (2004)
112. Kolokolnikov, T., Ward, M.J., Wei, J.: The existence and stability of spike equilibria in the one-dimensional Gray-Scott model: the low-feed regime. Stud. Appl. Math. **115**, 21–71 (2005)
113. Kolokolnikov, T., Ward, M.J., Wei, J.: The existence and stability of spike equilibria in the one-dimensional Gray-Scott model: the pulse-splitting regime. Physica D **202**, 258–293 (2005)
114. Kolokolnikov, T., Sun, W., Ward, M.J., Wei, J.: The stability of a stripe for the Gierer-Meinhardt model and the effect of saturation. SIAM J. Appl. Dyn. Syst. **5**, 313–363 (2006)
115. Kolokolnikov, T., Ward, M.J., Wei, J.: Spot self-replication and dynamics for the Schnakenberg model in a two-dimensional domain. J. Nonlinear Sci. **19**, 1–56 (2009)
116. Kolokolnikov, T., Wei, J., Yang, W.: On large ring solutions for Gierer-Meinhardt system in $\mathbb{R}^3$. J. Differ. Equ. **255**, 1408–1436 (2013)
117. Kolokonikov, T., Wei, J.: Stability of spiky solutions in a competition model with cross-diffusion. SIAM J. Appl. Math. **71**, 1428–1457 (2011)
118. Kolokonikov, T., Wei, J., Winter, M.: Existence and stability analysis of spiky solutions for the Gierer-Meinhardt system with large reaction rates. Physica D **238**, 1695–1710 (2009)
119. Kondo, S., Asai, R.: A reaction-diffusion wave on the skin of the marine angelfish *Pomacanthus*. Nature **376**, 765–768 (1995)

120. Kulesa, P.M., Cruywagen, G.C., Lubkin, S.R., Maini, P.K., Sneyd, J.S., Murray, J.D.: Modelling the spatial patterning of the teeth primordia in the lower jaw of *Alligator mississippiensis*. J. Biol. Syst. **3**, 975–985 (1995)

121. Kulesa, P.M., Cruywagen, G.C., Lubkin, S.R., Maini, P.K., Sneyd, J., Ferguson, M.W.J., Murray, J.D.: On a model mechanism for the spatial patterning of teeth primordia in the alligator. J. Theor. Biol. **180**, 287–296 (1996)

122. Kuramoto, Y., Tsuzuki, T.: On the formation of dissipative structures in reaction-diffusion systems—reductive perturbation approach. Prog. Theor. Phys. **54**, 687–699 (1975)

123. Kurata, K., Morimoto, K.: Construction and asymptotic behavior of multi-peak solutions to the Gierer-Meinhardt system with saturation. Commun. Pure Appl. Anal. **7**, 1443–1482 (2008)

124. Kwong, M.-K.: Uniqueness of positive solutions of $\Delta u - u + u^p = 0$ in $\mathbb{R}^n$. Arch. Ration. Mech. Anal. **105**, 243–266 (1989)

125. Kwong, M.-K., Yi, Y.: Uniqueness of radial solutions of semilinear elliptic equations. Trans. Am. Math. Soc. **333**, 339–363 (1992)

126. Kwong, M.-K., Zhang, L.: Uniqueness of positive solutions of $\Delta u + f(u) = 0$ in an annulus. Differ. Integral Equ. **4**, 583–599 (1991)

127. Lacalli, T.C.: Dissipative structures and morphogenetic pattern in unicellular algae. Philos. Trans. R. Soc. Lond. B, Biol. Sci. **294**, 547–588 (1981)

128. Lee, K.J., McCormick, W.D., Ouyang, Q., Swinney, H.L.: Pattern formation by interacting chemical fronts. Science **261**, 192–194 (1993)

129. Lee, K.J., McCormick, W.D., Pearson, J.E., Swinney, H.L.: Experimental observation of self-replicating spots in a reaction-diffusion system. Nature **369**, 215–218 (1994)

130. Lengyel, I., Epstein, I.R.: Modeling of Turing structures in the chlorite-iodide-malonic acid-starch reaction system. Science **251**, 650–652 (1991)

131. Levin, S.A.: The problem of pattern and scale in ecology. Ecology **73**, 1943–1967 (1992)

132. Li, Y.-Y.: On a singularly perturbed equation with Neumann boundary condition. Commun. Partial Differ. Equ. **23**, 487–545 (1998)

133. Lin, C.-S., Ni, W.-M.: On the diffusion coefficient of a semilinear Neumann problem. In: Calculus of Variations and Partial Differential Equations, Trento, 1986. Lecture Notes in Mathematics, vol. 1340, pp. 160–174. Springer, Berlin (1988)

134. Lin, C.-S., Ni, W.-M., Takagi, I.: Large amplitude stationary solutions to a chemotaxis systems. J. Differ. Equ. **72**, 1–27 (1988)

135. Lin, F.-H., Ni, W.-M., Wei, J.: On the number of interior peak solutions for a singularly perturbed Neumann problem. Commun. Pure Appl. Math. **60**, 252–281 (2007)

136. Lyons, M.J., Harrison, L.G.: A class of reaction-diffusion mechanisms which preferentially select striped patterns. Chem. Phys. Lett. **183**, 158–164 (1991)

137. Lyons, M.J., Harrison, L.G., Lakowski, B.C., Lacalli, T.C.: Reaction diffusion modelling of biological pattern formation: application to the embryogenesis of *Drosophila melanogaster*. Can. J. Phys. **68**, 772–777 (1990)

138. Madzvamuse, A., Wathen, A.J., Maini, P.K.: A moving grid finite element method applied to a model biological pattern generator. J. Comput. Phys. **190**, 478–500 (2003)

139. Madzvamuse, A., Maini, P.K., Wathen, A.J.: A moving grid finite element method for the simulation of pattern generation by Turing models on growing domains. J. Sci. Comput. **24**, 247–262 (2005)

140. Maini, P.K.: Turing patterns in fish skin? Nature **380**, 678 (1996)

141. Maini, P.K., Benson, D.L., Sherratt, J.A.: Pattern formation in reaction-diffusion models with spatially inhomogeneous diffusion coefficients. IMA J. Math. Appl. Med. Biol. **9**, 197–213 (1992)

142. Maini, P.K., Painter, K.J., Chau, H.: Spatial pattern formation in chemical and biological systems. J. Chem. Soc. Faraday Trans. **93**, 3601–3610 (1997)

143. Maini, P.K., Baker, R.E., Chuong, C.M.: The Turing model comes of molecular age. Science **314**, 1397–1398 (2006)

144. Maini, P.K., Wei, J., Winter, M.: Stability of spikes in the shadow Gierer-Meinhardt system with Robin boundary conditions. Chaos **17**, 037106 (2007)
145. Meinhardt, H.: Models of Biological Pattern Formation. Academic Press, London (1982)
146. Meinhardt, H.: Hierarchical inductions of cell states: a model for segmentation in *Drosophila*. J. Cell Sci., Suppl. **4**, 357–381 (1986)
147. Meinhardt, H.: A model for pattern-formation of hypostome, tentacles, and foot in hydra: how to form structures close to each other, how to form them at a distance. Dev. Biol. **157**, 321–333 (1993)
148. Meinhardt, H.: Growth and patterning—dynamics of stripe formation. Nature **376**, 722–723 (1995)
149. Meinhardt, H.: Organizer and axes formation as a self-organizing process. Int. J. Dev. Biol. **45**, 177–188 (2001)
150. Meinhardt, H.: Primary body axes of vertebrates: generation of a near-Cartesian coordinate system and the role of Spemann-type organizer. Dev. Dyn. **235**, 2907–2919 (2006)
151. Meinhardt, H.: Models of biological pattern formation: from elementary steps to the organization of embryonic axes. Curr. Top. Dev. Biol. **81**, 1–63 (2008)
152. Meinhardt, H.: The Algorithmic Beauty of Sea Shells, 4th edn. Springer, Berlin (2009)
153. Meinhardt, H.: Modeling pattern formation in hydra—a route to understanding essential steps in development. Int. J. Dev. Biol. **56**, 447–462 (2012)
154. Meinhardt, H., Gierer, A.: Generation and regeneration of sequences of structures during morphogenesis. J. Theor. Biol. **85**, 429–450 (1980)
155. Meinhardt, H., Klingler, M.: A model for pattern formation on the shells of molluscs. J. Theor. Biol. **126**, 63–69 (1987)
156. Mercader, I., Net, M., Knobloch, E.: Binary fluid convection in a cylinder. Phys. Rev. E **51**, 339–350 (1995)
157. Mielke, A.: The Ginzburg-Landau equation in its role as a modulation equation. In: Fiedler, B. (ed.) Handbook of Dynamical Systems, vol. 2, pp. 759–834. Elsevier, Amsterdam (2002)
158. Mielke, A., Schneider, G.: Attractors for modulation equations on unbounded domains: existence and comparison. Nonlinearity **8**, 743–768 (1995)
159. Mimura, M., Nishiura, Y.: Spatial patterns for an interaction-diffusion equation in morphogenesis. J. Math. Biol. **7**, 243–263 (1979)
160. Mimura, M., Tsujikawa, T.: Aggregating pattern dynamics in a chemotaxis model including growth. Physica A **230**, 499–543 (1996)
161. Moos, M., Wang, S.W., Krinks, M.: Anti-dorsalizing morphogenetic protein is a novel tgf-beta homolog expressed in the Spemann organizer. Development **121**, 4293–4301 (1995)
162. Morimoto, K.: Point-condensation phenomena and saturation effect for the one-dimensional Gierer-Meinhardt system. Ann. Inst. Henri Poincaré, Anal. Non Linéaire **27**, 973–995 (2010)
163. Muratov, C.B., Osipov, V.V.: Static spike autosolitons in the Gray-Scott model. J. Phys. A, Math. Gen. **33**, 8893–8916 (2000)
164. Muratov, C.B., Osipov, V.V.: Stability of the static spike autosolitons in the Gray-Scott model. SIAM J. Appl. Math. **62**, 1463–1487 (2002)
165. Murray, J.D.: Mathematical Biology I: An Introduction, 3rd edn. Interdisciplinary Applied Mathematics, vol. 17. Springer, Berlin (2002)
166. Murray, J.D.: Mathematical Biology II: Spatial Models and Biomedical Applications, 3rd edn. Interdisciplinary Applied Mathematics, vol. 18. Springer, Berlin (2003)
167. Nec, Y., Ward, M.J.: An explicitly solvable nonlocal eigenvalue problem and the stability of a spike for a sub-diffusive reaction-diffusion system. Math. Model. Nat. Phenom. **8**, 55–87 (2013)
168. Ni, W.-M.: Diffusion, cross-diffusion, and their spike-layer steady states. Not. Am. Math. Soc. **45**, 9–18 (1998)
169. Ni, W.-M.: The Mathematics of Diffusion. CBMS-NSF Regional Conference Series in Applied Mathematics, vol. 82. Society for Industrial and Applied Mathematics, Philadelphia (2011)

170. Ni, W.-M., Takagi, I.: On the Neumann problem for some semilinear elliptic equations and systems of activator-inhibitor type. Trans. Am. Math. Soc. **297**, 351–368 (1986)

171. Ni, W.-M., Takagi, I.: On the shape of least-energy solutions to a semilinear Neumann problem. Commun. Pure Appl. Math. **44**, 819–851 (1991)

172. Ni, W.-M., Takagi, I.: Locating the peaks of least-energy solutions to a semilinear Neumann problem. Duke Math. J. **70**, 247–281 (1993)

173. Ni, W.-M., Takagi, I.: Point condensation generated by a reaction-diffusion system in axially symmetric domains. Jpn. J. Ind. Appl. Math. **12**, 327–365 (1995)

174. Ni, W.-M., Wei, J.: On the location and profile of spike-layer solutions to singularly perturbed semilinear Dirichlet problems. Commun. Pure Appl. Math. **48**, 731–768 (1995)

175. Ni, W.-M., Wei, J.: On positive solutions concentrating on spheres for the Gierer-Meinhardt system. J. Differ. Equ. **221**, 158–189 (2006)

176. Ni, W.-M., Pan, X., Takagi, I.: Singular behavior of least-energy solutions of a semilinear Neumann problem involving critical Sobolev exponents. Duke Math. J. **67**, 1–20 (1992)

177. Ni, W.-M., Polacik, P., Yanagida, E.: Monotonicity of stable solutions in shadow systems. Trans. Am. Math. Soc. **353**, 5057–5069 (2001)

178. Ni, W.-M., Takagi, I., Yanagida, E.: Stability of least energy patterns of the shadow system for an activator-inhibitor model, recent topics in mathematics moving toward science and engineering. Jpn. J. Ind. Appl. Math. **18**, 259–272 (2001)

179. Ni, W.-M., Takagi, I., Yanagida, E.: Stability analysis of point-condensation solutions to a reaction-diffusion system proposed by Gierer and Meinhardt. Preprint (1999)

180. Nicolis, G., Prigogine, I.: Self-organization in Non-equilibrium Systems. Wiley, New York (1977)

181. Nishiura, Y.: Global structure of bifurcating solutions of some reaction-diffusion systems. SIAM J. Math. Anal. **13**, 555–593 (1982)

182. Nishiura, Y.: Far-from-Equilibrium-Dynamics. Translations of Mathematical Monographs, vol. 209. Am. Math. Soc., Providence (2002)

183. Nishiura, Y., Fujii, H.: Stability of singularly perturbed solutions to systems of reaction-diffusion equations. SIAM J. Math. Anal. **18**, 1726–1770 (1987)

184. Nishiura, Y., Ueyama, D.: A skeleton structure of self-replicating dynamics. Physica D **130**, 73–104 (1999)

185. Nishiura, Y., Ueyama, D.: Spatio-temporal chaos for the Gray-Scott model. Physica D **150**, 137–162 (2001)

186. Nishiura, Y., Teramoto, T., Ueda, K.: Scattering and separators in dissipative systems. Phys. Rev. E **67**, 056210 (2003)

187. Nishiura, Y., Teramoto, T., Yuan, X., Ueda, K.: Dynamics of traveling pulses in heterogeneous media. Chaos **17**, 037104 (2007)

188. Nüsslein-Volhard, C.: Determination of the embryonic axes of *Drosophila*. Development **1**(Suppl.), 1–10 (1991)

189. Nüsslein-Volhard, C.: Coming to Life: How Genes Drive Development. Yale University Press, New Haven (2006)

190. Nüsslein-Volhard, C., Wieschaus, E.: Mutations affecting segment number and polarity in *Drosophila*. Nature **287**, 795–801 (1980)

191. Oh, Y.G.: Existence of semiclassical bound states of nonlinear Schrödinger equations with potentials of the class $(V)_a$. Commun. Partial Differ. Equ. **13**, 1499–1519 (1988)

192. Oh, Y.G.: On positive multi-lump bound states of nonlinear Schrödinger equations under multiple well potential. Commun. Math. Phys. **131**, 223–253 (1990)

193. Okubo, A., Levin, S.A.: Diffusion and Ecological Problems: Modern Perspectives, 2nd edn. Interdisciplinary Applied Mathematics, vol. 14. Springer, Berlin (2001)

194. Ouyang, Q., Swinney, H.L.: Transition from a uniform state to hexagonal and striped Turing patterns. Nature **352**, 610–612 (1991)

195. Ouyang, Q., Swinney, H.L.: Transition to chemical turbulence. Chaos **1**, 411–420 (1991)

196. Page, K., Maini, P.K., Monk, N.A.M.: Pattern formation in spatially heterogeneous Turing reaction-diffusion models. Physica D **181**, 80–101 (2003)

197. Page, K., Maini, P.K., Monk, N.A.M.: Complex pattern formation in reaction-diffusion systems with spatially varying parameters. Physica D **202**, 95–115 (2005)
198. Painter, K.J., Maini, P.K., Othmer, H.G.: Stripe formation in juvenile *Pomacanthus* explained by a generalised Turing mechanism with chemotaxis. Proc. Natl. Acad. Sci. USA **96**, 5549–5554 (1999)
199. Pan, X.B.: Condensation of least-energy solutions of a semilinear Neumann problem. J. Partial Differ. Equ. **8**, 1–36 (1995)
200. Pan, X.B.: Condensation of least-energy solutions: the effect of boundary conditions. Nonlinear Anal., Theory Methods Appl. **24**, 195–222 (1995)
201. Pan, X.B.: Further study on the effect of boundary conditions. J. Differ. Equ. **117**, 446–468 (1995)
202. Pankratz, M.J., Jäckle, H.: Making stripes in the *Drosophila* embryo. Trends Genet. **6**, 287–292 (1990)
203. Pearson, J.E.: Complex patterns in a simple system. Science **261**, 189–192 (1993)
204. Pearson, J.E., Horsthemke, W.: Turing instabilities with nearly equal diffusion constants. J. Chem. Phys. **90**, 1588–1599 (1989)
205. Polya, G., Szego, G.: Isoperimetric Inequalities in Mathematical Physics. Annals of Mathematics Studies, vol. 27. Princeton University Press, Princeton (1952)
206. Prigogine, I., Lefever, R.: Symmetry breaking instabilities in dissipative systems. II. J. Chem. Phys. **48**, 1695 (1968)
207. Reynolds, J., Pearson, J., Ponce-Dawson, S.: Dynamics of self-replicating patterns in reaction diffusion systems. Phys. Rev. Lett. **72**, 2797–2800 (1994)
208. Reynolds, J., Pearson, J., Ponce-Dawson, S.: Dynamics of self-replicating spots in reaction-diffusion systems. Phys. Rev. E **56**, 185–198 (1997)
209. Riddihough, G.: Homing in on the homeobox. Nature **357**, 643–644 (1992)
210. Rothe, F.: Global Solutions of Reaction-Diffusion Systems. Lecture Notes in Mathematics, vol. 1072. Springer, Berlin (1984)
211. Sandstede, B.: Stability of traveling waves. In: Handbook of Dynamical Systems, vol. 2. North-Holland, Amsterdam (2002)
212. Sandstede, B., Scheel, A.: Absolute instabilities of standing pulses. Nonlinearity **18**, 331–378 (2005)
213. Satnoianu, R.A., Maini, P.K., Menzinger, M.: Parameter domains for Turing and stationary flow-distributed waves: I. The influence of nonlinearity. Physica D **160**, 79–102 (2001)
214. Schnakenberg, J.: Simple chemical reaction systems with limit cycle behaviour. J. Theor. Biol. **81**, 389–400 (1979)
215. Segel, L.A., Levin, S.A.: Applications of nonlinear stability theory to the study of the effects of dispersion on predator-prey interactions. In: Piccirelli, R. (ed.) Selected Topics in Statistical Mechanics and Biophysics. Conference Proceedings, vol. 27. American Inst. Physics, New York (1976)
216. Sick, S., Reinker, S., Timmer, J., Schlake, T.: WNT and DKK determine hair follicle spacing through a reaction-diffusion mechanism. Science **314**, 1447–1450 (2006)
217. Slater, L.J.: Generalized Hypergeometric Functions. Cambridge University Press, Cambridge (1966)
218. Sleeman, B.D., Ward, M.J., Wei, J.: The existence and stability of spike patterns in a chemotaxis model. SIAM J. Appl. Math. **65**, 790–817 (2005)
219. Spemann, H.: Embryonic Development and Induction. Yale University Press, New Haven (1938)
220. Spemann, H., Mangold, H.: Über Induktion von Embryonalanlagen durch Implantation artfremder Organisatoren. Wilhelm Roux' Arch. Entwicklungsmech. Org. **100**, 599–638 (1924)
221. Stevens, A.: The derivation of chemotaxis equations as limit dynamics of moderately interacting stochastic many-particle systems. SIAM J. Appl. Math. **61**, 183–212 (2000)
222. Strauss, W.A.: Existence of solitary waves in higher dimensions. Commun. Math. Phys. **55**, 149–162 (1977)

223. Sun, W., Ward, M.J., Russell, R.: The slow dynamics of two-spike solutions for the Gray-Scott and Gierer-Meinhardt systems: competition and oscillatory instabilities. SIAM J. Appl. Dyn. Syst. **4**, 904–953 (2005)

224. Suzuki, M., Ohta, T., Mimura, M., Sakaguchi, H.: Breathing and wiggling motions in three-species laterally inhibitory systems. Phys. Rev. E **52**, 3645–3655 (1995)

225. Takagi, I.: Stability of bifurcating solutions of the Gierer-Meinhardt system. Tohoku Math. J. **31**, 221–246 (1979)

226. Takagi, I.: Point-condensation for a reaction-diffusion system. J. Differ. Equ. **61**, 208–249 (1986)

227. Takagi, H., Kaneko, K.: Differentiation and replication of spots in a reaction-diffusion system with many chemicals. Europhys. Lett. **56**, 145–151 (2001)

228. Taniguchi, M.: A uniform convergence theorem for singular limit eigenvalue problems. Adv. Differ. Equ. **8**, 29–54 (2003)

229. Trembley, A.: Mémoires pour servir à l'histoire d'un genre de polypes d'eau douce, à bras en forme de cornes. Jean & Herman Verbeek, Leiden (1744)

230. Tse, W., Wei, J., Winter, M.: Spikes for the Gierer-Meinhardt system with many segments of different diffusivities. Bull. Inst. Math. Acad. Sin. (N.S.) **3**, 525–566 (2008)

231. Tse, W., Wei, J., Winter, M.: The Gierer-Meinhardt system on a compact two-dimensional Riemannian manifold: interaction of Gaussian curvature and Green's function. J. Math. Pures Appl. **94**, 366–397 (2010)

232. Turing, A.M.: The chemical basis of morphogenesis. Philos. Trans. R. Soc. Lond. B, Biol. Sci. **237**, 37–72 (1952)

233. Valeyev, N.V., Bates, D.G., Heslop-Harrison, P., Postlethwaite, I., Kotov, N.V.: Elucidating the mechanisms of cooperative calcium-calmodulin interactions: a structural systems biology approach. BMC Syst. Biol. **2**, 48 (2008)

234. Vastano, J.A., Pearson, J.E., Horsthemke, W., Swinney, H.L.: Chemical pattern formation with equal diffusion coefficients. Phys. Lett. A **124**, 320–324 (1987)

235. Vastano, J.A., Pearson, J.E., Horsthemke, W., Swinney, H.L.: Turing patterns in an open reactor. J. Chem. Phys. **88**, 6175–6181 (1988)

236. Volpert, A.I., Volpert, V.A., Volpert, V.A.: Traveling Wave Solutions of Parabolic Systems. Translations of Mathematical Monographs, vol. 140. Am. Math. Soc., Providence (1994)

237. Wang, Z.-A., Hillen, T.: Classical solutions and pattern formation for a volume filling chemotaxis model. Chaos **17**, 037108 (2007)

238. Wang, Z.-Q.: On the existence of multiple single-peaked solutions for a semilinear Neumann problem. Arch. Ration. Mech. Anal. **120**, 375–399 (1992)

239. Ward, M.J.: Asymptotic methods for reaction-diffusion systems: past and present. Bull. Math. Biol. **68**, 1151–1167 (2006)

240. Ward, M.J., Wei, J.: Asymmetric spike patterns for the one-dimensional Gierer-Meinhardt model: equilibria and stability. Eur. J. Appl. Math. **13**, 283–320 (2002)

241. Ward, M.J., Wei, J.: The existence and stability of asymmetric spike patterns for the Schnakenberg model. Stud. Appl. Math. **109**, 229–264 (2002)

242. Ward, M.J., Wei, J.: Hopf bifurcation of spike solutions for the shadow Gierer-Meinhardt system. Eur. J. Appl. Math. **14**, 677–711 (2003)

243. Ward, M.J., Wei, J.: Hopf bifurcations and oscillatory instabilities of solutions for the one-dimensional Gierer-Meinhardt model. J. Nonlinear Sci. **13**, 209–264 (2003)

244. Ward, M.J., McInerney, D., Houston, P., Gavaghan, D., Maini, P.K.: The dynamics and pinning of a spike for a reaction-diffusion system. SIAM J. Appl. Math. **62**, 1297–1328 (2002)

245. Wei, J.: On the construction of single-peaked solutions to a singularly perturbed Dirichlet problem. J. Differ. Equ. **129**, 315–333 (1996)

246. Wei, J.: On the boundary spike layer solutions of a singularly perturbed semilinear Neumann problem. J. Differ. Equ. **134**, 104–133 (1997)

247. Wei, J.: On the interior spike layer solutions of singularly perturbed semilinear Neumann problem. Tohoku Math. J. (2) **50**, 159–178 (1998)

248. Wei, J.: On the interior spike layer solutions for some singular perturbation problems. Proc. R. Soc. Edinb., Sect. A, Math. **128**, 849–874 (1998)
249. Wei, J.: On single interior spike solutions of the Gierer-Meinhardt system: uniqueness and spectrum estimates. Eur. J. Appl. Math. **10**, 353–378 (1999)
250. Wei, J.: Existence, stability and metastability of point condensation patterns generated by the Gray-Scott system. Nonlinearity **12**, 593–616 (1999)
251. Wei, J.: On the effect of domain geometry in a singularly perturbed Dirichlet problem. Differ. Integral Equ. **13**, 15–45 (2000)
252. Wei, J.: On a nonlocal eigenvalue problem and its applications to point-condensations in reaction-diffusion systems. Int. J. Bifurc. Chaos Appl. Sci. Eng. **10**, 1485–1496 (2000)
253. Wei, J.: Uniqueness and critical spectrum of boundary spike solutions. Proc. R. Soc. Edinb., Sect. A, Math. **131**, 1457–1480 (2001)
254. Wei, J.: Pattern formations in two-dimensional Gray-Scott model: existence of single-spot solutions and their stability. Physica D **148**, 20–48 (2001)
255. Wei, J.: Existence and stability of spikes for the Gierer-Meinhardt system. In: Chipot, M. (ed.) Handbook of Differential Equations: Stationary Partial Differential Equations, vol. 5, pp. 487–585. Elsevier, Amsterdam (2008)
256. Wei, J., Winter, M.: Stationary solutions for the Cahn-Hilliard equation. Ann. Inst. Henri Poincaré, Anal. Non Linéaire **15**, 459–492 (1998)
257. Wei, J., Winter, M.: On the Cahn-Hilliard equations II: interior spike layer solutions. J. Differ. Equ. **148**, 231–267 (1998)
258. Wei, J., Winter, M.: Multi-peak solutions for a wide class of singular perturbation problems. J. Lond. Math. Soc. **59**, 585–606 (1999)
259. Wei, J., Winter, M.: On the two-dimensional Gierer-Meinhardt system with strong coupling. SIAM J. Math. Anal. **30**, 1241–1263 (1999)
260. Wei, J., Winter, M.: On a two-dimensional reaction-diffusion system with hypercyclical structure. Nonlinearity **13**, 2005–2032 (2000)
261. Wei, J., Winter, M.: Spikes for the two-dimensional Gierer-Meinhardt system: the weak coupling case. J. Nonlinear Sci. **11**, 415–458 (2001)
262. Wei, J., Winter, M.: Spikes for the two-dimensional Gierer-Meinhardt system: the strong coupling case. J. Differ. Equ. **178**, 478–518 (2002)
263. Wei, J., Winter, M.: Critical threshold and stability of cluster solutions for large reaction-diffusion systems in $\mathbb{R}$. SIAM J. Math. Anal. **33**, 1058–1089 (2002)
264. Wei, J., Winter, M.: A nonlocal eigenvalue problem and the stability of spikes for reaction-diffusion systems with fractional reaction rates. Int. J. Bifurc. Chaos Appl. Sci. Eng. **13**, 1529–1543 (2003)
265. Wei, J., Winter, M.: Higher-order energy expansions for some singularly perturbed Neumann problems. C. R. Math. Acad. Sci. Paris **337**, 37–42 (2003)
266. Wei, J., Winter, M.: Stability of monotone solutions for the shadow Gierer-Meinhardt system with finite diffusivity. Differ. Integral Equ. **16**, 1153–1180 (2003)
267. Wei, J., Winter, M.: Existence and stability of multiple-spot solutions for the Gray-Scott model in $\mathbb{R}^2$. Physica D **176**, 147–180 (2003)
268. Wei, J., Winter, M.: Asymmetric spotty patterns for the Gray-Scott model in R^2. Stud. Appl. Math. **110**, 63–102 (2003)
269. Wei, J., Winter, M.: Existence and stability analysis of asymmetric patterns for the Gierer-Meinhardt system. J. Math. Pures Appl. **83**, 433–476 (2004)
270. Wei, J., Winter, M.: On the Gierer-Meinhardt system with saturation. Commun. Contemp. Math. **6**, 259–277 (2004)
271. Wei, J., Winter, M.: Higher-order energy expansions and spike locations. Calc. Var. Partial Differ. Equ. **20**, 403–430 (2004)
272. Wei, J., Winter, M.: Symmetric and asymmetric multiple clusters in a reaction-diffusion system. Nonlinear Differ. Equ. Appl. **14**, 787–823 (2007)
273. Wei, J., Winter, M.: Existence, classification and stability analysis of multiple-peaked solutions for the Gierer-Meinhardt system in $\mathbb{R}$. Methods Appl. Anal. **14**, 119–163 (2007)

274. Wei, J., Winter, M.: Stationary multiple spots for reaction-diffusion systems. J. Math. Biol. **57**, 53–89 (2008)

275. Wei, J., Winter, M.: Mutually exclusive spiky pattern and segmentation modeled by the five-component Meinhardt-Gierer system. SIAM J. Appl. Math. **69**, 419–452 (2008)

276. Wei, J., Winter, M.: Spikes for the Gierer-Meinhardt system with discontinuous diffusion coefficients. J. Nonlinear Sci. **19**, 301–339 (2009)

277. Wei, J., Winter, M.: On the Gierer-Meinhardt system with precursors. Discrete Contin. Dyn. Syst. **25**, 363–398 (2009)

278. Wei, J., Winter, M.: Stability of cluster solutions in a reaction-diffusion system with four morphogens on the real line. SIAM J. Math. Anal. **42**, 2818–2841 (2010)

279. Wei, J., Winter, M.: Flow-distributed spikes for Schnakenberg kinetics. J. Math. Biol. **64**, 211–254 (2012)

280. Wei, J., Winter, M.: Stability of cluster solutions in a cooperative consumer chain model. J. Math. Biol. (2012). doi:10.1007/s00285-012-0616-8

281. Wei, J., Zhang, L.: On a nonlocal eigenvalue problem. Ann. Sc. Norm. Super. Pisa, Cl. Sci. **30**, 41–61 (2001)

282. Wei, J., Winter, M., Yeung, W.-K.: A higher-order energy expansion to two-dimensional singularly perturbed Neumann problems. Asymptot. Anal. **43**, 75–110 (2005)

283. Weiss, J.D.: The Hill equation revisited: uses and misuses. FASEB J. **11**, 835–841 (1997)

284. Wilby, O.K., Webster, G.: Studies on the transmission of hypostome inhibition in hydra. J. Embryol. Exp. Morphol. **24**, 583–593 (1970)

285. Wilby, O.K., Webster, G.: Experimental studies on axial polarity in hydra. J. Embryol. Exp. Morphol. **24**, 595–613 (1970)

286. Wittenberg, R.W., Holmes, P.: The limited effectiveness of normal forms: a critical review and extension of local bifurcation studies of the Brusselator PDE. Physica D **100**, 1–40 (1997)

287. Wolpert, L.: Positional information and the spatial pattern of cellular differentiation. J. Theor. Biol. **25**, 1–47 (1969)

288. Wolpert, L.: Positional information and pattern regulation in regeneration of hydra. Symp. Soc. Exp. Biol. **25**, 391–415 (1971)

289. Wolpert, L., Hornbruch, A.: Double anterior chick limb buds and models for cartilage rudiment specification. Development **109**, 961–966 (1990)

290. Yuan, X., Teramoto, T., Nishiura, Y.: Heterogeneity-induced defect bifurcation and pulse dynamics for a three-component reaction-diffusion system. Phys. Rev. E **75**, 036220 (2007)

Index

Printed in the United States
By Bookmasters